WITHDRAWN FROM
KENT STATE UNIVERSITY LIBRARIES

Inland Fisheries
Ecology and Management

Inland Fisheries
Ecology and Management

Compiled by

R.L. Welcomme
Renewable Resources Assessment Group, Imperial College of Science, Technology and Medicine, London, UK

Published for
Food and Agriculture Organization of the United Nations
by Blackwell Science

Fishing News Books
An imprint of Blackwell Science

b
Blackwell Science

Copyright © 2001 by FAO

Fishing News Books
A division of Blackwell Science Ltd
Editorial Offices:
Osney Mead, Oxford OX2 0EL
25 John Street, London WC1N 2BS
23 Ainslie Place, Edinburgh EH3 6AJ
350 Main Street, Malden, MA 02148 5018, USA
54 University Street, Carlton, Victoria 3053, Australia
10, rue Casimir Delavigne, 75006 Paris, France

Other Editorial Offices:

Blackwell Wissenschafts-Verlag GmbH
Kurfürstendamm 57
10707 Berlin, Germany

Blackwell Science KK
MG Kodenmacho Building
7–10 Kodenmacho Nihombashi
Chuo-ku, Tokyo 104, Japan

Iowa State University Press
A Blackwell Science Company
2121 S. State Avenue
Ames, Iowa 50014-8300, USA

The right of the Author to be identified as the Author of this Work has been asserted in accordance with the Copyright, Designs and Patents Act 1988.

All rights reserved. No part of this publication may be reproduced, stored in a retrieval system, or transmitted, in any form or by any means, electronic, mechanical, photocopying, recording or otherwise, except as permitted by the UK Copyright, Designs and Patents Act 1988, without the prior permission of the publisher.

First published 2001

Set in Times by Gray Publishing, Tunbridge Wells, Kent
Printed and bound in Great Britain by
MPG Books Ltd, Bodmin, Cornwall

The Blackwell Science logo is a trade mark of Blackwell Science Ltd, registered at the United Kingdom Trade Marks Registry

DISTRIBUTORS
Marston Book Services Ltd
PO Box 269
Abingdon
Oxon OX14 4YN
(*Orders*: Tel: 01865 206206
Fax: 01865 721205
Telex: 83355 MEDBOK G)

USA and Canada
Iowa State University Press
A Blackwell Science Company
2121 S. State Avenue
Ames, Iowa 50014-8300
(*Orders*: Tel: 800-862-6657
Fax: 515-292-3348
Web: www.isupress.com
email: orders@isupress.com)

Australia
Blackwell Science Pty Ltd
54 University Street
Carlton, Victoria 3053
(*Orders*: Tel: 03 9347 0300
Fax: 03 9347 5001)

A catalogue record for this title
is available from the British Library

ISBN 0-85238-284-7 (FNB)
ISBN 92-5-104545-3 (FAO)

Library of Congress Cataloging-in-Publication Data
Welcomme, R. L.
  Inland fisheries: ecology and management/
compiled by R.L. Welcomme.
     p. cm.
  Includes bibliographical references (p.).
  ISBN 0-85238-284-7
  1. Fishery managment. 2. Freshwater fishes–
Ecology. I. Food and Agriculture Organization of the United Nations. II. Title.
SH328.W46 2001
333.95′6′153–dc21

For further information on Fishing News Books, visit our website: http://www.blacksci.co.uk/fnb/

The designations employed and the presentation of material in this publication do not imply the expression of any opinion whatsoever on the part of the Food and Agriculture Organization of the United Nations concerning the legal status of any country, territory, city or area or of its authorities, or concerning the delimitation of its frontiers or boundaries.

The designations 'developed' and 'developing' economies are intended for statistical convenience and do not necessarily express a judgement about the stage reached by a particular country, territory or area in the development process.

The views expressed herein are those of the author and do not necessarily represent those of the Food and Agriculture Organization of the United Nations.

Authorship

The author prepared most of the text. Particular sections were prepared as follows:

Dynamics of Fish Populations and Sections of Stock Assessment, by Ashley S. Halls, MRAG Ltd, 47 Princes Gate, London SW7 2QA, UK, and R.L. Welcomme

Fish Utilisation, by F. Teutcher, FIII, Fisheries Department, FAO, Viale delle Terme di Caracalla, Rome, Italy

Telemetry, by E. Baras, University of Liège, Aquaculture Research and Education Centre, 10 Chemin de la Justice, B-4500 Tihange, Belgium
Email: E.Baras@ulg.ac.be

Social and Economic Evaluation, by M. Aeron-Thomas and V. Cowan, MRAG, 47 Princes Gate, London, UK

Geographical Information Systems (GIS) in the Management of Inland Fisheries, by J.M. Kapetsky, 5410 Marina Club Drive, Wilmington, NC 28409, USA

Stocking, by K. Lorenzen, Renewable Resources Assessment Group, Imperial College of Science, Technology and Medicine, 8 Princes Gardens, London SW7 2QA, UK

Social and Policy Considerations, by M. Aeron-Thomas and V. Cowan, MRAG, 47 Princes Gate, London, UK

Biodiversity and Conservation Issues, by D.M. Bartley, FIRI, Fisheries Department, FAO, Viale delle Terme di Caracalla, Rome, Italy

Harvest Reserves, by D.D. Hoggarth, SCALES Inc., 6 Highgate Gardens, St Michael, Barbados

Legislation, by C. Leria, LEGA, Legal Office, FAO, Via delle Terme di Caracalla, Rome, Italy

Acknowledgements

I wish to thank the many colleagues who have participated in the development of current thinking on inland fisheries through their own efforts and through their participation in working groups of the FAO Regional Fishery Bodies as well as in numerous symposia and scientific meetings organised by a range of institutions. Particular thanks are due to my collaborators in FAO, in particular J. Kapetsky, D. Bartley, U. Barg, G. Marmulla and H. Naeve, whose enthusiasm and support have made the publication possible. I also wish to express my gratitude to my present colleagues at MRAG and RRAG, Imperial College, K. Lorenzen, A. Halls, V. Cowan, M. Aeron-Thomas and D. Hoggarth for their contributions and stimulus in developing many of the fresh approaches here discussed.

Contents

Authorship	v
Acknowledgements	vi
Guidelines	xvii

1. INTRODUCTION — 1

Resource availability — 1
- Human population — 1
- Water supply — 2
- Feed and other resources — 2
- Land availability — 3

Political and economic climate — 3
- National trends — 3
- International initiatives — 3

The changing situation of fisheries — 4
- A brief history of inland fisheries management. — 4
- Conservation as opposed to exploitation — 5
- Improved tools for management — 6

A new vision for management — 7

2. OBJECTIVES — 8

Objectives for the fishery — 8
- Extractive objectives — 8
  - Food fisheries — 8
  - Recreational fisheries — 10
  - Fisheries for ornamental species — 11
  - Bait fisheries — 11
  - Fry fisheries — 11
  - Fisheries for other purposes — 11
- Sustainability objectives — 12
  - Sustainability — 12
  - Ecological objectives — 12

Social objectives 12
   Income to fishers 13
   Equity/benefit distribution 13
   Conflict reduction 13
Government objectives 13
   Revenue to government 13
   Contribution to GDP 14
   Export income 14
Combinations of objectives 14
Aquaculture 14

Regional differences in management strategy 15

3. THE NATURE OF INLAND WATERS 17

Types of inland water 17
  Lakes 17
  Rivers 22
  Swamps, marshes and rice fields 28
  Reservoirs 28
  Coastal lagoons 29

4. THE NATURE OF FISH POPULATIONS 32

Fish populations in different types of inland water 32
  Lakes 32
  Rivers 36
  Swamps, marshes and rice fields 39
  Reservoirs 39
  Coastal lagoons 40

The size structure of fish populations 40
  The species structure of the assemblage 42

Dynamics of fish populations 42
  Reproduction 43
    Maturation 43
    Seasonality of reproduction 45
    Fecundity 47
  Recruitment 50
  Feeding 54
    Measuring feeding 55
    Fish condition 59
    Feeding behaviour 59
  Growth 60

Modelling growth	62
Density-dependent growth	63
Mortality	66

Biomass and production — 69
  Relationships between factors — 72

Migration and movements — 72
  Longitudinal migrations — 73
  Lateral migrations — 78
  Vertical migrations — 79

Responses of fish populations to stress — 79
  Responses to fishing and environmental change — 79
  Responses to exploitation under fluctuating water levels — 80
  Summary of changes occurring in response to stress — 82

5. THE FISHERMAN AND THE FISHERY — 84

The fisherman in society — 84
  The fishermen — 84
  Food fishermen — 86
    Full-time fishermen — 86
    Part-time fishermen — 87
    Subsistence fishermen — 88
  Recreational fishermen — 88
    Match — 88
    Specimen — 88
    Relaxation — 88
    Domestic consumption — 89
  Other stakeholders — 89
  Upstream — 89
    Gear manufacturers — 89
    Owners of water rights, boats and gear — 89
    Tourist industry — 90
  Downstream — 90
    Processors — 90
    Transporters — 91
    Retailers — 91

6. FISHING TECHNIQUES — 92

Types of fishing regime — 92
  Catch and remove — 92
    Natural — 92

Enhanced	92
Stocked	93
Other interventions	93
Catch and return	93

Fishing technology — **94**

Fishing gear	94
Factors influencing the choice of gear	94
Selectivity of gear	96
Principal types of gear	98
Associated technology	119
Echo sounding	121
Mobile telephones	121
Fishing craft	121

Social and policy implications of fishing technology — **122**

Seasonality of fishing — **123**

7. FISH UTILISATION — 125

Fish as food and nutrition — **125**

Fish preservation — **125**

Live fish	126
Icing and freezing	126
Smoking	128
Drying and salting	130
Canning	130
Fermentation	131
Fish meal, fish oil and animal feeds	131

Marketing — **131**

Collection of fish	131
Autoconsumption	132
Markets	132
Traders	133

8. RESOURCE EVALUATION — 134

Stock assessment — **137**

Models of potential yield and empirical regression models	137
Models of potential yield	137
Empirical regression models	138
Surplus production models	140
Age-structured (dynamic pool) models	140

Special considerations of multi-species stocks	141
Simulation models	141
Floodplain models	141
ECOPATH models	143

Catch assessment — 143
Recording catches — 143
Catch per unit effort — 144

Analysis of markets — 146

Analysis of consumption — 146

Area-catch studies — 147
Total removal methods — 147
Mark–recapture — 147
Catch depletion — 148
Fish counters and acoustic methods — 149
   Acoustic methods — 149
   Fish counters — 149
   Telemetry — 149

Environmental data — 152
Habitat Quality Indices — 153
Instream Flow Methodology — 154
Physical Habitat Simulation System — 154
Index of Biotic Integrity — 155

9. SOCIAL AND ECONOMIC EVALUATION — 160

Evaluation of the social and institutional context of a fishery — 160
Stakeholder analysis — 160
Gender awareness — 162
Participatory Rural Appraisal — 163
Participatory Monitoring and Evaluation — 164
Conflict management — 165
Traditional methods for quantitative data collection — 166
   Household surveys — 166
   Market surveys — 167
   Analysis of consumption — 168

Sampling — 169

xii Contents

10. INTEGRATING INFORMATION — 171

Resource mapping — 171

Geographical information systems — 171
- Practical considerations for the implementation of GIS — 171
- Overview of recent GIS applications in inland fisheries — 174
  - Approach — 174
  - Characterisation of recent GIS applications — 179
- Selected applications of GIS in inland fisheries — 180
  - Fishery resources — 181
  - Habitat approaches to biodiversity — 183
  - Environment — 189
- Multi-sectoral planning and management including fisheries — 191

Summary and conclusions — 192
- Sources of information on applications — 192
- Geographical distribution of applications — 192
- Present and future directions of applications — 193
- Integration of applications — 193

11. FISHERY MANAGEMENT — 195

Social and policy consideration — 195
- Identifying needs — 196
- Encouraging change — 197
- Responsibility: the possible actors — 198
  - The need for co-management — 199
- Developing a co-management structure — 200
  - Responsibilities, rights and relationships for co-management of inland fisheries — 200
  - Devolving responsibilities to communities and user groups — 200
  - Building local skills — 203
  - Defining an overarching management structure — 204

Strategies for regulation of fisheries — 207
- Goals for regulation — 207

Measures for regulation — 211
- Technical measures — 212
  - Mesh limitations — 212
  - Gear limitations — 212
  - Closed seasons — 213
  - Closed areas — 213
- Input controls: control of access and effort — 213
  - Licences — 214

State-regulated access	214
Ownership	214
Output controls	215
Quotas	215
Size limits on fish landed	215

12. ENVIRONMENTAL MANAGEMENT — 217

Other users of the inland water resource — 217

Fisheries in the context of multi-purpose use	217
Power generation	219
Flood control	219
Navigation	219
Domestic use	220
Agriculture	220
Forestry	220
Animal husbandry	221
Industry	221
Mining	221
Wildlife conservation	222

Impacts of other users — 222

Water quality	222
Eutrophication	223
Pollution	223
Sediment	224
Acidification	224
Interruptions to connectivity	225
Dam building	225
Levees	226
Channelisation	226
Water quantity and timing	226
Changes in discharge	226
Drawdown	227

Remedial measures — 227

Basin planning — 233

Valuation of the living aquatic resource	233
Definition and costing of methods	234
Mechanisms	234
Education and information	235

13. ENHANCEMENT — 236

Techniques for enhancement — 236
- Introductions — 237
  - Motives for introductions — 237
  - Risks from introductions — 238
  - Measures to reduce risk — 239
- Stocking — 241
  - Objectives of stocking — 241
  - Stocking versus natural reproduction — 241
  - Patterns of stocking — 242
  - Population dynamics of stocking — 245
  - Assessing management regimes in practice — 247
  - Development of stocked fisheries and decision making — 249
  - Risks from stocking — 252
- Fertilisation — 253
- Elimination of unwanted species — 253
- Constructing faunas of selected species — 254
- Engineering the environment — 255
- Management as intensive fishpond — 256
- Cage culture — 256
- Genetic modification — 256

Summary of enhancement strategies — 258

Cost effectiveness — 258
- Economics of individual fisheries — 258
- Economics in the wider context — 260

14. MITIGATION AND REHABILITATION — 262

Objectives of rehabilitation — 263

Habitat requirements of fish — 266

Protection of water quality — 266

Control and use of vegetation — 267
- Vegetation within the drainage basin — 267
- Riparian vegetation — 267
- Aquatic vegetation — 268
  - Emergent — 268
  - Submersed and submerged — 268
  - Floating — 268
- Control of riparian and aquatic vegetation — 269
  - Mechanical — 269

Chemical	269
Biological	269

Rehabilitation of lakes and reservoirs — 269
    Control of nutrient balance — 270
        Eutrophy — 270
        Oligotrophy — 272
    Acidification — 273
    Physical modification — 273
        Shoreline development — 273
        Siltation — 273
    Water-level control — 274
    Spatial usage within the lake — 274

Rehabilitation of rivers — 274
    Rehabilitation of channels — 275
        Small-scale interventions — 276
        Larger scale interventions — 277
    Rehabilitation of floodplains — 279
        Setting back levees — 280
        Reconnection of relic channels and floodplain water bodies — 280
        Creation of new floodplain features — 283
        Submersible dams — 285
    Protection of fish movement — 286
        Fish passes — 286
        Fish ramps — 287
        Bypass canals — 288
        Sluice gates — 290
        Removal of dams — 290
    Water regime management — 291
    Management of aquatic vegetation — 294

15. BIODIVERSITY AND CONSERVATION ISSUES — 297

Importance of biodiversity — 297
    The theory — 297
        Productivity — 297
        Stability: resistance and resilience — 299
        Aesthetics and the existence value of biological diversity — 300
    Some measures of biodiversity — 301
        Genetic diversity — 302
        Species diversity — 303

Management for biodiversity — **304**

Indicators of change 305
Management of the gene pool 305
Management of the fishery and the fishers 306
Rehabilitation of the environment 306
Protocols, guidelines and codes of practice 307
Parks and reserves 307
    Conservation reserves 308
    Harvest reserves 308
    Size, location and area 312

16. LEGISLATION   313

International instruments   313
    Agenda 21 313
    Convention on Biological Diversity 314
    FAO Code of Conduct for Responsible Fisheries 316
    UNESCO Convention 317
    Ramsar Convention 317

National legislation   318
    Defining inland fisheries and their placement within the legislative framework ... 318
    Creating institutional linkages between inland fisheries and water management: the case of the United Kingdom ... 319
    The problem of inland fisheries in a federal system: the case of Sudan ... 320
    Creating a legal framework for co-management of inland fisheries: the case of Zambia ... 322

Regional legislation   324

17. CONCLUSION   326

Changing patterns for the resource   326

Monitoring   328

Needs for participatory management   329

References   332

Index   353

Guidelines

This guideline is in the form of a brief checklist of the major questions that need to be answered in formulating fisheries management policy. The list also includes procedures for gaining necessary information by referring to appropriate sections of the following text.

Guideline 1 Framework for management of inland fisheries

	Comment		Go to	Page No.

Decisions at National Level

1	Is the overall place of inland fisheries in national economies understood?	Yes No	2 1.1	8
1.1	Carry out surveys to determine the importance of the inland fisheries sector		2	134 160
2	Formulate policy for the inland fisheries sector		3	195
3	Legislate to reflect overall policy regarding the sector		4	313

Decisions at the Level of the Fishery

4	Are the local objectives of the fishery clearly formulated?	Yes No	5 4.1	8
4.1	Involve stakeholders in initial discussions about local problems and objectives		5	195
5	Establish policy and allocate fishery to clearly assigned user groups		6	211
6	Is the fishery is seriously constrained by external environmental impacts?	Yes No	6.1 7	
6.1	Identify, quantify and cost impacts from other users on the fishery		6.2	217
6.2	Negotiate with other users to establish which long-term management scenario applies: • greater damage and modification to the system • further intensification of use but with less risk • pressures on the aquatic environment will ease • accept the damage and adapt management practice to its constraints		6.2.1 6.2.2 6.2.3 7	*(continued)*

Guideline 1 (*Continued*)

	Comment		Go to	Page No.
6.2.1	Formulate policies to minimise the damage to the fishery		7	227
6.2.2	Negotiate measures to protect the fishery from further damage		7	233
6.2.3	Consider measure to rehabilitate the environment and restore the fish stocks		7	262
7	Is the potential of the water-body known?	Yes	9	
		No	8	
8	Assess the fishery potential of the water-body and the local community		9	134
9	Is the actual state of the fishery known?	Yes	11	
		No	10	
10	Assess stocks and evaluate the fishery		10.1 10.2	134
10.1	In fisheries based on one species and one type of gear, apply standard surplus yield models to the fishery		11	137
10.2	In multispecies, multigear fishery adopt an assemblage-based approach to management		11	141
11	Is the fishery: (A) Underexploited? (B) Equitably exploited? (C) Overexploited?		12 13 14	
12	Determine and resolve the constraints on expansion of the fishery through social and economic studies		23	160
13	Adopt a conservative approach to management		23	297
14	Can you carry out projects to enhance production?	Yes	19	
		No	15	
15	Identify the main causes of overfishing. Is this due to: (A) Demography or investment overcapacity? (B) Policy? (C) Bad fishing practice?		16 17 18	
16	Control fishing capacity		18	211
17	Rectify inadequate or inappropriate policy decisions		18	
18	Set control measures for the fishery		23	211
19	Is production limited by: (A) An unfilled niche or externally induced environmental change? (B) Recruitment overfishing or lack of breeding success? (C) Low fertility of the water? (D) Excessive diversion of productivity into unwanted species? (E) Limitations in the form of the water-body?		19.1 19.2 19.3 19.4 19.5	
			(*continued*)	

Guideline 1 (*Continued*)

	Comment		Go to	Page No.
19.1	Investigate the possibility of introduction of new species		20	237
19.2	Investigate the possibility of stocking		20	241
19.3	Investigate the possibility supplemental fertilisation		20	253
19.4	Investigate the possibility elimination of predatory or pest species		20	253
19.5	Investigate possibility of rehabilitation or engineering solution		20	262
20	Negotiate and establish legal frameworks for protection of the practitioner		21	318
21	Encourage establishment of financing institutions and Infrastructure		22	
22	Establish extension and education programmes		23	326
23	Establish a regular programme to monitor the fishery and provide feedback on the outcomes of management		24	326
24	Modify policy and regulations to improve the performance of the fishery			

Chapter 1
Introduction

Fishing in inland waters is among the most ancient of human practices and fishing tools have been found among the earliest human remains. Fish from inland waters has continued to provide a major source of animal protein to the present day, particularly in areas far removed from the sea. The current high demand for fish and increased awareness of the role of the environment in supporting human well-being have led to a situation whereby attitudes to inland water resources are changing rapidly. This change is part of a larger preoccupation for the long-term stability of ecosystems as well as a concern for the capacity of existing freshwater resources to meet human needs. Humanity is moving away from old exploitative models of development towards priorities for agriculture, forestry and fisheries that aim to secure the sustainability of the world's resources in the interest of long-term food security. Policies for conservation and sustainability are being pursued against increasing pressures on natural resources in general. Nowhere are these trends in resource use and environmental impact more evident than in inland waters, which are particularly vulnerable as they act as collectors of all the activities occurring in their basins and rank as some of the most endangered ecosystems in the world. The principal changes influencing the evolution of the aquatic resource for fisheries are described in this book.

Inland fisheries and aquaculture are strongly conditioned by the social and geographical context in which they are being pursued and respond to the demographic, economic and political evolution of the world at the present time. Several factors may be singled out from many as particularly important as conditioning policies and strategies for management and in determining the patterns in resource use in the future.

Resource availability

Human population

The most important factor affecting natural resources on a global scale is human population growth. The United Nations (UN, 1993) proposes three scenarios for human population growth into the twenty-first century. Even the lowest of these predicts a population of around 8 billion people by 2040 and the medium projection predicts 10 billion people by the middle of the century. The growing population places increasing demands on natural resources, not only as a life-support system through the services that a healthy ecosystem can provide but also as a source of food and recreation.

The impact of the growing population is also influenced by changes in human distribution. The major trend of the past decades has been towards urbanisation and demographic shifts to live along lakes, rivers and the marine fringes of the continents. These shifts result in a move away from subsistence agriculture at the rural level with a growing demand for transfer of food through commerce to urban markets. Urban markets tend to be more varied than rural ones and to require a corresponding diversification of species and products. Urbanisation and shifts to riparian areas imply greater environmental pressure on the resource as the amounts of waste grow, and other population-induced effects concentrate more around the available water.

Water supply

Water is now becoming one of the primary limiting factors to the growth of societies as supplies struggle to keep up with steadily increasing demand. By the middle of the twenty-first century the medium population growth prediction indicates that nearly half of the world's population will live in the 58 countries experiencing severely restricted access to water. In other parts of the world per caput water supply will be substantially decreased. Against this background of increasing pressure on a fundamental resource it is clear that the competing users of that resource, agriculture, domestic consumption, industry, transport, power generation, recreation and fisheries, will have to become more efficient and focused in the way in which they use water. Furthermore, the need to conserve water quality for human consumption means that intensive fisheries and aquaculture practices either will have to produce less pollution and eutrophicating nutrients or will have to be associated with downstream water-treatment systems.

Feed and other resources

Much of the production from aquaculture and enhanced inland fisheries is based on low-input systems relying on low-protein, agricultural byproducts and pond fertilisers. Agriculture, animal husbandry and poultry rearers also use the same resources. As the general demand for food rises the intensification of agricultural activities in parallel with aquaculture will lead to competition. Fewer resources may be available to the fisheries sector. About 1.26 million tonnes of finfish and 0.8 million tonnes of shrimp are totally dependent on fishmeal. This source of fish feed is finite and its limited supply means that the culture of all the carnivorous species that depend on it can only occur if a substitute high-protein feed is found. In the mean time, if efforts to market the small pelagics that are used for fishmeal directly for human consumption are successful, the availability of high-protein feeds may decline. Both these trends, together with the rapid increases in other farm operating costs, point to the need for improvements in the efficiency of aquaculture and intensively managed systems in general.

Land availability

As populations rise the proportion of land needed to produce food will increase, although much of the land already under culture may deteriorate through bad practices. There will, therefore, be increased pressure on land for all uses including agriculture, aquaculture and human occupancy. Extra land will be obtained by the draining of swamps and lakes and the isolation of floodplains, thereby diminishing the area of water available for capture and culture fisheries. This implies competition for available land, which means that fisheries and aquaculture practices must use the available areas more efficiently. It also implies limits to the spread of traditional freshwater aquaculture systems, which will eventually direct future expansion to the marine ecosystem, or to highly efficient water recycling systems. The search for efficiency is reinforced by the current conservation-oriented attitude which places a premium on maintaining some ecosystems intact and by limiting practices which negatively affect the gene pool such as introductions and stocking.

Political and economic climate

National trends

Recent years have seen a shift in political and economic orientation from socialist models, involving central planning, open access to natural resources and subsidies to the fishing, forestry and agriculture sectors, to capitalist or open-market ones which imply private ownership and financial self-sufficiency of capture and culture sectors. This transition involves a breakdown in strong central government in favour of more devolved regional models. At the same time there has been a growth in concern about the future of natural resources, which encourages countries, groups and individuals to develop more sustainable approaches to management that are sometimes inconsistent with the short-term, market approach.

International initiatives

Concerns were being expressed world-wide as to the sustainability of all natural resources. In response the United Nations convened a Conference on Environment and Development in 1992 whose Chapter 21 set out a number of principles for sustainable development. At the same time a Convention on Biological Diversity was being adopted by most of the world's nations which legally bound its signatories to protect and conserve the diversity of living organisms around the world. These initiatives have been fundamental in changing the old exploitative vision towards a more conservationist one.

In the field of fisheries the adoption in 1982 of the United Nations Convention on the Law of the Sea provided a new framework for the better management of marine resources. The new legal regime of the oceans gave coastal states rights and responsibilities for the management and use of fishery resources within their Exclusive Economic Zones (EEZs), which embrace some 90% of the world's marine fisheries.

Nevertheless, overexploitation of important fish stocks, modifications of ecosystems, significant economic losses, and international conflicts on management and fish trade continued to threaten the long-term sustainability of fisheries and the contribution of fisheries to food supply. A Code of Conduct, specifically addressing the issue of the responsible fisheries, was prepared by the Food and Agriculture Organisation (FAO) as a means of containing these threats.

Unlike the Convention on Biological Diversity, the Code is voluntary. However, certain parts of it are based on relevant rules of international law, as reflected in the United Nations Convention on the Law of the Sea of 10 December 1982. The Code also contains provisions that may be or have already been given binding effect by means of other obligatory legal instruments amongst the Parties, such as the Agreement to Promote Compliance with Conservation and Management Measures by Fishing Vessels on the High Seas, 1993. Although the Code was primarily elaborated to meet the needs of marine capture fisheries and in particular industrial fisheries, its provisions are applicable in principle to the world's inland waters and to aquaculture.

Additional initiatives are forthcoming. Ramsar (Convention on Wetlands of International Importance especially as Waterfowl Habitat) has now inserted fish into the list of organisms justifying the dedication of a wetland conservation area. The World Conservation Union (IUCN) has, for a long time, run a freshwater conservation programme encouraging the participation of independent universities from all over the world.

The changing situation of fisheries

A brief history of inland fisheries management

Attempts at managing inland fisheries are very ancient. The Capitoline museum in Rome houses a bas-relief of a sturgeon that is reputedly the standard for the minimum size of fish landed in the port of the ancient city. In the middle ages in Europe inland fishermen were grouped into guilds which were charged with the exploitation and management of the resource. In seventeenth century France, Colbert, Finance Minister to Louis IV, regulated inland fisheries by controlling landings. In other parts of the world inland fisheries were regulated by local tradition often reinforced by religious sanctions, many of which persist today. Fishing pressure at these times may be supposed to have been less than that today but the very existence of regulations of this type shows that there was an awareness of the limited nature of the inland fishery resource and the need to control the fishery.

Modern governments have attempted to impose centralised regulations to control inland fisheries at national level through limitations on mesh and fish sizes, season or gear type. In general, however, inland fisheries have been considered open-access, commons type resources and thus tended to attract the more disadvantaged in the communities. However, centralised systems have proved impractical for several reasons. Firstly, the inconsistencies inherent in trying to impose blanket legislation over very diverse resources and large geographical areas; secondly, the immense difficulties experienced in policing and enforcing the regulations; and thirdly, the overcapacity that has been generated by the open-access nature of the resource. These deficiencies have caused a crisis in management of inland waters that parallels that of the marine sector and, indeed, many other natural resources.

Old models of fisheries management were based on assumptions as to the apparent inexhaustibility of living aquatic resources and the low environmental impact of fishing. It quickly became apparent that these assumptions were flawed as increasing numbers of marine and inland fisheries were classified as overexploited. This culminated in the estimation by Garcia & Newton (1997) that 75% of all marine stocks were either fully or excessively fished. No comparable estimate was made for inland waters but observation indicates that most rivers, reservoirs and lakes are fished to levels beyond their optimum (FAO, 1999a). Living aquatic resources, although renewable, are not infinite and need to be properly managed if their contribution to the nutritional, economic and social well-being of the world's growing population is to be sustained.

An alternative system for management is to seek systems for participatory management involving power sharing between the government and the fishing communities. In this way benefits can be drawn from modern scientific approaches as well as traditional management systems. These involve much greater flexibility in the management of individual resources as well as provisions for limitation of access or even of redefining ownership.

Conservation as opposed to exploitation

One of the major issues facing the world today is how to balance the increasing demand for food against the need to conserve natural resources for the future. The debate on this topic has already produced shifts in the way in which natural resources are perceived by society and has provoked international and national initiatives for the protection of such resources. This perception has moved through society at an uneven rate. In most rich, temperate-zone countries strongly conservationist attitudes have emerged which have tended to be against the exploitative use of resources such as inland fisheries. In the more food-hungry countries of the tropics pressures are to maintain high levels of off-take that are still inconsistent with sustainability. Nevertheless, even here there is a realisation that there is a need to protect resources from overharvesting and conservationist policies are being adopted to an increasing degree.

Improved tools for management

Experience with inland fisheries in the tropics and the temperate zone over the past few decades has resulted in the development of a range of new tools for the evaluation and management of inland water environments and their fish stocks. It has become apparent that traditional methods developed for individual fish species were of limited value and confined in their use mainly to certain large lakes where the fisheries depended on only one or two species. In most other environments fish stocks consisted of numerous species and were not amenable to the methods of population dynamics applied to marine fisheries. A more holistic ecological approach has been developed which is more appropriate to the types of fish community found in most rivers and lakes.

At the same time it has been realised that considerations of the biology and dynamics of the stock alone were not adequate fully to describe the fishery. In recent years there has been a greater concentration on the social, economic and policy aspects of fisheries management. These studies, together with the general failure of centralised management, have led to the exploration of new techniques for co-management involving the devolution of many of the previously centralised powers to local government or even the fishing communities. The new attitudes to management often involve transfers of ownership and responsibility from central government or large landowners to the fishermen themselves.

A range of techniques has emerged to assist the manager and the fishermen better to manage their fishery by giving them choices for action depending on their priorities and those of the societies in which they operate. Thus, tools exist to improve the productivity of the fishery, such as the enhancement of natural fish stocks, or they may be more conservation oriented, such as for the rehabilitation of damaged environments. At the same time better understanding has been acquired of the consequences of human activities on the environment within the river or lake basin and on the distribution of benefits between different members of the community. More recently, too, the implications for the genetic health of fish stocks, particularly with respect to the introduction of species and the impacts of fishing and aquaculture, have become apparent. Such tools are not necessarily new but sit within a new more integrated approach to the management of the environment and the fishery.

A new vision for management

Present trends in the development of natural resources emphasise the need for approaches that improve the quality of aquatic ecosystems and the sustainability of the fisheries that operate within them. The shift in political and economic orientation from central planning, open access to natural resources and subsidies to the fishery to open markets, private ownership and financial self-sufficiency of capture and culture sectors involves a redefinition of the role of the various levels within a national hierarchy. Whereas, previously, paternalistic central governments would retain most of the scientific expertise and the responsibility to make laws and enforce them, modern practice suggests that many of these responsibilities will now fall upon the local governments and fisher communities. This, in turn, implies that local officials and fishermen have to operate at a higher level of understanding of the resource and build institutions to negotiate together and with outside interests. New systems are therefore necessary to enable these stakeholders to discharge their newly acquired obligations. To a certain extent such systems already exist but further research is required on how to develop local management capacities and how to resolve conflicts between these competing interests.

This manual examines the issues underlying present-day inland fisheries management. Management is defined in the *Oxford English Dictionary* as the process of controlling or administering and here the process is applied to a natural resource, inland fisheries. Fisheries management is defined by Lackey (1979) as 'the practice of analysing, making and implementing decisions to maintain or alter the structure, dynamics, and interaction of habitat, aquatic biota, and man to achieve human goals and objectives through the aquatic resources'. The thrust of the manual is therefore aimed at human goals and is anthropocentric rather than adopting the more traditional biocentric focus that considers ecological integrity and preservation as its goal.

Chapter 2
Objectives

Objectives for the fishery

Three major principles have emerged from the UNCED process, The Convention on Biological Diversity and the FAO Code of Conduct for Responsible Fisheries as governing modern fishery management:

- Conservation of the diversity of living aquatic resources
- Sustainability of the fishery
- Equitable distribution of the benefits of the fishery and of aquatic ecosystems

Within these overall principles inland waters are used for a number of fishery-related purposes. These include extractive outputs such as food, recreation, provision of ornamental or aquarium species and, in some parts of the world, medicinal products. In addition, aquatic organisms can be managed for non-extractive purposes such as control of disease vectors, control of water quality, and aesthetic or moral intangibles.

Extractive objectives

Food fisheries

Fishing for food is still the dominant use for inland waters over much of the world. Catches of fish for food are reported to FAO yearly by countries throughout the world (FAO, 1999a). Historically, catches from inland fisheries have not been reported separately but have been included with aquaculture in a global figure for inland waters. Any estimate has had to be reached on the basis of subtraction of the aquaculture production from the total. This method has had some limitations.

Firstly, the distinction between aquaculture and inland capture fisheries is becoming increasingly blurred as enhancement techniques become more widespread and there is no agreed definition to distinguish between the two. As a result, countries report their production in different ways.

Secondly, distinction between species is poor and the 'general finfish category' is generally too large. This means that catches are poorly defined and trends in catch are not readily discernible.

Table 2.1 World catches from inland waters by continent for 1996 (estimated from FAO statistics)

Continent	Catch (t)	% of total
Asia	4 681 788	62.0
Africa	1 819 887	24.1
South America	326 250	4.3
Countries of the former USSR	314 124	4.2
North America	209 845	2.8
Europe	183 384	2.4
Oceania	19 485	0.3
Total	7 554 763	100.0

Thirdly, inland fisheries statistics are normally collected at relatively few major fish landings. Diffuse and low-level sources of fish are often not reported. Thus the statistics may exclude large portions of the subsistence and artisanal fisheries, production from agriculture-related sources such as rice fields and the recreational fishery.

This situation is being rectified with the separate reporting of inland catch statistics by member nations of FAO.

Total reported yields from inland waters world-wide, including brackish water environments, were 22.2 million tonnes in 1996. Of that production, 13.2 million tonnes came from freshwater aquaculture, 1.5 million tonnes from brackish waters (mainly shrimp cultured in coastal ponds) and 7.5 million tonnes were from capture fisheries.

Most of this production comes from Asia and particularly from China (see Table 2.1). Catches from Asia increased considerably during the 1990s and have been of such magnitude as to conceal trends in almost all other continents. Catches from Africa, Latin America and North and Central America have remained stable over the past few years. Those in Europe and the countries of the former Soviet Union declined through the early 1990s, mainly because of the transition from centrally planned to free-market economies. Catches are now rising again in many areas of Eastern Europe, although environmental degradation continues to be a problem in Central Asia and Russia.

The species composition world-wide (Table 2.2) is heavily biased towards cyprinids and unspecified fishes. Thirteen taxa (groupings at family level or above) make up 96% of the cumulated catch and the remaining 4% are distributed among 35 other taxa.

Catches throughout most of the world have shown a drift towards smaller species, indicating that the fish populations are being exploited at levels which are possibly not sustainable in the long term. The widespread decline in quantity and quality of catch can be traced variously to environmental degradation arising from the pressures on water as a resource, excessive levels of exploitation and, in some cases, use of fish-

Table 2.2 Composition of world catches for 1996 (estimated from FAO statistics)

Taxon	Catch (t)	% of total
General (not specified)	397 678	45.0
Cyprinidae	830 411	11.0
Mollusca	534 515	7.1
Crustacea	529 421	7.0
Cichlidae	515 385	6.8
Clupeidae	415 431	5.5
Centropomidae	342 467	4.5
Characidae	179 411	2.4
Ophicephalidae	166 044	2.2
Siluroidea	114 210	1.5
Clariidae	89 132	1.2
Salmonidae	72 688	1.0
Percidae	51 690	0.7
35 other taxa	314 752	4.2

eries for recreation rather than food production. Rises in yield in the face of the general decline in environmental quality can usually be traced to increasing use of enhancement techniques to improve productivity.

Recreational fisheries

Recreational fisheries have long represented the major use of living aquatic resources in temperate countries. However, recreational fishing is now becoming common in some tropical countries. Brazil, for instance, has witnessed an explosive expansion in interest in recreational fishing to satisfy the needs of the growing urban populations.

Recreational fisheries have high social and amenity value and are often the main participant sport of a country. They are far more ecologically friendly than capture fisheries for food as the catch is often returned unharmed to the water. This catch-and-return policy is now the norm for many European and North American countries. Nevertheless, fishers and their families may eat catches from recreational fisheries, and there is evidence that in some poorer areas such catches are subsistence in nature. Cumulatively, catches from recreational fisheries may represent a considerable tonnage that only rarely appears in the national catch statistics.

Recreational fisheries are also far more valuable in financial terms than food fisheries. It has been estimated that the value per kilogram of fish caught by the recreational fishery is some ten times greater than the same fish sold purely as food. Such costings include all expenses incurred by the fisherman, such as travel, accommodation and gear.

Fisheries for ornamental species

Large numbers of small, brightly coloured fishes are caught mainly from tropical waters for export as part of the aquarium fish trade. In many cases the more common species are cultured, but capture from the wild remains the major source of supply. Several thousand species are caught from inland and marine environments. Although large numbers of some species may be removed locally, there is little evidence that there is any great risk to any one species by extraction alone. However, in the marine environment and in some inland situations capture by poisons may pose a risk to fish in surrounding areas.

Fisheries for ornamental species are frequently the only source of income in isolated and impoverished areas where few alternative sources of income exist. Governments also favour them as a source of foreign currency.

Bait fisheries

In some countries there are specialised fisheries for small species to be used as bait by recreational and, less frequently, commercial fishermen. The major impact of such fisheries is to serve as a vehicle for introductions of small species to other waters where they may prove a nuisance. Many invasions, such as that of *Pseudorasbora parva* throughout Europe, owe their success to the indiscriminate disposal of live bait-fish at the end of a day's fishing.

Fry fisheries

Before the development of techniques for the artificial breeding of culture species, most aquaculture was based on extraction of fish seed from the wild. In Asia large-scale extractions, mainly of cyprinids, persisted over may decades for stocking into ponds. This tradition persists principally for enhancement stocking into reservoirs in India and some parts of China. Some cage culture is also based on wild-caught fry. For example, cage culture of *Pangasius* in South-East Asian rivers depends on fry caught from the Mekong. Recent legislation is seeking to suppress this.

Fisheries for other purposes

Fish are used for a range of other purposes which are not strictly extractive but which benefit communities. For example, some species of fish are used to control the vectors of water-borne diseases, such as schistosomiasis, malaria and onchocerciasis. They may also be used to control insect pests such as the stem borers of rice. Here there may be benefits in allowing fish into the rice paddies not only because of the additional crop that they provide but also because the rice crop might benefit. In some areas fish are used for medicinal purposes or for rituals and rites of passage. Ritual uses of this kind usually involve eating the fish or some portion of it but here the purpose of consumption is not nutrition.

In addition to these extractive objectives, the benefits of which are immediately apparent as food, money or some health or ceremonial output, there is a range of other objectives, examined below, the value of which is not always immediately apparent.

Sustainability objectives

Sustainability

The overriding objective for natural resource management has become sustainability. Sustainability is defined by FAO (1995a) as:

> the management and conservation of the natural resource base and the orientation of technological and institutional change in such a manner as to ensure the attainment and continued satisfaction of human needs for present and future generations. Such sustainable development (in the agricultural, forestry and fisheries sectors) conserves land, water, plant and animal genetic resources, is environmentally non-degrading, technically appropriate, economically viable and socially acceptable.

Policies for fisheries development and management should therefore aim at extracting only that amount of fish from the aquatic system that is consistent with the continuity of supplies at similar levels into the future.

Ecological objectives

Several objectives are related to the more general one of sustainability. These include biodiversity and conservation. Signatories to the Convention on Biological Diversity are obliged to preserve their biological diversity to the maximum possible extent. As inland aquatic systems and their living components are among the most endangered world-wide, special efforts are needed to secure such protection. Conservation of individual aquatic species, species assemblages and ecosystems therefore become objectives for management which call for limitations on exploitation and prohibition of damaging fishing practices.

Conservation objectives also have strong implications outside the fishery. The gravest threats to the sustainability of the aquatic ecosystem do not arise from within the fishery but from outside it. Conservation therefore becomes a question of landscape or basin management where all users of the water contribute to policies for sustainability of the aquatic ecosystem as a whole.

Social objectives

A range of social objectives reflects government and public expectations of the fishery, as detailed below.

Income to fishers

As well as being a source of food, fisheries provide an important source of income to a diverse sector. Fishermen and other associated trades such as fish merchants will always seek to maximise their benefit. In capture fisheries benefits are usually viewed in the short term and there is, therefore, a strong tendency to fish with methods or at intensities that are damaging to the stock. Managers should demonstrate to fishermen that improved and more secure incomes depend on more sustainable levels of fishing effort and the renunciation of damaging fishing practices.

Equity/benefit distribution

This objective groups a number of goals such as employment and poverty reduction. In many parts of the world inland fisheries have been treated as a sort of occupation of the last resort whereby governments can deflect the landless poor towards an alternative activity and thus avoid them migrating to urban situations. In some cases this objective remains valid, but it does tend to add to fishing capacity at a time when other priorities would call for a lessening of fishing pressure and a restriction of access.

The employment created by a fishery is not confined to the fishermen. An active fishery creates a number of ancillary jobs in the form of boat builders, fishing gear manufacturers and suppliers, traders, etc.

Managers are also called to address imbalances to established fisheries that are caused by management decisions. Current trends to restrict access to the fishery in the interest of controlling effort will clearly leave some elements of the fishing community dispossessed of previous rights. Similarly, decisions may favour certain types of gear and their users over others. In many cases it is the poorer elements of the fishing community that suffer from such actions. Efforts should be made to keep such disruptions to a minimum.

Conflict reduction

Inland aquatic systems are subject to a wide range of fishery and non-fishery interests. Inevitably conflicts will arise between groups of fishermen, such as users of different types of gear, or recreational and commercial fishermen, or between fishermen and some interest outside the fishery. One objective for managers will be to minimise potential conflicts by clearly assigning access and user rights to water and to fish stocks among the various interested parties.

Government objectives

Revenue to government

Governments at all levels have traditionally viewed fisheries as a source of revenue.

Revenue is usually obtained either through taxes on fishermen's income, through taxes on their product at market or through licence fees on boats, gear or fishing places. For example, in some parts of South Sumatra, Indonesia, over 50% of local government revenues are raised by the annual auction of fishing rights in floodplain water-bodies.

Contribution to GDP

At a slightly more abstract level the products and benefits of fisheries can be seen as a contribution to the gross domestic product (GDP). In some cases, such as major commercial fisheries or recreational fishing, a whole commercial sector is created around the fishery to service the fishery directly and to supply those involved in the fishery with the necessities of life. In this way the economic development of whole areas can be built up around a successful fishery. Sustainable management then becomes essential to avoid major economic recessions in such areas following a collapse of the stock.

Export income

Many countries view their fisheries as important sources of foreign currency. This can be either through the culture, capture and sale of valuable commodities such as shrimp or high-grade finfish, or through tourism associated with recreational fisheries. Unfortunately, a strong export orientation in capture fisheries or aquaculture risks being detrimental to local communities by depriving them of locally produced food and is a potential source of conflict.

Combinations of objectives

Unfortunately, not all objectives can be achieved at the same time and many of them are mutually exclusive. Managers must attempt to define a panel of objectives that most closely approximates to the goals of society. This will inevitably involve compromise and those in authority must recognise that their own goals for the fishery, such as maintaining biodiversity or raising revenues, may not all be shared by other stakeholders such as fishing communities. Where communities are expected to participate in management and enforcement they should be allowed a dominant voice in the selection of objectives.

Aquaculture

Aquaculture is defined by FAO for statistical purposes as follows:

Aquaculture is the farming of aquatic organisms including fish, molluscs, crustaceans and aquatic plants. Farming implies some sort of intervention in the rearing process to enhance production, such as regular stocking, feeding, protection from predators, etc. Farming also implies individual or corporate ownership of the stock being cultivated. For statistical purposes, aquatic organisms which are harvested by an individual or corporate body which has owned them throughout their rearing period contribute to aqua culture while aquatic organisms which exploited by the public as a common property resource, with or without appropriate licences, are the harvest of fisheries.

Aquaculture is not so much an objective as a tool for the controlled production of fish and other aquatic organisms. Inland aquaculture contributed over 13.4 million tonnes to world fisheries production in 1996. This represents about 75% of total world inland fisheries yields. The demarcation between capture fisheries and aquaculture is becoming increasingly difficult to define as fisheries managers and fishermen assume greater control of the fishery through stocking, fertilisation, environmental manipulation and other practices. Aquaculture systems range from extremely intensive to extensive. Intensive aquaculture is usually based on specific installations, whereas extensive methods may simply be applied as a method for intensifying existing capture fisheries. The differences between capture and culture fisheries and between extensive and intensive aquaculture are important ecologically. Sociologically too, aquaculture and its practitioners have a different status under law from simple capture fishermen, as aquaculture implies ownership of the production facility and the organisms being reared. There are many interactions between aquaculture and capture fisheries, including competition for water, risks of introduction of exotic species and genetic degradation of native fishes through escapes from aquaculture installations, damage to wild stocks from overfishing of young fish for rearing and pollution from pond effluents.

Regional differences in management strategy

Two main strategies for the management of inland waters for fisheries are being adopted based on different societal views of natural resources and their use. Conservation approaches tend to be dominant in the more developed economies of the temperate zone where food surpluses allow for alternative uses of natural resources. In less developed economies, usually in the tropics, food shortages are such that a more production-oriented approach is needed (Table 2.3).

In developed countries management is oriented towards satisfying the recreational fishing community, although in recent years an increasingly vocal protectionist lobby enjoys growing influence. Management has tended to concentrate on mitigating the effects of environmental degradation through stocking coupled with habitat maintenance or rehabilitation. More recently, aquatic resource use is becoming increasingly subordinated to conservation. Commercial food fisheries have largely disappeared and

Table 2.3 Differing strategies for management of inland waters for fisheries in developed and developing countries

	Developed (temperate)	Developing (tropical)
Objectives	Conservation/preservation Recreation	Provision of food Income
Mechanisms	Sport fisheries Habitat restoration Environmentally sound stocking Intensive, discrete, industrialized aquaculture	Food fisheries Habitat modification Enhancement through intensive stocking and management of ecosystem Extensive, integrated, rural aquaculture
Economic	Capital intensive Profit	Labour intensive Production

production facilities are generally isolated in carefully controlled fish farms. Inland waters are destined mainly for aesthetic and recreational uses.

In developing countries the high demand for food is forcing an intensification of production systems. In many cases this has led to overfishing and management has to concentrate on containing effort and access within levels of sustainability. Efforts at increasing production through enhancement are widespread and extensive aquaculture is practised alongside other forms of fishery. Inland waters are still used mainly for food production.

Chapter 3
The nature of inland waters

Types of inland water

Inland waters are distributed throughout the continents and are represented everywhere except in the main desert areas. Several schemes have been devised to classify inland waters, the most basic of which is into standing waters (lakes and swamps) and running waters (rivers). Some systems (reservoirs and coastal lagoons) are intermediate between the two. The nature of the fishery and the way in which fish respond to fishing depend greatly on the type of water body, its size and morphology, as well as its accessibility.

Lakes

Lakes are bodies of water enclosed by land. They can be regarded as relatively closed systems as most of their hydrology is internal, although they may have substantial inflowing and outflowing rivers. They range in size from the Caspian Sea, which covers 371 795 km^2, to small ponds of a few hectares. Their depth ranges from 1742 m (Lake Baikal) to less than 1 m.

Lakes may be classified according to their origin (Hutchinson, 1975), with principal categories being as follows.

- Glacial lakes are found throughout the cold temperate zones where ice sheets during the last glaciation scoured large shallow basins out of the underlying rock. Glaciation also produced long lakes at the lower end of valleys in all of the major temperate mountain ranges. Such lakes frequently appear grouped in districts such as the alpine lakes or the lakes of southern Chile and Argentina (Fig. 3.1). Large numbers of smaller depression lakes formed by glacial pressure and scour are found throughout northern North America, northern Russia and Scandinavia. Smaller lakes, such as corries or cirque lakes, are also found as scour features high up in mountain ranges (Fig. 3.2).
- Rift valley lakes are found along the great fault lines and include lakes such as Baikal and Balkhash in Russia or the East African Great Lakes. Rift valley lakes are usually deep with little shoreline development (Figs 3.3 and 3.4) although some smaller lakes such as Lake Naivasha may be shallow.

18   *Inland fisheries*

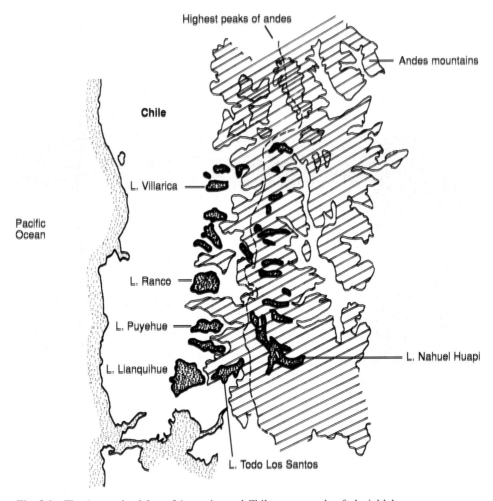

Fig. 3.1   The Araucarian lakes of Argentina and Chile, an example of glacial lakes.

- Depression lakes are often small and shallow, but may also cover great areas, such as Lakes Chad and Victoria in Africa. Depression lakes have various origins. They may persist as a relic of a larger body of water that has been largely filled by siltation as in the case of Lake Chad. They may originate from uplifting of the Earth's crust as in the case of Lake Victoria. They may also be formed in groups in depressions left by ice sheets during the last glaciation, as is the case with most North American and Scandinavian lakes, or in the hollows of wind-driven sand dunes as in the Pampas wetlands of Argentina. Depression lakes often have drainage basins that do not discharge to the sea (endorhoeic) (Fig. 3.5). As a result they tend to become highly saline over time, to such an extent that they do not support fish.
- Volcanic lakes are associated with volcanic activity and form in calderas and craters. They are normally relatively small and isolated.

Fig. 3.2 Twin lakes, Colorado, an example of a high mountain lake.

- River lakes are located on floodplains, where they form integral parts of the river system (see below). Such lakes are usually small and relatively transient features of the landscape as they are subject to constant filling by depositional processes of the river. New lakes arise through river migration and scour. Some river lakes may be large, especially where the river occupies an incompletely silted depression.

Lakes may also be classified by their richness.

- Oligotrophic (nutrient-poor) lakes of glacial origins are usually oligotrophic because the underlying rocks tend to be granitic.
- Mesotrophic (balanced nutrient status) rift valley lakes tend to have a balanced nutrient status, although their depth usually causes them to stratify thermally for much of the year, producing seasonal fluctuations in the abundance of nutrients, phytoplankton and zooplankton.
- Eutrophic (nutrient-enriched) depression lakes are often eutrophicated because they are associated with fertile plains and intensive agriculture. In addition, the surface area relative to depth is such that evaporation causes a rapid build-up of nutrients and salts in their water.
- Dystrophic lakes are formed in bogs and marshes where the water may be poor in nutrients but carry a heavy load of tannic acids from the associated vegetation. Dystrophic lake water is the equivalent of riverine black waters.

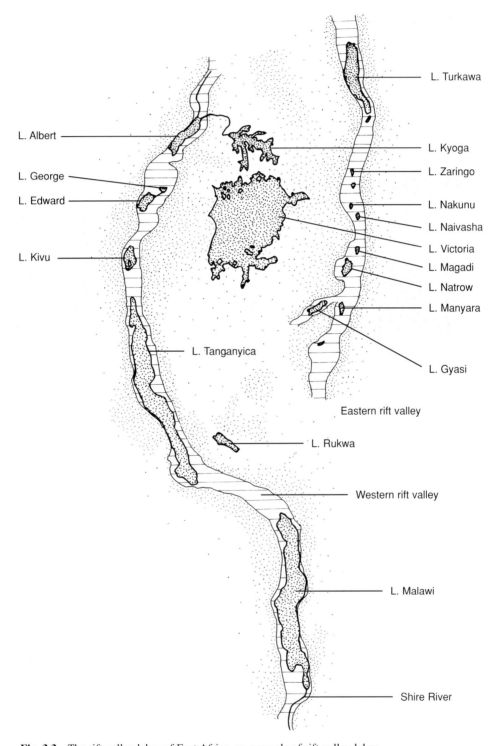

Fig. 3.3 The rift valley lakes of East Africa, an example of rift valley lakes.

The nature of inland waters 21

Fig. 3.4 A fish landing at Lake Chilwa, Malawi – a tropical rift valley lake.

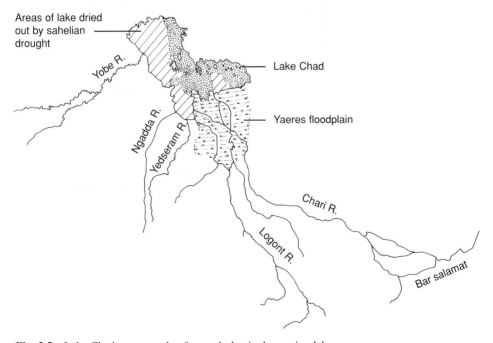

Fig. 3.5 Lake Chad, an example of an endorhoeic depression lake.

Lakes are generally chemically and physically stable environments on a year-to-year basis but may undergo considerable seasonal change within any one year. In shallow lakes the water is mixed thoroughly by wind and wave action, but deep lakes usually stratify in response to temperature differences between the surface and the deeper waters in the summer. In this case the bottom water (hypolimnion) may become deoxygenated to a degree that it cannot support fish. In any eventuality nutrients accrue in this layer during the period of stratification. The stratified condition breaks down as temperatures cool in the autumn and an overturn occurs, mixing the water and producing a pulse of nutrients. Lakes in the cool temperate zone and Arctic usually freeze over in winter, with a risk of reduced oxygen below the ice. Smaller lakes in the tropics may dry out in the dry season. Some degree of seasonality is therefore a feature of lakes at nearly all latitudes.

Trophic richness is not the only indicator of productive potential in lakes. Productivity may also be associated with physical features. For example, deep lakes are usually less productive than shallow ones of the same area, as the impact of lessened light and temperature at greater depths together with the deoxygenated hypolimnion reduces the amount of life that the deeper waters are able to support. Shoreline development is also considered important as deeply indented shorelines increase the amount of riparian vegetation and the number of shallow embayments.

Rivers

Rivers are linear features of the landscape that transfer the water falling on the land to the sea. As such, they are open systems with considerable connectivity (Welcomme, 1985). They have a hierarchical structure from small tributary streams to large rivers. This hierarchy is used as a system of river classification under which rivers are assigned orders. Under the standard classification rivers of the first order have no tributaries. Rivers of the second order are formed at the junction of two first-order streams. Third-order rivers form at the junction of two second order rivers, and so on (Fig. 3.6).

Geomorphological studies show that there is a clear relationship between river basin area and the length of the main river channel. The relationship varies according to continent but typically takes the form:

$$A = 0.362 L^{1.804}$$

as derived for Europe by Petrere *et al.* (1998).

Furthermore, the number of streams in any river basin decreases logarithmically as order increases, and the length of streams of any order increases as order increases (Table 3.1).

The hierarchy of orders is closely linked to gradient as small, low-order, headwater streams tend to have high gradients, whereas larger, high-order rivers are found on low-gradient plains areas.

The nature of inland waters   23

Fig. 3.6 Diagram illustrating river order.

- High-gradient streams tend to be segmented into pools and rapids that recur as a regular sequence along the channel (Fig. 3.7). They also have waterfalls, which segment the stream, and glides, which are featureless stretches of rapid flowing water.
- Braided systems with numerous channels interspersed among islands are characteristic of intermediate gradients (Fig. 3.8).
- Low-gradient, plains rivers usually consist of two components:
  ○ The main channel or channels which meander through the landscape with a sequence of bends at which the inner side forms a shallow point bar and the outer a deep under an undercut bank. Many channels retain water throughout the year, but in some systems even the main channels may sun dry or separate into a series of stagnant pools. The channel is flanked by natural levees where silt is deposited as flow slows during overbank discharge (Figs 3.9 and 3.10).

24 Inland fisheries

Table 3.1 Number and lengths of streams of different orders as exemplified by Africa

Order	Number	Average length (km)	Combined length (km)	Example
1	4 166 000	1.6	6 667 150	
2	870 615	3.7	3 203 865	
3	181 900	8.5	1 540 693	
4	38 005	19.5	741 097	
5	7 940	44.8	356 347	
6	1 659	133.3	171 358	
7	347	237.4	82 492	Moa
8	72	547.1	39 392	Ouémé
9	15	1 259.1	19 013	Volta
10	3	2 897.8	8 691	Niger
11	1	6 669.0	6 669	Nile

After Welcomme (1976a).
The order shown here is that of Horton (1945), whereby the order of the highest order stream is extended back through the longest tributary, as opposed to the more commonly used Strahler (1957) index whereby all streams of the same rank have the same order number.

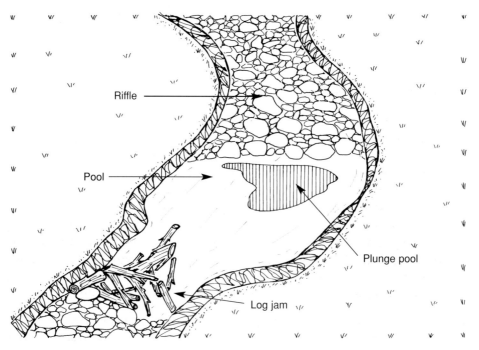

Fig. 3.7 Diagram of a rhithronic reach.

- The floodplain, which is a complex of seasonally flooded land, interspersed with seasonal and permanent lakes and channels (Fig. 3.11).

Two main nutrient flows are found in rivers.

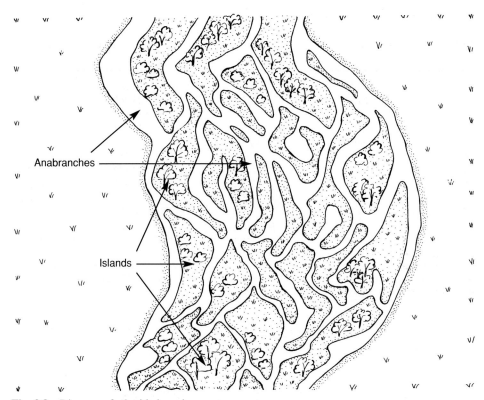

Fig. 3.8 Diagram of a braided reach.

- A longitudinal transfer of coarse organic material and dissolved nutrients washed into the river channel from the surrounding basin. This degrades as it proceeds downstream and is connected with a characteristic succession of invertebrates that feed on the material. The transformation and the organisms responsible for it are described by the River Continuum Concept (Vannote *et al.*, 1980).
- A lateral transfer of water onto the plains that brings with it some nutrient-laden silt. As the plains are flooded there is a release of locally generated material originating from decaying vegetation, dung and other deposits on the floodplain. This seasonal enrichment of the floodwaters favours the growth of vegetation on the plain and is used as the basis for agriculture. The process of nutrient and energy transfer during flooding is described by the Flood Pulse Concept (Junk *et al.*, 1989).

In their natural state most rivers are highly seasonal, alternating between periods of high and low flow. This cycle gives rise to four phases that are important in regulating the ecology of such systems.

- Low water (dry season) when the river is confined to the main channels and permanent floodplain lakes. At this time floodplain water-bodies are often disconnected from the main channel and the channel may break down into a series of disconnected pools.

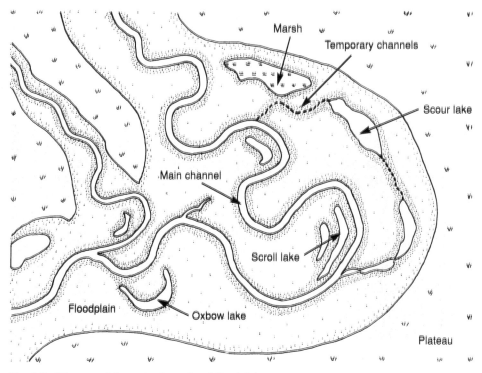

Fig. 3.9 Diagram of the main channel and floodplain.

Fig. 3.10 Main channel of the Parana River at the level of Parana, Argentina, showing the braided nature of the channel.

Fig. 3.11 The Colastine River, a channel of the Parana system showing a lateral lagoon separated from the main channel by a levee.

- Rising water when the water begins to fill the main channels until it reaches a point just before it begins to spill over the levees. Prior to the overtopping of the levee water may already be moving onto the plain through connecting channels.
- High water (floods) when the water overtops the bank and spreads out over the floodplain. At this time the floodplain water-bodies are connected to the channel and to each other, and the low-lying land between them is submerged.
- Falling water when the water retreats from the plain towards the main channels, leading to increasing isolation of the floodplain water-bodies and desiccation of the plain.

The origin and timing of flooding varies in different climatic zones. In the Arctic major floods occur in north-flowing rivers when the spring snow melt releases large volumes of water from the more temperate upstream area. This is then penned back by the continuing ice dams nearer the mouth and spreads laterally to fill extensive marshes. In the temperate zone rivers may flood as a result of spring snowmelt on the mountain chains at the source producing a regular pattern of spring flooding. Floods may also occur in response to heavy rainfall at all times of the year, although this is more likely to occur in the spring and early summer when rainfall and snowmelt combine. In the tropics a regular pattern of unimodal flooding is associated with the monsoon or major rainy season. In the equatorial zone two floods may occur each year in response to the biannual rains.

28 *Inland fisheries*

Fig. 3.12   Flooded rice fields on the Red River system near Hanoi, Vietnam.

Many river systems, particularly in the temperate zones, are now controlled to the point where normal flooding of the lateral plains no longer occurs.

Swamps, marshes and rice fields

Swamps and marshes are often associated with rivers as extensions of their flooded area. Many of the world's greatest wetlands are riverine features either as internal or terminal deltas. In some cases large swamp areas may also be associated with depression lakes or exist as independent features of the landscape. Irrigated agriculture, particularly for rice, increases the area of seasonally flooded land (Fig. 3.12).

Reservoirs

Impoundment of streams and rivers has created a vast number of artificial waterbodies world-wide. These are mostly small, taking the form of dams for agriculture,

The nature of inland waters 29

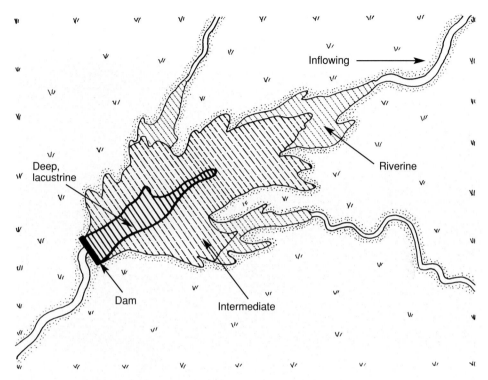

Fig. 3.13 Diagram of a reservoir.

flood control, small-scale power generation and drinking water supply. Larger, sometimes very large, reservoirs are also a feature of most large river systems. Reservoirs combine many of the features of lakes and rivers. Water transit times are generally short so the nutrient cycle is more riverine than lacustrine. Marked seasonality and year-to-year variations in production have been observed in reservoirs as large as Kariba (5364 km^2) as responses to differences in discharge by inflowing rivers (Karenge & Kolding, 1994). At the same time the lower end of the reservoir may be sufficiently deep for stratification to occur. In general, reservoirs are regarded as having a lacustrine part near the dam and a riverine portion at the upstream end (Figs 3.13 and 3.14).

Coastal lagoons

Coastal lagoons take the form of lakes separated from the sea by sandbars and are often associated with estuaries (Figs 3.15 and 3.16). They show great seasonal variation in salinity, being fed from associated freshwater rivers for part of the year and from the sea for the remainder.

30 Inland fisheries

Fig. 3.14 Ubolratana reservoir with the dam wall to the right of the picture.

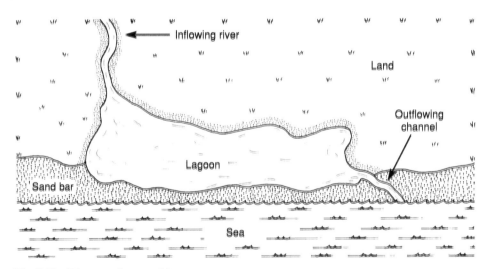

Fig. 3.15 Diagram of a coastal lagoon.

Fig. 3.16 A coastal lagoon: the channel at the mouth of Lake Nokoue, Republic of Benin, showing the large numbers of fish fences for trapping migrating shrimp.

Chapter 4
The nature of fish populations

Fish populations in different types of inland water

The fisheries ecology of inland waters is very different for lakes and reservoirs, rivers and coastal lagoons. The fish populations of each of these environments are described below and representative species are illustrated in Fig. 4.1.

Lakes

Fish in lakes can be divided into two main types:

- Demersal fish that are adapted to the large range of habitats found along the shores and near the bottom. Demersal species are usually numerous and divided into many different communities which are associated with different substrates such as rock reefs, sandy shores and muddy bays. Species in these areas feed on a wide range of foods from mud and detritus, benthic (bottom-living) invertebrates, plants and organisms attached to rocks and other substrates, to other fish. Demersal species show a wide range of breeding habit, often with advanced systems of parental care such as nest building or mouth brooding.
- Pelagic communities are found at the surface and in the water column in the open water. These usually consist of far fewer species but are greater in total biomass. Feeding habits are more restricted being confined to zooplankton and phytoplankton, insects and other fish. Breeding tends to be confined to egg scattering, although some species migrate inshore to spawn over rock or sand bottoms.

In large lakes the pelagic and demersal communities are usually distinct. In smaller lakes the differences in the use of the water column between demersal and pelagic fishes is less marked but some stratification among species still exists.

The number of species forming the populations of lakes is linked to lake area. For example, a sample of over 160 tropical lakes and reservoirs gave a log–log relationship:

$$N = 5.9A^{0.2684}$$

where N is the number of species and A is the lake area (km^2). The low coefficient of correlation of $r^2 = 0.37$ indicates a high degree of variability. This arises because the

The nature of fish populations 33

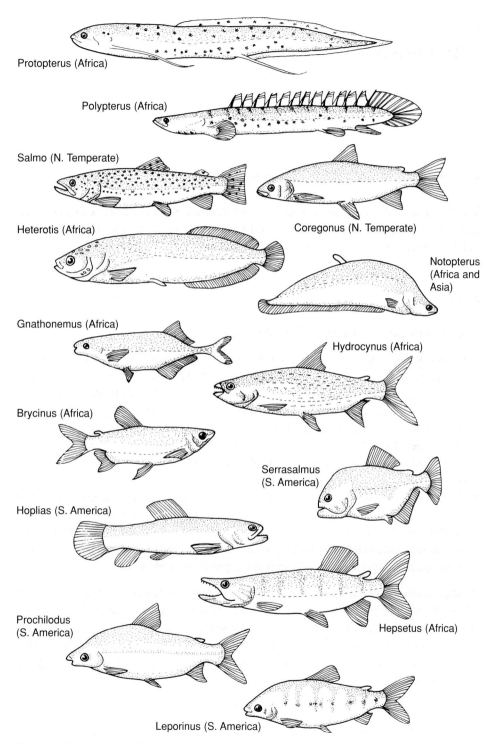

Fig. 4.1 Representative fishes from inland waters.

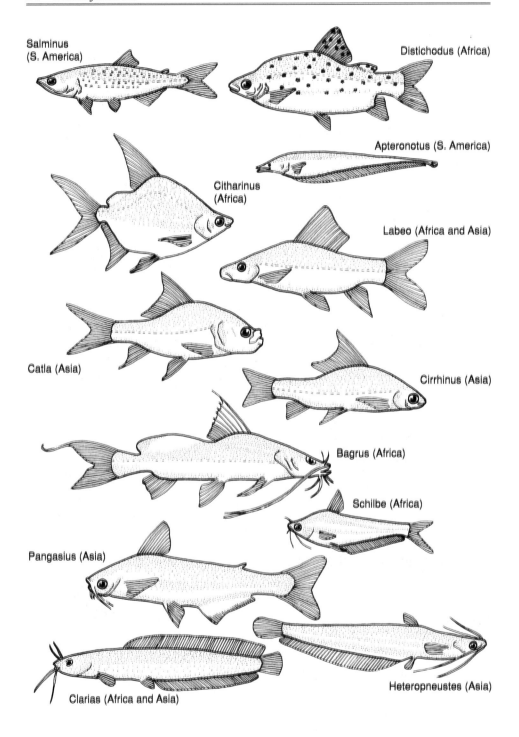

Fig. 4.1 Representative fishes from inland waters (*continued*).

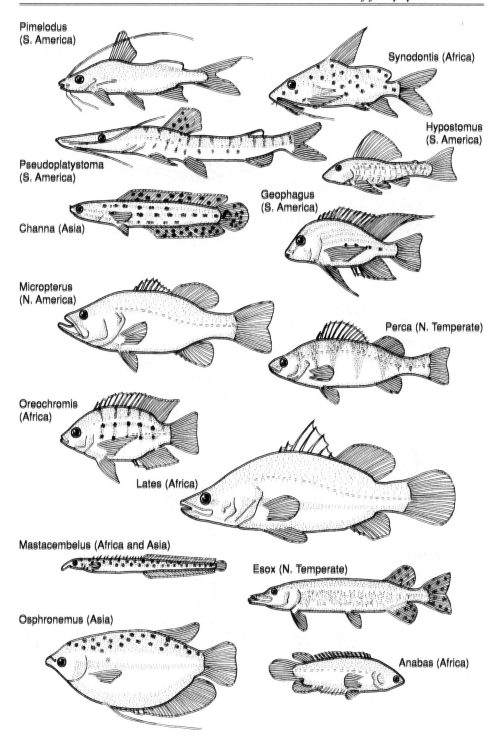

Fig. 4.1 Representative fishes from inland waters (*continued*).

number of species in lakes of any given area is also a function of latitude, altitude, and the trophic and morphological nature of the lake. In sets of lakes from individual areas the correlation is better. For example, in an African series of 29 non-riverine lakes:

$$N = 22.5 + 0.005A$$

with an $r^2 = 0.88$ (figures derived from Vanden Bosch & Bernacsek, 1990).

Rivers

Gradient is one of the major factors influencing the fish fauna in rivers. Rivers may be divided into segments or zones depending on the type of fish fauna present. Zonation is more readily demonstrated in temperate rivers (Fig. 4.2), where successions are established by substitution of species, than in the tropics, where successions tend to be more by accretion.

High-gradient streams are usually segmented into pool and riffle reaches, each of which attracts specific species complexes. Riffles are rich in invertebrate food organisms and their waters are well aerated. They are occupied by flow-loving species that show a variety of adaptations that enable them to survive in strong currents. Such species often have hooks, spines or suckers that enable them to fasten themselves to

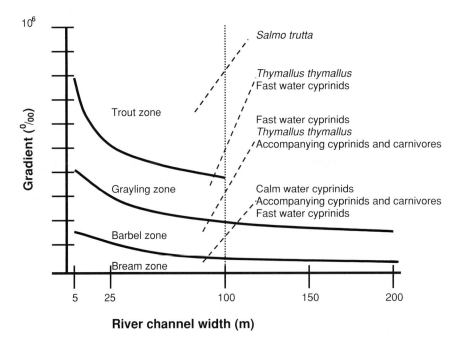

Fig. 4.2 Zonation of fish in European rivers (after Huet, 1949).

rocks and vegetation. Small species or juveniles of larger ones tend to live in the interstices of rock and gravel bottoms. The quieter waters of the pools are inhabited by less energetic swimmers that can take refuge in areas of slack flow, as well as by larger fish which feed on the drift of organisms dislodged from the riffles. Some large, migratory species move upstream to the headwater tributaries from the lower reaches of the river to spawn and to place their young on the food-rich riffles. Fish in this zone may bury their eggs in pockets in the gravel of the riffles or may attach their eggs to rocks or submersed vegetation. Some species of up-river spawner scatter their semi-pelagic eggs, which drift downstream with the current until they reach suitable nursery sites on the floodplain.

Fish assemblages from the downstream regions are much more complex as a function of the increased complexity and extent of the environment. They fall into three principal behavioural guilds, which are often categorised according to the nomenclature originally developed by traditional peoples in the Mekong River.

- *Whitefish*: This guild groups large, strongly migratory fishes from several families that move large distances within the river channels between feeding and breeding habitats. The fish may pass their whole life history in the main channel or may move onto the floodplains to feed. They are generally intolerant of low dissolved oxygen concentrations, preferring migration as a means to escape the adverse conditions downstream during the dry season. Whitefish are generally one-shot spawners, scattering numerous eggs, which may remain to hatch *in situ* or may be pelagic or semi-pelagic, being swept downstream with the current.
- *Blackfish*: This guild consists of fish that move only locally from floodplain waterbodies to the surrounding plain when flooded and return to the pools during the dry season. They have are adapted to remain on the floodplain at all times, often having auxiliary respiratory organs that enable them to breathe atmospheric air or behaviours that give them access to the well-oxygenated surface film. They have a wide range of elaborate breeding behaviours that allows them to maintain the eggs and newly hatched fry in relatively well-oxygenated localities.
- *Greyfish*: These species are intermediate between the floodplain-resident and the long-distance migrants guilds. Greyfish generally execute short migrations between the floodplain, where they reside at high water for breeding and feeding, to the main river channel, where they shelter in marginal vegetation or in the deeper pools of the channel over the dry season. These species are less capable of surviving extremely low oxygen levels but do have elaborate reproductive behaviours, which enable them to use the floodplain for breeding.

The main differences between the various behavioural guilds are summarised in Table 4.1.

Many rivers in climatically unstable areas have parallel faunas adapted to different climatic regimes. In general there is a fauna adapted to periods of drought, which spawns and lives within the main channel of the river, and one adapted to more normal flood regimes, which spawns and feeds on the floodplain. In some rivers, such

Table 4.1 Some contrasting features of three behavioural guilds of reverine fishes

Feature	Whitefish	Greyfish	Blackfish
Respiratory organs	Gills	Gills with some physiological adaptations to low dissolved oxygen	Gills; supplementary air breathing organs; physiological adaptations to low dissolved oxygen
Respiratory tolerance	Highly oxygenated waters	Medium to low oxygen tensions	Low dissolved oxygen to anoxic
Dominant sense	Eyes	Eyes; electrosensory	Tactile; olfactory electrosensory
Muscle fibre type	Red	Red/white	White
Migratory behaviour	Long distance longitudinal	Short distance longitudinal; lateral; often complex local movements	Local movements
Reproductive guild	Non-guarders; open substrate spawners; lithophils pelagophils	Guarders; nest spawners; open substrate spawners; phytophils	Guarders; external and internal bearers; complex nest builders
Body form	Round, streamlined fusiform	Laterally compressed, spiny often heavily scaled	Vertically compressed; soft and flabby; elongated scales often absent or heavily armoured and spiny
Colour	Silvery or light	Dark, frequently ornamented or coloured	Very dark often black
Dry season habitat	Main channel, lake or sea	Backwaters and fringes of main channel	Floodplain, waterbodies
Wet season habitat	Main channel or flooded plain	Floodplain	Floodplain or marshy fringes

as the Niger, the two faunas are represented by species homologues (see, for example, Dansoko, 1975), one of which prefers main channels and the other the floodplain. In other areas different behavioural groups within the same species ensure the diversity.

In addition to the permanent residents of rivers, diadromous fish occupy the inland water system for only part of their life cycle. Anadromous species such as the salmonids and sturgeons spawn and pass their early life stages in freshwaters and their adult, feeding stages in the sea. Catadromous species such as the eels pass their adult stages in rivers and lakes and return to specific areas in the sea to breed. Both of these behaviours require long migrations and highly specific physiological adaptations to enable the fish to return to their natal waters for breeding. In general, anadromy is more common in the Arctic and temperate zones where the seas are richer than rivers, and catadromy predominates in the tropics where the rivers are richer than the marine environments (Gross et al.,1988).

The number of species in rivers is highly correlated with the size of the river system as indicated by main channel length or basin area. A sample of 292 rivers from around the world (Oberdorff et al., 1995) gives a relationship:

$$\log N = 0.478 + 0.266 \log D$$

with an r^2 of 0.439, where N is the number of species and D is the drainage basin area (km^2). The variance in this relationship is caused by differences between continents, and relationships calculated for rivers in each continent give higher values of r^2. Little effect of latitude has been detected on species richness in rivers, and temperate rivers such as the Danube or Mississippi fit well within the scatter for tropical systems. The only exception to this occurs in rivers from previously glaciated areas where repopulation of the rivers is still incomplete.

Swamps, marshes and rice fields

These generally poorly oxygenated habitats tend to attract fish of the black- and greyfish guilds with a predominance of air-breathing species. Some such habitats are strongly seasonal, holding water during and after the rains but eventually drying out completely. In such cases new fish may migrate in from adjacent watercourse. In completely isolated systems, however, some species have developed the capacity to survive the dry phase either buried in mud, in cocoons or as dormant eggs. Some 'annual' species appear spontaneously shortly after the temporary water-body is filled.

Reservoirs

Fish occupying reservoirs are drawn from the fauna of the impounded river. However, many riverine species are unable to adapt to the new regime and disappear from the main body of the reservoir. Some migratory species may persist in the shallow upper end of the reservoir as that remains accessible to the river. Other species may adapt well to the new ecosystem, often by changing their breeding and feeding habits. Reservoir faunas often consist of the minority elements of the previous river fauna, which take on a new prominence by occupying the new habitats. After impoundment there is a regular succession of dominant species that may last for more than a decade before a more or less stable fauna is established. There is also a pulse of productivity during which levels of harvest are increased considerably as nutrients freed by the flooding are used up. Where native species have proved unable to adapt to the new conditions it has proved necessary to introduce exotic lacustrine species such as the tilapias and carps into the basin to compensate.

Coastal lagoons

Coastal lagoons are transitional systems that are occupied by three main blocks of species.

- Freshwater species move into the lagoon from the inflowing rivers during rainy and flood seasons to feed when the water is mainly fresh to slightly brackish. During periods of low flow these species withdraw into the rivers and are replaced by marine species.
- Marine species migrate in from the sea, often to reproduce during dry seasons when the water is primarily saline.
- A few species are permanently resident in the lagoon, being adapted to the fluctuating salt concentrations. Some species such as peneid prawns, mullets and milkfish (*Chanos chanos*) have a larval phase that remains in the lagoon for at least one freshwater season.

The size structure of fish populations

The size and species characteristics of fish populations are very important for management. Comparison of the numbers of species at different sizes (where size is defined as the maximum length to which a species grows) in any water-body shows that small species are far more abundant than big ones. These types of relationship are illustrated by cumulated values for a number of major temperate and tropical rivers (Welcomme, 1999) (Fig. 4.3). These show that about 50% of the number of species present in any system do not grow larger than 15 cm standard length and that 90% of the species never exceed 50 cm. Larger fishes of up to 300 cm are present in many inland water systems but are represented by one or two species only.

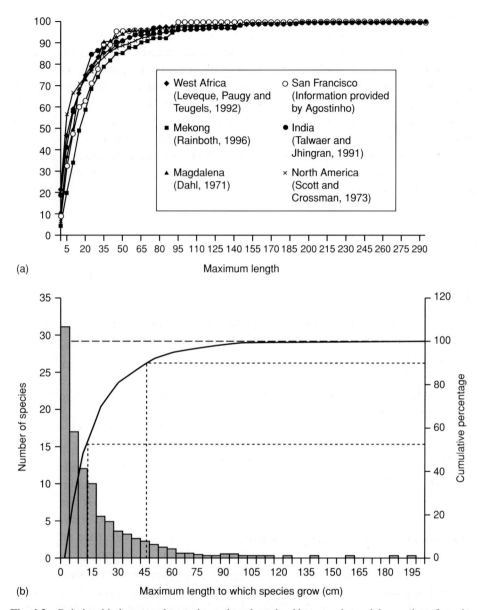

Fig. 4.3 Relationship between the maximum length attained by a species and the number of species attaining that length for various river systems (numbers expressed as cumulative percentage): (a) examples from various river systems; (b) generalised model extrapolated from the above data.

The species structure of the assemblage

The relative abundance of species in any environment has been shown to follow a pattern whereby the logarithm of the species' abundance is normally distributed (Preston, 1962a, b). Fish conform to this type of relationship, which is described by a simple exponential:

$$N = ab^R$$

between species ranked in order of abundance, where R is rank and N is number. This canonical distribution suggests that fish assemblages are always dominated by very few species. A large number of species will be moderately abundant and a few other species will be rare.

Dynamics of fish populations[1]

Fish are influenced by four major factors that operate at the population level to determine their absolute abundance at any one time in the environment (Fig. 4.4). These factors are regulated in turn by external factors, some natural, such as rainfall and temperature, some caused by humans, such as pollution. Each species responds differently to these factors, resulting in a constant shifting in relative abundance within the assemblage of species present.

Management of fisheries and fish stocks usually attempts to exercise some control over one or other of these factors. Management of the fishery aims at regulating **F** (fishing mortality). Aquaculture and enhancement tend to ignore **F** but regulate **N** (numbers accruing through recruitment) through stocking, **G** (growth) through feeding and **M** (natural mortality) through full control of the rearing environment. Efforts to rehabilitate seek to improve **N** through restoration of spawning grounds and

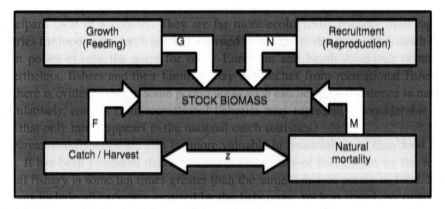

Fig. 4.4 Factors affecting the biomass of fish populations.

nursery habitats and reduce **M** by reducing pollution and creating environments favourable to the fish.

In general, the dynamics of individual species in lake fish populations are similar to those described for marine species and conform well to conventional models in that the environment is stable and does not influence the processes described. In river fisheries the converse is true as most dynamic processes are influenced heavily by abiotic environmental factors. Therefore, this section explores in some detail divergences from conventional models in rivers. In this context most rivers were probably flooded regularly in their pristine condition but many are now regulated to the point where they no longer inundate their floodplains. In such systems, as in canals, the dynamics of the fish populations more closely resemble those of lakes.

Reproduction

'The perpetuation and evolution of species is dependent upon the process of reproduction, the success of which depends upon resource allocation and the location and the timing of reproduction defined by the reproductive strategy of the species' (Lagler *et al.*, 1977). Reproductive strategies are shaped largely by the abiotic environment, food availability, presence of predators and the habitat of parental fish (Wootton, 1990). Lowe-McConnell (1975) and Welcomme (1985) review spawning sites, reproductive behaviour and adaptations, and classifications of reproductive guilds of floodplain fish (Table 4.2).

From a fisheries perspective, particular elements of the reproductive strategy are of interest since they, coupled with mortality rates, determine how many fish survive to recruit to the fishery. These recruits ultimately reproduce to form further cohorts and therefore have a direct bearing on the sustainability of production. These elements include length (age) at maturity, seasonal timing of reproduction, and fecundity.

Maturation

The length at maturity may be defined as the length at which 50% of all individuals within a population are sexually mature (Lm_{50}), where mature individuals are characterised by the presence of spermatophores or ova in the gonads (Bagenal, 1978). Because maturation is age or size dependent it is strongly linked to growth. This, in turn, is regulated by the temperature regime of the water in which the fish live. Fishes exhibit significant interspecific and intraspecific variation in age or length at maturity. Small tropical cyprinodonts can reach maturity at the age of a few weeks (Wootton, 1990; Halls, 1998). Longer lived temperate species such as white sturgeon (*Acipenser transmontanus*) may not reach sexual maturity until they are 25 years of age (Beamesderfer *et al.*, 1995). It is generally assumed that the size at maturation follows a normal distribution, and therefore a plot of the percentage (or fraction) mature at length $M(L)$ will follow a cumulative normal distribution (King, 1995). This can be

Table 4.2 Examples of main types of reproductive behaviour in fishes

Type of fecundity	Seasonality	Examples	Movement and parental care
Big bang	Once in a lifetime	*Anguilla*	Very long catadromous migrations, no parental care
Total spawners (very high fecundity)	Highly seasonal, concentrated on annual or biannual floods	Characins, e.g. *Prochilodus, Salminus, Alestes*	Long-distance migrants, open substratum spawners
		Cyprinids, e.g. *Labeo, Barbus, Cirrhinus*	
		Siluroids, e.g. *Schilbe*	
		Heteropneustes, Catla catla, Labeo rohita	Local lateral migrants, open substratum spawners
		Mormyrids	
Partial spawners	Throughout flood season(s)	Some cyprinids, characins and siluroids, e.g. *Clarias, Microalestes acutidens*	
			Mainly lateral migrants, open substratum spawners
Grades into		*Protopterus, Arapaima, Serrasalmus, Hoplias, Heterotis*	Bottom nest constructors and guarders
		Ophicephalus, Gymnarchus	Floating nest builders
		Hepsetus, Hoplosternum, Anabantids	Bubble nest builders
Small brood spawners (low fecundity)	High water but may start during low water or may continue throughout the year	*Tilapia, Hypostomus*	Nest constructors with various behavioural patterns
		Aspredo, Loricaria sp.	Egg carriers
		Osteoglossum, Sarotherodon spp.	Mouth brooders
		Potamotrygon, Poeciliids	Live bearers
	End of rains	Some cyprinodonts	Annual species with resting eggs

Adapted from Lowe-McConnell (1975).

well described by a logistic function fitted using non-linear least squares to provide the parameter estimates with 95% confidence intervals according to the following relationship (Fig. 4.5):

$$M(L) = \frac{1}{1 + e^{\alpha(Lm_{50}-L)}}$$

Intraspecific variation in age at maturity may be induced by environmental change. Typically, under favourable conditions, when food resources are abundant, age at maturity declines, but length at maturity remains unchanged (Wootton, 1990). Length at maturity in some species may decline in response to intensive fishing (various authors in Lowe-McConnell, 1987) and in others, such as the tilapias, in response to adverse environmental conditions – a condition known as stunting. Several studies, including Aass *et al.* (1989), Crisp *et al.* (1990), Crisp & Mann (1991) and Beamesderfer *et al.* (1995), have also demonstrated changes in age at maturity associated with hydraulic engineering and other sources of environmental degradation.

Fish species in tropical river systems are particularly noted for very rapid maturation, which is believed to be an adaptive response to the ephemeral conditions (Lowe-McConnell, 1987). Most small tropical species reach sexual maturity within their first year of life, ready to spawn at the onset of the next rainy season. Larger, migratory species such as the Indian Major carps, the larger South American pimelodid catfishes and prochilodontids, mature after 2–4 years (Welcomme, 1985; MRAG, 1994a). Maturation is generally slower in temperate waters and even small species may require 2 years to reach maturity.

Seasonality of reproduction

Intra-annual variations in the suitability of environmental conditions for hatching, growth and survival of the young largely dictate the time of year at which fish reproduce. Reproduction is generally timed to coincide with a seasonal abundance of food resources, shelter from predators and benign abiotic conditions (Wootton, 1990). In temperate latitudes, these periods are correlated with changes in temperature and daylength associated with the seasons. In tropical latitudes, where seasonal changes in temperature and daylength are less pronounced, suitable environmental conditions are governed more by changes in rainfall and water level forming the sequence of wet and dry seasons. Thus, many species in equatorial lakes breed throughout the year. By contrast, species inhabiting river systems often spawn just prior to, or during, the period of flooding, enabling developing progeny to take advantage of abundant food resources and shelter from predators upon recently flooded plains (Welcomme, 1985; Wootton, 1990). Spawning during falling floods is rare, although some species of cichlid and of several other families are known to spawn throughout the year, albeit at a lower intensity in riverine environments. Many annual cyprinodonts spawn towards

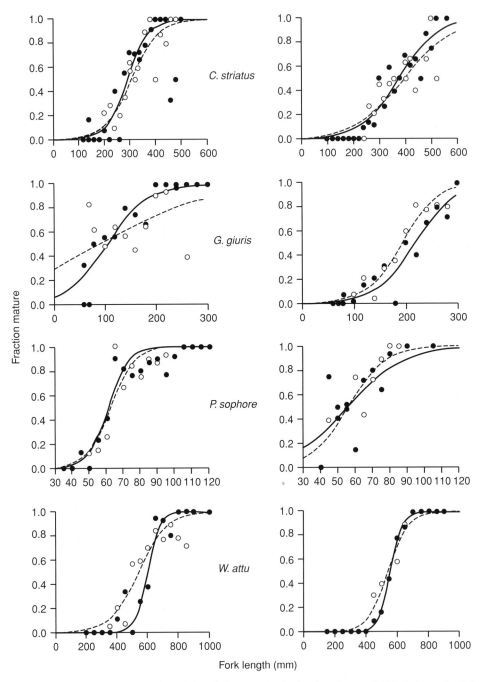

Fig. 4.5 Fraction of mature male and female key species in the size classes. Solid circles and solid lines, and open circles and dashed lines denote data and corresponding logistic model fits for populations sampled from inside and outside the Pabna Irrigation and Rural Development Project (PIRDP) in north-west Bangladesh, respectively. (From Halls *et al.*, 1999.)

the end of the rainy season, especially in seasonal lakes and pools (see Welcomme, 1985, for a review). In equatorial areas where there are two floods each year, fish are known to spawn during each flood phase, but it is uncertain whether individual fish spawn twice each year (Lowe-McConnell, 1987). Pulse spawning, with repeated batches of eggs being released by the same individual, has been recorded in some species of *Prochilodus* in reservoirs in the Parana River, despite the species' normal behaviour as a total spawner once a year under riverine conditions (Espinach, pers. comm.). Maturation and spawning stimuli are less well understood in tropical species than in those inhabiting temperate latitudes. Important cues may include changes in food abundance, water chemistry (conductivity), rainfall, flow, water depth, small changes in temperature and photoperiod or combinations thereof (Welcomme, 1985; Lowe-McConnell, 1987; Wootton, 1990). In total spawners in particular, hydrological conditions are seen as particularly important stimuli. Reproduction may be prolonged with several peaks, deferred or fail altogether if floods are delayed or inadequate to trigger maturation (see Welcomme, 1985; Lowe-McConnell, 1987, for reviews). Partial or batch spawners are capable of breeding throughout the year or whenever conditions are suitable. Harris (1988), Cambray (1990) and Liu and Yu (1992), among others, have shown that spawning behaviour can be significantly disrupted by the effects of engineering structures on the hydrological regime.

The seasonal timing of reproduction (spawning time) is often identified from changes in the gonadosomatic index (GSI) (Fig. 4.6), which is a measure of the relative weight of the gonad (W) with respect to total or somatic weight (Tw) (Welcomme, 1985; Wootton, 1990; King, 1995).

$$GSI(\%) = \frac{W}{Tw} \times 100$$

Increases in the GSI during the reproductive cycle reflect the growth of developing oocytes during vitellogenesis. In total spawners, the GSI is highest just prior to spawning, after which the GSI declines as increasingly more of the population contains spent females. In batch spawners, several peaks may be present during the breeding season. Peak GSI values vary among species from less than 5% up to 30%. These differences are governed largely by the temporal pattern of egg development and spawning. Total spawners often have higher GSI than batch spawners, since the latter may produce several batches of egg in a season. The GSI may be as high as 30% for total spawners but much lower (4%) for multiple (batch) spawners (Wootton, 1990).

Fecundity

Fecundity, defined as the number of vitellogenic oocytes (yolked eggs) in the female prior to the next spawning, varies intraspecifically and interspecifically but, in general, is a function of somatic weight (or body length) (Bagenal, 1978; Lowe-McConnell, 1987; Wootton, 1990). Total spawners produce large numbers of small eggs that are deposited over a short period. This reproductive strategy is geared towards producing large numbers of eggs, and therefore females are generally larger than males. Migratory whitefish are conspicuous members of this category. In contrast, batch or multiple spawners usually produce small numbers of much larger eggs and often provide parental care to the offspring. They have a more protracted breeding period which sometimes lasts throughout the year, with only a proportion of eggs within the gonad becoming ripe at any one spawning (Welcomme, 1985; Lowe-McConnell, 1987). Males are often larger than females, reflecting the lower bioenergetic cost of spermatozoa compared with the massive cytoplasmic investment associated with ova production (Wootton, 1990). Where eggs or progeny are guarded by one sex, often the male, the guarding sex is the larger (Lowe-McConnell, 1975). Lowe-McConnell (1987) and Welcomme (1979) suggest that multiple spawning behaviour is an adaptive response to fluctuations in water level. In multiple spawners, batch fecundity can vary interannually or differ between populations of the same species. Batch fecundity

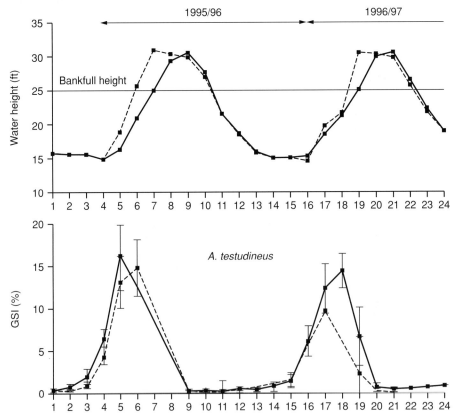

Fig. 4.6 Changes in the mean monthly GSI with 95% confidence intervals for populations of *Anabas testudineus*, *Catla catla* and *Channa striatus* sampled from inside (solid line) and outside (dashed line) the Pabna Irrigation and Rural Development Project (PIRDP) in north-west Bangladesh during a 24-month sampling period. The uppermost plot shows changes in mean monthly water height (ft) inside (solid line) and outside (dashed line) the scheme. (From Halls *et al.*, 1999.)

and the number of spawnings during a breeding season have been shown to be positively correlated with fish size, food availability and temperature. Fecundity of both types of spawners may also be reduced by acidic water, abrupt changes in water level and pollutants (Wootton, 1990).

Methods for estimating fecundity are described by Bagenal (1978). These include total counts or estimations of total egg numbers by subsampling using gravimetric or volumetric methods. Fecundity, F, is often expressed as a function of length, L, by:

$$F = aL^b$$

The parameters of the model are easily estimated using simple linear regression after $\log_e$ transformation of the length data and the corresponding fecundity estimates before back-transformation (see Fig. 4.7 for example).

Fecundity in fishes that show a high degree of parental care may be modified by the capacity of the parent to care for the brood. Thus, mouthbrooders typified by many cichlid species have a linear relationship between body length and the number of fry brooded (Welcomme, 1967).

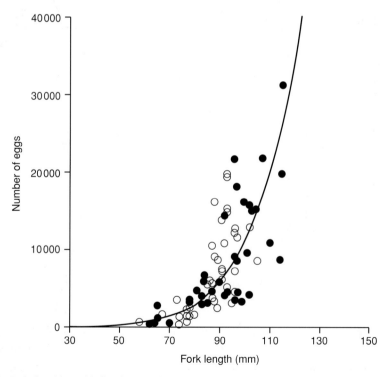

Fig. 4.7 Relationship (with fitted regression model) between fecundity (numbers of eggs) or gonad weight (g) and fork length (mm) for populations of *Puntius sophore* sampled from inside (solid circles, solid line) and outside (open circles, dashed line) the Pabna Irrigation and Rural Development Project (PIRDP) in north-west Bangladesh. (From Halls *et al.*, 1999.)

Recruitment

The stock–recruitment process is the main mechanism controlling and maintaining fish populations and therefore one of the most basic issues in fishery science and management (Gulland, 1983; Salojarvi & Mutenia, 1994). It has been explained as a response which has developed over evolutionary time to buffer the effects of varying food and spatial resources brought about by changes in stock density (biotic or density-dependent factors) and environmental perturbation (abiotic or density-independent factors) (Pitcher & Hart, 1982; Beyer & Sparre, 1982; Gulland, 1983). Since recruitment ensures the continuity of the stock and the fishery, knowledge of the process is crucial, particularly in heavily exploited populations where a significant proportion of the stock's natural 'buffer' may have been removed (Pitcher & Hart, 1982). The simulation models of Welcomme & Hagborg (1977) and Halls (1998) identified recruitment to be the most important factor affecting the dynamics of a floodplain fish community. Failure to recognise the relationship has, in the past, led to recruitment overfishing and eventual collapse of several stocks. However, quantifying the relationship between stock size and recruitment is the most difficult problem in the biological assessment of fisheries (Cushing, 1988; Hilborn & Walters, 1992).

The relationship between stock and recruitment is most often explored by examining the empirical relationship between spawning stock size and subsequent recruitment, the latter resulting from a complex chain of events through spawning, egg deposition, hatching, larval growth, metamorphosis, growth and survival, and often migration to adult feeding grounds. Many stock–recruitment relationships show a great deal of scatter. Recruitment is generally higher at higher stock sizes, although it often stops increasing and may start decreasing above some stock size. The scatter shown in many stock–recruitment relationships may be an artefact of environmental perturbations affecting one of the many stages in the recruitment process. It may equally originate from bad data or simply the fact that the stock and recruitment are not closely related (Hilborn & Walters, 1992; King, 1995).

The mechanisms determining the relationship between stock and subsequent recruitment may be categorised as density independent or density dependent.

Density-independent mechanisms imply that egg survival, and therefore recruitment, is independent of stock size or the number of eggs deposited, giving rise to some form of linear stock–recruitment relationship across all stock sizes. This model must have limits since no population can increase indefinitely given that resources (food and space) are finite. Density-dependent models describe compensatory mechanisms that ultimately limit population size by maintaining some ceiling on the level of recruitment, independent of stock size, or act to reduce recruitment at high stock sizes. This process, coupled with the scatter inherent in many data sets, often leads to the erroneous conclusion that 'there is no relationship between stock and recruitment' (Pitcher & Hart, 1982; Hilborn & Walters, 1992; King, 1995).

Density-dependent mortality at the pre-recruit stage may arise from (1) starvation, (2) a numerical or functional response of predator populations, (3) an extended period of vulnerability to predation arising from density-dependent growth due to intercohort or intracohort competition, and (4) cannibalism (Jones, 1973; Cushing, 1974; Pitcher & Hart, 1982; Gulland, 1983; Mann & Mills, 1985; Hilborn & Walters, 1992). Density-dependent egg mortality may also arise from a numerical or functional response of predator populations (Jones, 1973). Overcutting of redds (in salmonids), oxygen limitation and disease transmission may also be important factors if egg deposition is very heavy or concentrated, as is often observed in salmonid populations (Pitcher & Hart, 1982; Wootton, 1990; Hilborn & Walters, 1992). Density-independent egg mortality may arise from abiotic conditions, for example, extreme temperature, turbulence and desiccation. Estimates of egg mortality for marine pelagic and demersal eggs range from 4% to 63% per day (Wootton, 1990), although Pitcher & Hart (1982) quote a range of 2% to 5% for most species. Under favourable conditions, the mortality of salmonid eggs (3%) is well within this range (Le Cren, 1965). Compensation, arising from a reduction in egg quality (and therefore probability of survival) or fecundity of individual spawners, may also occur, as a consequence of density-dependent growth, as length at age, and condition, decline with increasing stock size (Le Cren, 1965; Wootton, 1990, for a review). Depensation, an increase in recruits-per-spawner (egg survival) with increasing stock size, is also a feature of density-dependent models. This process arises when the numerical or functional response of predators is limited beyond some prey stock size or results from the Allee effect (reduced probability of finding a mate, or reduced fertilisation success at low densities) (Hilborn & Walters, 1992).

Several models have been used to describe the relationship between stock and recruitment (Hilborn & Walters, 1992). The two models described below are most commonly used. Both assume that compensatory mechanisms are operating, although in different ways.

The Beverton & Holt (1957) model describes a family of asymptotic curves with constant recruitment above some stock size, where R is the recruitment, α is the maximum number of recruits and β is the spawning stock size, S, required to produce recruitment equal to $\alpha/2$ (Hilborn & Walters, 1992):

$$R = \frac{\alpha S}{\beta + S}$$

The Ricker (1954) model describes a family of humped curves with declining recruitment above some maximum level of recruitment, where α is a constant and β is the coefficient describing the magnitude of the effect of stock size on subsequent recruitment.

$$R = \alpha S e^{-\beta S}$$

Both models are derived from the logistic growth model (Gulland, 1983; Hilborn & Walters, 1992), incorporating density-dependent mortality at the pre-recruit stage. The Beverton and Holt model assumes that mortality is dependent upon the density of the pre-recruits, whereas the Ricker model assumes that mortality is a function of the spawning stock, S.

The Beverton and Holt model is applicable when a 'ceiling' of recruitment is imposed by available food or habitat resources or when the numerical or functional response of predators continually changes in response to prey (pre-recruit) abundance. The Ricker model is more applicable to situations where adults prey on pre-recruits of the same species (cannibalism) or where adults can compete more successfully for the same resources as those used by the pre-recruits. The latter may lead to mortalities directly due to starvation or indirectly through density-dependent growth when pre-recruits remain vulnerable to predation for an extended period. 'Scramble competition' for limited resources such as food and space may also give rise to this response. In this situation, each individual's share is reduced to the extent that mortality is very high. This response is also expected when a time lag exists between a functional or numerical response of predators to prey (pre-recruit) abundance (Pitcher & Hart, 1982; Jones, 1984; Hilborn & Walters, 1992; King, 1995).

The importance of chance environmental factors on recruitment success is widely recognised (Pitcher & Hart, 1982; Welcomme, 1985; Le Cren, 1987; Cushing, 1988; Eckmann *et al.*, 1988) and has been responsible for obscuring density-dependent effects (King, 1995). Mann & Mills (1985) used environmental variables alone to explain variation in year class strength. However, recruitment cannot be entirely independent of spawning stock biomass (Salojarvi, 1991; Hilborn & Walters, 1992) and therefore several workers, including Craig & Kipling (1983), Penn & Caputi (1984b), Stocker *et al.*, (1985), Lorda & Creeco (1987) and Halls (1998), have incorporated environmental factors into extended stock–recruitment models.

The majority of stock–recruitment relationships that have been documented for freshwater fish populations have been derived for lake-dwelling coregonid populations (e.g. Henderson *et al.*, 1983; Salojarvi, 1991; Salojarvi & Mutenia, 1994) and salmonids (e.g. Le Cren, 1973; Chadwick, 1982; Elliott, 1985; Crozier & Kennedy, 1995). Stock–recruitment relationships have been fitted to populations of pike (*Esox lucius*), perch (*Perca fluviatilis*), roach (*Rutilus rutilus*) (e.g. Craig & Kipling, 1983) and vimba (*Vimba vimba*), a migratory cyprinid (Backiel & Le Cren, 1978).

The relationship between stock and recruitment in tropical rivers is less well understood, although Halls (in press) fitted a Ricker stock–recruitment relationship to discrete populations of the Bangladesh swamp-barb *Puntius sophore*. Recruitment was found to be highly density dependent but also strongly affected by biolimiting nutrient concentrations (Fig. 4.8).

Welcomme (1985), Fremling *et al.* (1989), Harris (1988), Underwood & Bennett (1992) and Merron *et al.* (1993) have shown that spawning success is dependent upon the hydrological regime in the same year. Moreover, because many floodplain fisheries are based almost entirely upon the exploitation of 0+ and 1+ fish, variations in year class strength (YCS) have a very strong influence upon yield and therefore have been the subject of many studies. Typically, these studies employ linear models of the form:

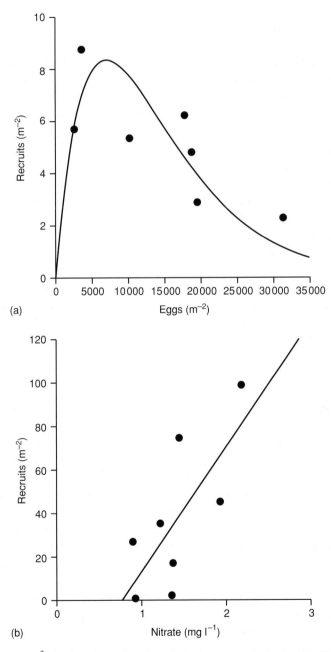

Fig. 4.8 Recruits m^{-2} (R) plotted as a function of egg (spawning stock) density (S) with: (a) fitted Ricker model $R = \alpha S \exp(\beta S)$ where $\alpha = 0.003195$ and $\beta = 0.0001409$; and (b) recruits m^{-2} plotted as a function of nitrate concentration with fitted regression (intercept = –42.72, slope = 56.87, $r^2 = 0.54$).

$$\text{Yield}_y = \alpha + \sum_i \beta_i(\text{FI}_{y-i})$$

to relate yield in year y to some flood index FI in previous years, where $i = 0, 1, 2 \ldots n$.

Significant positive correlations have been demonstrated for several floodplain populations where values of i appear closely related to the age of recruitment or the age of recruitment +1 and the age composition of the exploitable stock. This implies that hydrological conditions during the year of spawning or during the year prior to spawning has a strong influence upon recruitment (de Merona & Gascuel, 1993).

It is uncertain which component of the hydrological regime is most influential, since flooding intensity is often highly correlated with dry-season water levels. However, Welcomme (1985) proposes that the more stringent the drawdown, measured as the proportion of residual water remaining during the dry season, the greater the influence of the low water regime on catch (recruitment) in the following year.

Although the causal mechanisms behind these empirical relationships remain poorly understood, it is clear that during extreme dry seasons both fishing and natural mortality are likely to be higher because fish will be more vulnerable to capture by both fishermen and other predators. Natural mortalities arising from stranding and severe abiotic factors will also be greater and therefore the number of fish surviving to form the spawning stock in the next spawning season will be lower. Egg survival and pre-recruit growth are also likely to be less favourable in drier years because fish density and thus competition for limited resources, including food, shelter from predators and suitable spawning substrate, will be more intense. Total biolimiting nutrient input (nitrate and phosphate) is also likely to be lower during shorter, shallower floods, leading to lower primary and secondary production, thereby reducing feeding opportunities for recently hatched larvae (Junk *et al.*, 1989; Gehrke, 1992; Kolding, 1993).

One major drawback with these empirical models is that they assume that spawning stock biomass is independent of fishing effort and that any changes in fishing mortality depend solely on the hydrological regime through its effect on catchability.

Feeding

The type of food taken by any fish depends on the availability of food items in the environment and on the physical adaptations of the fish in terms of gut length, nature and composition of digestive physiology, tooth and pharyngeal bone form, body shape and behaviour. The food type taken is usually judged by the adult behaviour, although younger stages may have very different food requirements. Most larval and juvenile fish feed on periphyton (aufwuchs), phytoplankton and zooplankton or fine detritus irrespective of the final adult diet. Some of the more specialised species, however, begin to show adult feeding patterns at a remarkably early post-larval stage (Rossi, 1989, 1992). In general, specialisations have developed so that individual species of fish are more competent at the capture and assimilation of particular groups of food (Table 4.3). These specialisations have led to the elaboration of complex food webs, particularly in lakes where the relative stability of the water body allows full expression of trophic specialisations (Fig. 4.9). The amount of any food type available may be regarded as limiting to the abundance of individual species, which leads to partitioning of the resources. This degree of specialisation can create problems for the fish when conditions change. For example, the disappearance of *Oreochromis esculentus* from Lake Victoria has been attributed in part to the shift from a diatom-based plankton to a phytoplankton composed of blue–green algae because of eutrophication. *Oreochromis esculentus* was specialised through its gill raker structure and gastric juices to a diet of diatoms, whereas an introduced competitor, *O. niloticus*, was more competent at digesting blue–greens. In rivers many species appear more flexible, shifting food items according to availability. There has been much debate on the role of specialisations under these conditions. Some authors consider that specialised trophic behaviour confers a competitive advantage during low water when food resources are short; others that the competitive advantage arises at times of abundance. In any eventuality it is hypothesised that, under normal flood regimes the amount of food available at high water exceeds the demand by the fish and is not limiting to their abundance. Welcomme (1985) has extensively reviewed studies on the sources of food, partitioning among individual species and seasonality in feeding patterns of floodplain fish populations.

Measuring feeding

Methods for describing diet composition have been reviewed by Bagenal (1978), Hyslop (1980), Hoggarth (1985) and Wootton (1990). They can be broadly categorised as numerical, volumetric or gravimetric. Numerical methods are the simplest and measure either 'frequency of occurrence' or 'percentage composition by number' (Bagenal, 1978). Frequency of occurrence, defined as the number of stomach samples in which a given food item is found, has been criticised for only providing presence and absence information and not the relative number or bulk of individual food items (Wootton, 1990). Percentage composition by number, defined by Bagenal (1978) as 'the number of food items of a given type that are found in all specimens examined … expressed as a percentage of all food items' has also been criticised for overemphasising the importance of small and numerous items (e.g. zooplankton). It also underemphasises the importance of large less common items (e.g. fish) (Hoggarth, 1985; Wootton, 1990). Moreover, it is also only applicable when the diet comprises discrete food items and therefore is not suitable for those which contain significant proportions of detritus or plant material (Bagenal, 1978; Wootton, 1990).

Volumetric and gravimetric methods emphasise the relative bulk of the food items, expressed as the percentage of total volume or weight of all items present in the samples, respectively, but are generally labour intensive and time consuming (Wootton, 1990).

Table 4.3 Main feeding guilds of freshwater fishes

Bottom feeders		
Mud feeders	Finely divided silt and attached organic molecules	*Prochilodus, Citharinus, Phractolaemus Pterigoplichthys*
Detritus feeders	Coarse decaying plant and animal matter, small crustacea, fungi, diatoms and other microrganisms	Many cyprinids e.g. *Cyprinus carpo, Labeo sp.* Siluroids e.g. *Clarias bathupogon, Plecostomus*, catastomids
Herbivores		
Phytoplankton filterers	Filter floating or decanted phytoplankton from the water column, usually diatoms and green algae, more rarely blue-greens	Cichlids e.g. *Oreochromis niloticus*, several pelagic haplochromine species
Phytoplankton grazers	Graze over rocks and plants for attached algae and fungi, and associated 'Aufwuchs' communities (rotifers, micro-crustacea and other micro-organisms)	Many cyprinids, cichlids, siluroids, catastomids
Higher plant feeders	Graze and eat higher plants	*Tilapia zillii, Ctenopharyngodon, Leporinus maculatus, Cichlasoma, Distichodus*
Fruit, seed and nut eaters	Penetrate flooded forest and floodplains to eat seeds, falling fruit etc.	*Colossoma bidens, Puntius, Leptobarbus, Brycinus*
Omnivores		
Omnivores	Unspecialised feeders that eat insects, zooplankton, detritus, plant matter according to abundance	Many species, even highly specialised forms will occasionally eat other foods when their main food item is scarce
Micro-predators		
Zooplankton feeders	Pelagic species feeding on zooplankton	*Limnothrissa miodon, Rasrtineobola argentaeus, Astyanax, Hypopthalmus*
Insectivores	Surface feeders taking insects falling on surface of water	*Toxotes chatareus,* pelagic haplochromine cichlids
	Bottom feeders feeding over rock, mud or sandy bottoms	Many species, characins, cyprinids, siluroids, cichlids, *Synodontis*
Meso-predators		
Crustacean and large insect feeders	Feed on large bottom-living arthropods	Young of many fish-eating predators, Siluroids, *Schilbe*
Molluscivores	Pickers – remove bivalve and gastropods from their shells with special pick-like teeth	Haplochromine cichlids
	Crushers – crush whole gasteropods	Astatoreus, Siluroids, haplochromie cishlids

Macro-predators		
Fish eaters	Lurking predators, hide in vegetation or adopt disguise	*Esox, Channa*
	Pursuing predators, rapidly pursue prey	*Salminus, Lates, Hoplias, Hepsetus odoe*
Commensals		
Cleaner fish	Species feeding on insect and other surface and gill parasites in freshwaters found especially in highly evolved species flocks	Haplochromine species from lakes Victoria, Malawi and Tanganyika
Parasites		
Fin nippers	Some fin nippers are found in most fish assemblages in the tropics. They tend to be lurking predators that creep up on other fish to remove part of their fins	*Phago, Serasalmus*
Flesh biters	Predators that bite lumps out of prey species. Thes are often relatively small fishes	Piranhas
Scale eaters	Scale eaters have similar behaviour to fin nippers but have special dentition that enables them to rasp scales from the caudal peduncle	Specialized individuals in cichlid species flocks
Egg robbers	Species that prey especially on mouth-brooding cichlids, persuading the brooding fish to surrender her brood	Specialized individuals in cichlid species flocks
Blood suckers	Small species that attach themselves to the gills of other species to suck blood	Candiru

The 'points' method is a semi-quantitative approach based upon the volumetric method. Because of its greater simplicity and speed, this method has been recommended by Hyslop (1980), Hoggarth (1985) and Wootton (1990). The method is based upon subjective allocation of points to each food category in proportion to its visually estimated contribution to stomach volume.

In addition to the nutritional value of ingested food, the quantity of food consumed will largely determine growth rates until the 'maximum ration' is reached. The Stomach Fullness Index (SFI) is a useful indicator of relative quantities of consumed food. Plots of mean SFI with time can provide information on seasonal patterns of feeding.

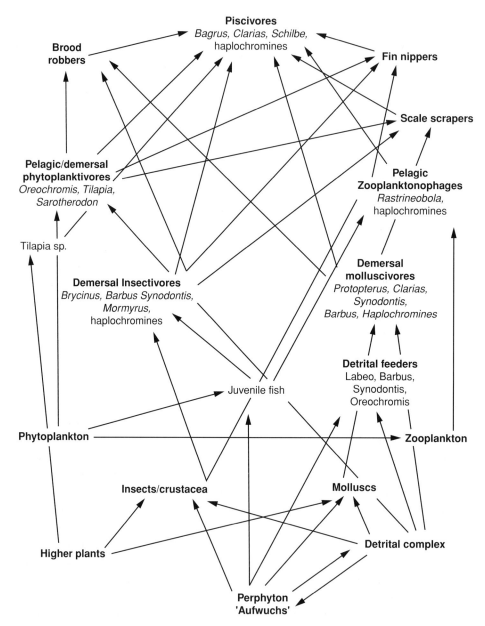

Fig. 4.9 Example of a complex food web from a tropical lake (Lake Victoria before the introduction of the Nile perch). (Adapted from Witte & van Densen, 1995.)

Fish condition

Fish condition (Tesch, 1971) reflects the state of well-being of a fish or population. Fish condition is often expressed in terms of the length (L) and weight (W) by the condition factor index, K, where:

$$K = \frac{W}{L^3}$$

Condition factor can be used to indicate other processes. For example, K increases radically during maturation, especially in female fish, only to fall after spawning. K is also unduly high in fish leaving floodplains because of the heavy load of fat, which is used up during the dry season when feeding levels are low. To illustrate better the long-term population trends length–weight relationships can be expressed graphically by plotting weight against length. The resulting graph can demonstrate age-dependent effects such as reproduction-related slowing of growth after maturity. Direct comparison of the parameters (intercept and slope) of regressions of plots of length and weight is regarded by Wootton (1990) as preferable to comparison of condition factor indices because they can be compared statistically and because the effects of short-term processes are minimised.

Feeding behaviour

Feeding behaviour is tightly linked to growth, as it is the primary motor for increase in body weight. The rate of feeding is determined by the abundance of food, the temperature, the levels of dissolved oxygen and the presence or absence of predators. It is also influenced by temperature through its affect on metabolism, rate of gastric evacuation and swimming performance. Dissolved oxygen concentrations will limit the maximum rate of aerobic respiration and therefore can have the same effects as temperature. Because maximum dissolved oxygen concentrations are dependent upon temperature, it is often difficult to separate these effects. Feeding rates may also be influenced by the physiological state of the fish. Feeding rate or consumption may decrease or cease completely as fish become reproductively active. Predator density is also likely to affect feeding rates. When predator densities are high, prey species must be more attentive, reducing the time available for foraging (Bayley, 1988; Belk, 1993). Interspecific and intraspecific competition for food will also determine feeding rates, although interspecific competition is believed to be less important among river fishes because of the inherent flexibility in trophic organisation among floodplain communities (Welcomme, 1985). Intraspecific competition is likely to be more important and will be largely governed by the density of fish. Because biomass density is known to vary significantly both interannually and intra-annually, the existence of intraspecific density-dependent growth has important implications for determining the dynamics of floodplain fish populations (Chapter 8, p. 141). Feeding rates can also be determined by factors that influence the abundance, type and quality of food. These include human-induced environmental effects such as siltation, change of feeding substrate or eutrophication. Such alterations to the environment will find expression in the changes in the relative abundance of species as the trophic web adjusts to changes in the basic food types.

Seasonal changes in these environmental attributes impose a seasonality in feeding in all environments but equatorial lakes. The cessation of feeding of fish during the winter in subarctic and cold temperate water-bodies is well established. An analogous process occurs in tropical rivers where feeding ceases almost completely in most species in the dry season, a phenomenon called a physiological winter by Lowe McConnell (1975). Much of this effect may be traced to the need for oxygen in the capture and digestion of food which may exceed the concentration of oxygen available in the water. Other factors such as crowding may also play a part. For example, in the Mekong system, feeding is regarded to be highly seasonal, linked to changes in food supply and population density (Chevey & Le Poulain, 1940). During the high water period, food resources become abundant and fish density is at its lowest, favouring intensive feeding. Conversely, during the dry season, densities are at their highest, food resources in discrete water-bodies are finite and therefore may become depleted, and dissolved oxygen levels are generally low. These patterns of feeding intensity may be detected by changes in fish condition, measured as a function of weight in relation to length (see above). The pattern of feeding described above is likely to describe the general pattern for the majority of species in most rivers. It is not, however, universal among either species or river systems where air-breathing predators such as *Hoplias malabaricus* in Latin America or the snakeheads in southern Asia and Africa can continue feeding throughout the dry season because of their capacity to utilise atmospheric oxygen.

Situations do arise where existing population structures are unable to utilise completely all food resources in a body of water. This occurs where there are insufficient species to occupy all habitats and build food webs capable of efficiently exploiting all types of food, as in the species impoverished inland waters east of the Wallace line in South-east Asia. It can also occur where existing populations of fish are unable to adjust to environmental change, as in reservoirs or rivers in which the flow and morphology have been altered (Fernando & Holcik, 1982). In such cases the concept of the 'empty niche' is frequently advanced as a reason for introduction of new species capable of using the underutilised resource. Strictly speaking, the idea of a vacant niche is incorrect ecologically as the niche is a property of the species not the environment. However, the idea is extremely useful for management, as the efficient utilisation of all food resources is an important factor in maximising yield, should this be the objective.

Growth

Growth is an important component of biological production. As well as affecting overall production directly, negative changes to growth rates may also result in decreased individual health, decreased reproductive success and an increased risk of predation, and thereby mortality (Wootton, 1990). Growth-rate estimates also provide information required to estimate mortality rates.

Factors influencing growth of fish are discussed, among others, by Lagler *et al.* (1977), Lowe-McConnell (1987), Welcomme (1985), Weatherly & Gill (1987), Wootton (1990), Mann (1991), Pauly (1994) and McDowall (1994). These factors may be broadly categorised as either endogenous or exogenous.

Endogenous factors include the genetic component of fish growth that ultimately limits the maximum size of a given species. The other main intrinsic factor controlling fish growth is body size. Growth rates generally decline with increasing fish size (age). Processes that account for this response are not fully understood, but may be related to the allometric growth of the gill surface area (Wootton, 1990; Pauly, 1994).

Exogenous factors affecting growth are better understood. Growth will depend largely on food availability and feeding rate and the nutritional value of the food ingested. Above the minimum ration required for maintenance, specific growth rate increases as a logarithmic function of ration. Specific growth rate reaches an asymptote corresponding to the maximum ration that can be consumed by the species. Fish generally forage to maximise growth rates rather than growth efficiency. The protein and energy contents of the ingested food also largely determine growth rates. The size of food items may also be important, particularly in predator species. The profitability of a given prey item is a function of its energy content and the amount of energy required for its capture and handling. Growth rates will therefore be largely determined by the availability of different prey size, allowing the predator to 'select prey which will yield enough energy to support further increases in size' (Wootton, 1990).

The time of spawning (hatching) also has the potential to affect growth, particularly in annual species or in populations which comprise only one cohort for most of the year (Crisp & Mann, 1991).

Abiotic factors also affect growth. Salinity levels outside the tolerable limits of the species will demand energy costs associated with osmotic and ion regulation. High levels of ammonia and pH have also been shown to slow growth (Wootton, 1990).

The growth of floodplain species is rapid, particularly for the 0+ fish, postulated to be an adaptation to avoid intense predation by rapidly outgrowing the gape of predators before shelter on the floodplain disappears (Lowe-McConnell, 1967; MRAG, 1994a). For example, up to 75% of the first year's growth of *Oreochromis* species inhabiting the Kafue system was achieved within the first 6 weeks of life (Dudley, 1972).

Growth is also seasonal, being fastest during the flood or high-water season and slowest, often ceasing completely, during the dry season. In the Amazon floodplain, Bayley (1988) found that mean weight increments of 12 species were 60% higher during the rising-water period than during the remainder of the year. Seasonality in growth has been explained in terms of changes in food availability, temperature, competition for food, fasting prior to spawning activity and 'deep physiological rhythms'

(Welcomme, 1985). It is difficult to distinguish the relative importance of these factors since they often occur at the same time. The rapid growth during the high-water period generally correspond to periods when food resources are at their most abundant, temperature is at its highest and density (competition) is at its lowest. These conditions are reversed during the dry season when growth is at its slowest. However, explanations for the seasonality of growth are often inconsistent or contradictory among different species and systems (see Welcomme, 1985, for reviews).

In addition to the inherent intra-annual variation described above, the growth of floodplain fish often exhibits significant interannual variation. These variations have been correlated with flooding intensity and duration, and temperature during the dry season (Dudley, 1972; Kapetsky, 1974). More extensive flooding is likely to promote greater primary and secondary production upon the floodplain (Junk et al., 1989) and bring greater quantities of allocthonous food inputs into the system, thereby improving feeding opportunities. At the same time, fish densities, and therefore interspecific and intraspecific competition for resources, are likely to decline, making it difficult to separate the effects of the two factors. During exceptionally poor flood years in the Senegal river system, *Citharinus citharinus* failed to reach sexual maturity, achieving less than 50% of its normal size, resulting in recruitment failure (Welcomme, 1985). Many fish species in Bangladesh comprise almost entirely 0+ age fish, reaching maturity at the end of their first year. Failure to reach sexual maturity owing to poor growth could have terminal implications for these populations (Halls, 1999).

Modelling growth

A general growth model was proposed by von Bertalanffy (1934) as:

$$L_t = L_\infty[1 - \exp(-K(t - t_0))]$$

This equation provides an adequate description of the interannual growth of most species of fish. However, the growth of most species fluctuates during the year in all but the most equatorial lacustrine systems. Several models exist to describe these intra-annual variations. Perhaps the most commonly applied and flexible model is the seasonal version of the von Bertalanffy growth function (VBGF) after Pitcher & Macdonald (1973), which allows for sinusoidal variation in growth rates throughout the year.

$$L_t = L_\infty \left(1 - e^{-K\left[(t-t_0) + \frac{C \sin(2\pi(t-t_s))}{2\pi}\right]}\right)$$

where L_t is length at time t (decimal years), L_∞ is asymptotic length (the maximum size the fish would grow to in the absence of predation or disease, K is a growth coefficient (describes the rate at which L_∞ is reached), t_0 is age when length is zero, C is the amplitude of seasonal growth rates oscillation (range 0–1), and t_s is the 'summer point' (time within the year when growth is fastest; range = –0.5 to +0.5). Growth is slowest at 'winter point' (t_s+0.5).

In spite of its considerable flexibility, the VBGF is often inadequate for describing floodplain fish growth that commonly exhibits significant interannual variation (Halls, 1998; Bayley, pers. comm.). To take account of this growth variation, Welcomme & Hagborg (1977) adopted a logarithmic growth model:

$$L_{t,i} = L_{52,i-1} + G_i \mathrm{Ln}(t)$$

where $L_{t,i}$ is the length at week t, for age group i, and G_i is the growth coefficient of age group i.

The model predicts rapid initial growth at the start of each year followed by a period of slower growth. Successive years of growth approximate the non-seasonal VBGF (Fig. 4.10). The model has the advantage that the age-specific growth coefficient (G) can be adjusted to take account of interannual variations in growth according to empirical relationships with factors such as flooding intensity which introduce density dependence into the growth model. One drawback is that the model predicts very abrupt changes in growth rate from one year to the next.

More conventional models such as the seasonal VBGF provide a better description of the seasonal pattern of growth and can be easily modified to account for interannual variability in growth rates simply by introducing year-dependent values for either K or L_∞.

This approach was adopted by Halls (1998) to model the growth of the Bangladeshi floodplain swamp-barb *P. sophore*. In this study, the year-dependent values for L_∞ were varied according to an inverse linear relationship with fish density (see below).

Density-dependent growth

The dependence of growth on population density in marine and freshwater fish populations has been described by several workers (Beverton & Holt, 1957; Le Cren, 1958, 1987; Backiel & Le Cren, 1978; Hanson & Legget, 1985; Ross & Almeida, 1986; Overholtz, 1989; Wooton, 1990; Mann, 1991; McDowell, 1994), where it is mainly the result of intraspecific competition for food (Weatherley & Gill, 1985; Walters & Post, 1993; Lorenzen, 1996a). In rivers where fish densities are in a state of flux, driven by the dynamic hydrological regime, and under conditions where water volume and hence fish densities can be manipulated (e.g. FCDI schemes), density-dependent process have potentially important implications for determining the pattern of fish growth both intra-annually and interannually.

Examples of density-dependent growth studies in the floodplain environment are scarce, reflecting the difficulty of studies of this type where the population density is continually changing. However, Dudley (1972) found a negative correlation at the $P < 0.1$ level between annual fish yield (an index of density) and mean annual growth

64 Inland fisheries

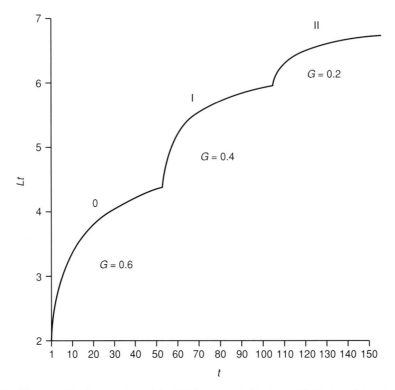

Fig. 4.10 The logarithmic growth model of Welcomme & Hagborg (1977) describing changes in length, L_t, over time, t (weeks), for three cohorts with declining growth coefficient (G).

increment of age 3+ *Tilapia macrochir* in the Kafue floodplain. Several quite strong, but statistically insignificant ($P = 0.2$) correlations for other tilapia species and age classes were also established. Dudley (1972) and Kapetsky (1974) also found significant correlations between the mean annual growth increment of several cichlid species on the same floodplain and the annual flood index (FI) across a number of years. Welcomme (1979, 1985) cites several other examples where workers have found strong correlations between annual growth performance and flood extent and duration. If differences in population biomass between the years examined were small relative to the differences in FI, these observations would be indicative of density-dependent processes. Indeed, Dudley (1972) states that 'the apparent decrease in growth at higher fish densities agrees with the correlation of growth with high water levels'. He speculates that improved growth performance during high floods arises from more abundant food resources and reduced intraspecific and interspecific competition for them. These results may reflect the same effects as those obtained from other aquatic environments by Le Cren (1958, 1987), Kennedy & Fitzmaurice (1969), Ross & Almeida (1986) and Overholtz (1989) who, among others, found increases in the width of growth checks in scales and bones from fish following reductions in their abundance and hence biomass density (see Backiel & Le Cren, 1978; Weatherley & Gill, 1987; Mann, 1991, for reviews). These comparisons are, however, complicated by the fact that other growth-influencing factors, such as temperature and dissolved oxygen, may also be linked with FI.

Bayley (1988) sought evidence for interspecific competition mediated density-dependent growth among 12 species of fish inhabiting the central Amazon floodplain. Species were divided into guilds – detritivores, piscivores and omnivores – based upon their diet overlap. Density-dependent growth of each species was tested with respect to guild biomass. Piscivorous species could not be tested due to insufficient data. None of the detritivores and only two of the four species of omnivore showed evidence of density-dependent growth during the falling-water period. No evidence of density-dependent growth was found in any of the 12 species during the rising-water period. The apparent general absence of interspecific competition found by Bayley implies either that food resources were not limited or that the species could mitigate the effects of competition by shifting to less intensely exploited food resources (Weatherly & Gill, 1987). Omnivores are wide-spectrum feeders and have considerable scope for varying their diet in an opportunistic manner. Detritivores have less scope for dietary shift, but Bayley (1988) states that 'the lack of density-dependent effects on the growth of young detritivores is not surprising when one considers the abundance of fine detrital matter in the floodplain'. Piscivorous and planktivorous species guilds, not examined by Bayley, often possess specialised feeding structures and dentition which restrict their scope for diet change and are therefore more likely to show evidence of interspecific density-dependent growth.

Density-dependent growth has been modelled empirically in terms of growth increment (Le Cren, 1958; Dudley, 1972), percentage change in body weight (Hanson & Legget, 1985), specific growth rate (Le Cren, 1965), weight at age (Ross & Almeida, 1986; Overholtz, 1989) and length at age (Salojarvi & Mutenia, 1994) expressed as a function of (log) population density or abundance indices, e.g. CPUE.

An alternative density-dependent growth model has been formulated by Lorenzen (1996a) based on the work of Beverton & Holt (1957). The foundations of the model descend from the von Bertalanffy growth function which, in its differential form, is expressed as:

$$\frac{dL}{dt} = -K(L - L_\infty)$$

where L_∞ is the maximum size the fish would grow to in the absence of predation or disease and is a measure of anabolic activity, i.e. the renewal of body materials by metabolic processes. The parameter K is a measure of catabolic activity, the breakdown of existing body materials, and describes the rate at which L_∞ is reached. Catabolism is independent of food consumption, although anabolism and hence L_∞ are dependent on the rate at which nutrients are brought into contact with absorbing surfaces. Competition for food through increased population density is therefore expected to influence L_∞ but not K (Beverton & Holt, 1957; Le Cren 1965; Walters & Post,

1993; Lorenzen, 1996a). The asymptotic length is therefore expressed as a linear function of population density, B, by:

$$L_{\infty B} = L_{\infty L} - gB$$

where $L_{\infty B}$ is the asymptotic length at biomass B, and $L_{\infty L}$ the limiting asymptotic length of the growth curve in the absence of competition. Because, in the absence of competition, anabolism is still dependent on the available food resources, the limiting asymptotic length $L_{\infty L}$ is dependent only on the productivity of the water in which the population lives. The parameter g is termed the 'competition coefficient' and describes the amount by which $L_{\infty B}$ decreases per unit of biomass density. The value of g reflects the degree of dietary overlap (competition) among individuals in the population. Gape-limited predators, omnivorous bottom feeders and planktivorous species are likely to exhibit increasing overlap and therefore stability in the value of g (Weatherley & Gill, 1987; Lorenzen, 1996a).

For mixed-age populations characterised by approximately constant biomass, growth may be described by the standard or seasonal integrated form of the VBGF where L_∞ is replaced by $L_{\infty B}$.

Lorenzen (1996a) found that this density-dependent VBGF model provided an excellent description of the growth of carp (*Cyprinus carpio*) stocked at a broad range of densities in ponds with different productivities. He also found that the competition coefficient g to be very similar among populations stocked in ponds of different productivities. Halls (1998) also successfully applied the model to discrete populations of the swamp barb *P. sophore* subject to a range of biomass densities, under controlled environmental conditions.

Mortality

Mortality in exploited fish populations is the combination of fishing and natural mortality Factors affecting mortality in exploited lake and floodplain river fish populations are discussed by Welcomme (1985, 1989) and summarised in Fig. 4.11.

In lakes natural mortality is due mainly to predation by fish, birds and mammals, although disease may also play a role, particularly among dense, stunted populations. Mortalities may also be due to high temperature and low dissolved oxygen, and sudden 'fish kills' have been linked to rapid changes in abiotic conditions, particularly dissolved oxygen, following a breakdown in the stratification of the water column causing sudden overturn (Das & Pande, 1980; Welcomme, 1985).

Stranding in isolated pools is a major cause of mortality in some floodplain river systems. Direct losses arising from stranding may be as much as four times the fish actually caught (Hoggarth *et al.*, 1999a). Stranding is also likely to make fish more vulnerable to fishing gear, other predators and starvation. Moreover, extreme abiotic conditions (temperature and dissolved oxygen concentrations) within small pools or discrete water-bodies may eventually become intolerable, causing mortalities directly,

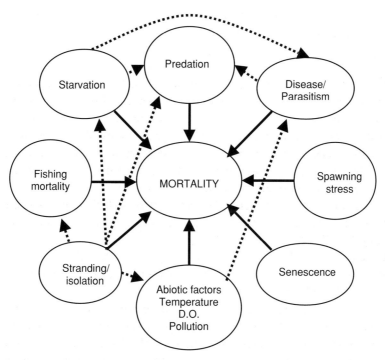

Fig. 4.11 Diagrammatic representation of the main direct (solid line with arrow) and indirect causes (dashed line with arrow) of mortality in floodplain systems. (From Halls, 1998.)

or indirectly as the result of increased vulnerability to disease and parasitism arising from physiological stress. Predation by fish, birds, mammals, reptiles, amphibians and some aquatic insects is also believed to be a major case of natural mortality in floodplain fish populations (Welcomme, 1985, 1989).

The plasticity of fish growth means that fish are unlikely to die from starvation, although fish in poor condition are likely to be more susceptible to predation and disease. The importance of disease as a cause of mortality is uncertain.

In river systems the major factors influencing mortality rates – fishing, stranding, abiotic factors and predation – are assumed to vary with the flood cycle, intensifying during the drawdown and with the progression of the dry season (Welcomme, 1985, 1989). During this time fish become concentrated in channels or pools as water recedes from the floodplain. These periods of high concentrations or densities of fish, with high catchability (q), are exploited by both fishermen and other predators (Welcomme, 1985; MRAG, 1994a). Abiotic pressures are also likely to reach extremes during this period, leading directly, or indirectly, through greater vulnerability to disease (and thereby predation), to increased mortality rates. Conversely, during the high-water period, the floodplain offers abundant food resources, shelter from predators, tolerable abiotic conditions and low fish densities. Mortality rates during this period are therefore assumed to be low. Mortality rates of floodplain fish are therefore anticipated to be density dependent.

Density-dependent mortality processes may be categorised as either competition or predation mediated (MRAG, 1994b). Competition-mediated density-dependent mortality results from interspecific and intraspecific competition for limited resources and would encompass those mortalities arising from stranding or isolation within temporary or permanent water bodies. Predation-mediated density-dependent mortality occurs when the mortality rate changes in response to prey density. This is brought about by a functional or numerical response of the predator population, that is, by a change in predator fishing/feeding activity or by a change in abundance, respectively (Begon & Mortimer, 1986; Wootton, 1990).

Evidence of competition-mediated density-dependent mortality in the form of stranding and isolation is well documented (Welcomme, 1985). Evidence supporting predation-mediated density-dependent mortality is scarce. Welcomme (1979) observed predation-mediated density-dependent mortality on the Oueme and Niger rivers as a functional response to large numbers of small and juvenile fish present during the drawdown when their densities are at their highest. Predation during the dry season is known to occur as predators generally gain condition during this period and therefore must be feeding (Welcomme, 1985). Furthermore, Rodriguez & Lewis (1994) observed that piscivory continues throughout dry season in the Orinoco River. However, the intensity of predation is believed to be lower during this period than during the drawdown, perhaps because of the lowered dissolved oxygen tensions in some waters.

Lowe-McConnell (1964) and Williams (1971) made observations on numerical responses of predator populations that support the process of predation-mediated density-dependent mortality. They noted the presence of large numbers of water birds preying on fish stranded in pools and leaving the floodplain through channels. Several workers have made estimates of the amount of fish taken by birds. For example, Reizer (1974) calculated that birds took some 70 000 t year^{-1} from the Senegal River, compared with a fishery catch of 50 000 t. There is little equivalent published evidence of a numerical response to prey density by predatory fish in rivers, although Holcik & Kmet (1986) found that populations of predators in the Danube depended in part on the density of forage fish.

Because fish densities in floodplain–river systems are in a state of flux, driven by the dynamic hydrological conditions, mortality rates are likely to change throughout the year. Simple exponential models of mortality are therefore inadequate for modelling floodplain populations. More realistic models must take population density into account, and Welcomme & Hagborg (1977) and Halls (1998) employed empirical density-dependent mortality models to describe this process.

The simple exponential model (Gulland, 1983) is, however, adequate for describing and comparing annual rates of mortality, assuming that interannual variations in density are not significant:

$$N_1 = N_0 e^{-z}$$

where N_0 is the initial number and N_1 the number after one unit of time, and Z is the total annual instantaneous mortality rate.

Pauly (1980) examined data on the relationships between natural mortality, maximum length, growth and mean environmental temperature in 175 stocks of 84 species, including many from inland waters. He concluded that mortality rates were determined by the size of the fish, its growth rate and the environmental temperature, according to the following regression:

$$\log M = -0.0066 - 0.279 \log L_\infty + 0.6543 \log K + 0.4634 \log T$$

An alternative relationship between mortality and body weight takes the form:

$$M_w = M_u W^b$$

where M_w is natural mortality at weight w, M_u is natural mortality at unit weight and b is the allometric exponent. Empirical studies (McGurk, 1986; Lorenzen, 1996b) show that the allometric weight exponent of fish natural mortality is generally in the range −0.37 to −0.29. Therefore, mortality is approximately inversely proportional to body length (corresponding to a weight exponent of 0.33). Explicit consideration of the mortality–size relationships is particularly important in several areas of management, for example, estimating the optimal release size in fish stocking.

It should be noted that estimates of natural mortality are generally difficult to derive because most inland fish populations are already heavily exploited. Total mortality is easier to estimate (Quinn & Deriso, 1999). Natural mortality rates are often estimated by subtracting fishing mortality from total mortality. Total mortality rate estimates for a number of river fish populations are shown in Table 4.4.

The few published estimates of Z for tropical river fish populations relate almost exclusively to populations inhabiting river systems in Africa (Welcomme, 1985). Only two estimates have been made for systems outside Africa, both for *Colossoma maropomum* in the Amazon River. Petrere (1983) found $Z = 0.45$ in 1983, whereas Isaac & Ruffino (1996) estimated $Z = 1.37$ year^{-1} (1992) and 1.28 year^{-1} (1993). In 1992 this was composed of $M = 0.45$ year^{-1} and $F = 0.95$ year^{-1}. The difference between the two estimates is possibly due to increases in fishing pressure over the intervening decade. Unusually high estimates of Z often reflect intensive fishing pressure on the stock. For example, in Bangladesh estimates of Z for fish age 0+ range from $Z = 1.2$ year^{-1} to 4.61 year^{-1} (mean $Z = 2.5$ year^{-1}) and are equivalent to survival rates of between 30% and 1% (mean survival = 8%). Where survival rates appear to improve with age (size), it is assumed that fish become less susceptible to predation with increasing size (Welcomme, 1985).

Biomass and production[2]

Biomass (B) is an instantaneous measure of the weight of living organic material in the environment. In the case of fisheries this usually represents the weight of all fish

Table 4.4 Annual mortality coefficients (Z) for certain fish species from some tropical and temperate rivers

Species	River	Year								Reference
		1	2	3	4	5	6	7	8	
Tropical										
Polypterus senegalus	Chari		0.54	0.53						Daget & Ecoutin (1976)
Brycinus leuciscus	Niger	1.20								Daget & Ecoutin (1976)
Hydrocynus brevis	Niger	3.08	0.98							Dansoko (1975)
Hydrocynus forskahlii	Niger	2.49	2.67							Dansoko (1975)
Oreochromis andersoni	Kafue	2.47	0.65	0.65	0.65	0.65	1.70	0.58	0.58	Kapetsky (1974)
Oreochromis macrochir	Kafue	3.98	0.70	0.70	0.70	0.70	0.70			Kapetsky (1974)
Tilapia rendalli	Kafue	4.61	1.40	1.40	1.40	1.40				Kapetsky (1974)
Serranochromis angusticeps	Kafue	3.12	2.20							Kapetsky (1974)
Hepsetus odoe	Kafue	2.74	1.84	1.84	1.84					Kapetsky (1974)
Labeo ulangensis	Rufigi				1.90					Payne & McCarton (1985)
Citharinus congicus	Rufigi				1.11					Payne & McCarton (1985)
Oreochromis urolepis	Rufigi				1.16					Payne & McCarton (1985)
Tilapia brevimanus	Taia				0.53					Payne & McCarton (1985)
Hemichromis fasciatus	Taia				0.82					Payne & McCarton (1985)
Total community[a]	Bandama				2.67					Daget et al. (1973)
Prochilodus platensis	Pilcomayo				1.2					Payne (1987)
Colossoma macropomum	Amazon	0.45	0.45							Petrere (1983)
Colossoma macropomum	Amazon				1.37					Isaac & Ruffino (1996)
Petenia splendida	San Pedro				1.6					Noiset (1994)
Cichlasoma spilura	San Pedro				0.91					Noiset (1994)
Cichlasoma managuense	San Pedro				2.85					Noiset (1994)
Labeo rohita	Meghna				0.6					MRAG 1994
Anabas testudineus	Meghna				3.5					
Catla catla	Meghna				4.7					
Channa striatus	Meghna				3.9					
Wallago attu	Meghna				3.7					

(*continued*)

Table 4.4 (Continued)

Species	River	Year								Reference
		1	2	3	4	5	6	7	8	
Puntius sophore	Meghna				3.2					Halls (1998)
Temperate										
Salmo trutta	Bere stream	2.10	1.34	1.81						Mann (1971)
Cottus gobio	Bere stream	1.66	1.33							Mann (1971)
Rutilus rutilus	Thames	0.42	0.42	0.68	0.70	0.92	1.00	1.26	1.86	Mann (1971)
Rutilus rutilus	Danube	0.69	0.98	0.65	0.70	0.56	0.61	0.68	0.75	Chitravadivelu (1972)
Alburnus alburnus	Thames	1.29	1.29	1.29	3.90	4.80	4.10			Mann (1971)
Alburnus alburnus	Danube	0.88	0.86	0.82	0.84	0.73	0.63			Chitravadivelu (1972)
Leuciscus leuciscus	Thames	0.60	0.60	0.60	0.60	0.56	1.15	1.28	1.46	Mann (1971)
Perca fluviatilis	Thames	0.98	0.98	0.98	1.61	1.49	1.39	1.45	1.47	Mann (1971)
Gobio gobio	Thames	0.88	0.88	0.88	0.88	2.52	4.24			Mann (1971)

[a]Dominant species *Labeo coubie* and *Alestes rutilus*.

72 Inland fisheries

in a specified environment or habitat. It may also be derived for a specified species, in which case it is equivalent to standing stock. Biomass is usually expressed as weight per unit area or volume.

Production (P) is the total weight of organic material that is produced over a specified time. This includes fish that have died as well as gonadal products such as eggs and sperm. It is expressed as increase in weight over the specified time.

The standing stock represents the amount of fish that is available to capture, although clearly total removal would stop any further exploitation and the amount removed is usually a percentage of the biomass. Production represents the rate at which the biomass is renewed and affects the percentage. For example, high standing stocks are sometimes present in blackwater lakes and rivers, but because the production in such environments is low only a small percentage of the fish can be removed without producing a decline in catches. Conversely, in eutrophicated lakes production is usually high and a much greater proportion of the standing stock can be taken without causing a decline in the fishery.

An indicator of the production potential is the productivity biomass ratio (P/B), which indicates the rate at which the fish population in any environment renews itself. Allen (1971) determined that mean age or mean life span is equal to the reciprocal of the P/B ratio. This means that smaller fish and species have higher production rates and are thus able to support greater fishing effort, a fact of great significance in fisheries management.

Typical values for biomass, production and P/B ratios are given in Table 4.5 for river systems and for biomass in floodplain lakes in Table 4.6. Further values for a range of North American and European species are given by Mann & Penczak (1986).

Relationships between factors

The parameters of fish dynamics (growth, mortality, production and biomass) are strongly related to each other and to length (Allen, 1971; Pauly, 1980; MRAG, 1994a). As a rough generalisation:

$$L_\infty \approx \text{age} \approx \frac{1}{K} \approx \frac{1}{M} \approx \frac{1}{\frac{P}{B}}$$

The correlation between length and these parameters has important implications for the behaviour of multi-species fish communities (see Fig. 4.13 and Chapter 11, p. 207).

Migration and movements

Fish in inland waters move, sometimes over great distances, between breeding, feeding and refuge areas. These patterns of movement render fish especially vulnerable to fisheries and to certain types of environmental disturbance whereby the pathways are interrupted. Knowledge of migration patterns is therefore essential in carrying out any assessment of impact for river modification through damming and channelisation.

The degree to which fish depend on migration for the completion of their life cycle is also crucial to management of fisheries and biological diversity. Many riverine species appear to be obligate migrants and disappear should the migration pathway be interrupted for any reason. In such cases where the species is sufficiently valuable, such as in the salmonids and sturgeons of Europe and North America, every attempt is made to maintain populations through artificial rearing programmes. More normally, however, the species disappears from the fish community and as a result obligate migrant freshwater fish species are among the vertebrates most threatened with extinction. Consequently, in many such cases there are now attempts to re-establish stocks in areas from which they have been eliminated.

Other species appear to be able to adapt to lacustrine conditions when rivers are regulated or impounded. It would appear that many species are behaviourally heterogeneous, with part of the stock being static and part mobile. In rivers this may have been an adaptation to climatic variability so that some fish could always breed whether there was a strong flood or a failed one. This behaviour appears to have persisted in species such as roach, which were migrants in the original European rivers but have adopted a more sedentary existence in the regulated systems of today. A part of the stock still shown signs of wanting to migrate even after some 200 years of regulation (Bouvet et al., 1985; Linfield, 1985), which would indicate sufficient resilience in the genotype should more pristine conditions be restored.

Migrations are basically longitudinal, lateral and vertical (Fig. 4.12).

Longitudinal migrations

Longitudinal migrations take place within the main channels of rivers. In reservoirs the original patterns of riverine movement are often conserved, with fish migrating primarily along the drowned course of the original riverbed (Poddubnyi, 1979). Fish may also migrate from lakes and reservoirs up tributary streams. Longitudinal migrations are of three main types:

- *Anadromous*: Movements from the sea to rivers to breed in the river and feed in the sea, e.g. shads, salmonids
- *Catadromous*: Movements from river to the sea to breed in the sea and feed in rivers, e.g. eels
- *Potamodromous*: Movements within the river, e.g. many cyprinids, catastomids, characins and siluroids.

Longitudinal migrations frequently cover enormous distances. Salmonids in temperate rivers have long marine migrations before ascending the rivers. Before the damming of the Volga River the large sturgeon species such as *Huso huso* used to

74 Inland fisheries

Table 4.5 Typical values for biomass (B), production (P) and P/B from inland waters

River	Country	Biomass (kg ha^{-1})	Production (kg ha^{-1} year^{-1})	P/B	Reference
Amazon Manaus	Brazil	1600.0	2800.0	1.75	Bayley (1983)
Thames					
(0+fish)	UK	1315.0	2000.0	1.52	Mann (1972)
(1+fish)	UK	659.0	426.0	0.65	Mann (1965)
Hinaki	New Zealand	880.8	735.4	0.82	Hopkins (1971)
Kafue	Zambia	520.0	618.0	1.19	Kapetsky (1974)
Ellis	Canada	376.0	375.4	1.0	Mahon (1981)
Warkosc	Poland	307.5	161.4	0.5	Mahon (1981)
Carrol	Canada	274.8	280.3	1.0	Mahon (1981)
Hopewell	Canada	228.4	263.5	1.0	Mahon (1981)
Hinau II	New Zealand	217.7	241.7	1.11	Hopkins (1971)
Lemki	USA	212.0	136.0	0.64	Goodnight & Bjorn (1971)
Jaruma	Spain	178–221	221–583	1.2–2.6	Lobon-Cervia & Penczak (1984)
Tarrant	UK	198.0	596.0	3.01	Mann (1971)
Kejin	Borneo	173.1	261.5	1.50	Watson & Balon (1984)
Bere	UK	161.7	270.0	1.69	Mann (1971)
Irvine	Canada	149.9	226.4	1.50	Mahon (1981)
Hall	UK	129.0	52.0	0.40	Le Cren (1969)
Swann	Canada	124.4	174.5	1.40	Mahon (1981)
Struga	Poland	111.2	110.3	1.0	Mahon (1981)
Deer	USA	84.7	160.0	1.89	Chapman (1967)
Utrata	Poland	84.5	289.3	3.4	Penczak (1981)
Big Springs	USA	84.2	118.0	1.4	Goodnight & Bjornn (1971)
Mesta	Bulgaria	80+82	80.0	1.0	Penczak et al. (1982)
Dockens	UK	75.0	140.0	1.87	Mann (1971)
Clemons Fork (3rd order)	USA	71.5	77.2	1.08	Lotrich (1973)
Kejin	Borneo	71.0	98.3	1.4	Watson & Balon (1984)
Clemons Fork (2nd order)	USA	63.6	105.5	1.66	Lotrich (1973)

(continued)

Table 4.5 (*Continued*)

River	Country	Biomass (kg ha^{-1})	Production (kg ha^{-1} year^{-1})	P/B	Reference
Appletreeworth	UK	62.0	30.0	0.48	Le Cren (1969)
Black Brows	UK	59.0	100.0	1.69	Le Cren (1969)
Clemons Fork (1st order)	USA	54.9	83.5	1.52	Lotrich (1973)
Bobrza	Poland	50.2	83.4	1.7	Mahon (1981)
Needle Branch	USA	45.9	90.0	1.96	Chapman (1967)
Zalewka	Poland	43.9	54.0	1.23	Penczak (1981)
Kaka	Borneo	38.5	47.5	1.2	Watson & Balon (1984)
Wolborka	Poland	37.4	35.3	0.94	Penczak et al. (1982)
Speed (3rd order)	Canada	32.3	54.0	1.67	Mahon et al. (1979)
Lava	Borneo	30.5	39.3	1.3	Watson & Balon (1984)
Payau	Borneo	27.1	31.2	1.2	Watson & Balon (1984)
Lawa	Borneo	21.3	33.3	1.6	Watson & Balon (1984)
Bulu	Borneo	21.5	26.0	1.2	Watson & Balon (1984)
Hinau I	New Zealand	19.6	42.8	2.18	Hopkins (1971)
Speed (4th order)	Canada	7.0	16.2	2.31	Mahon et al. (1979)

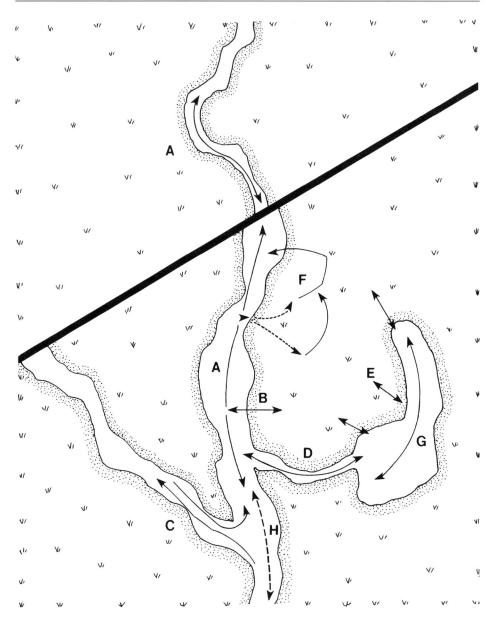

Fig. 4.12 Various types of fish migration in rivers and lakes: A, longitudinal migration in river channel with downstream drift of fry (whitefish pattern); B, lateral migration of adults and fry to floodplain; C, 'piracema' type migrations of adult fish into and out of tributaries; D, lateral migrations between main river channel and floodplain lakes; E, lateral migrations from floodplain lakes onto the floodplain (blackfish pattern); F, migrations of adult to spawn at river margin with movements of fry onto floodplain and return movement of young of the year to main channel (greyfish pattern); G, internal movements within lakes; H, anadromous/catadromous migrations into the system from the sea.

Table 4.6 Estimates of dry-season ichthyomass in pools and lagoons of the floodplains of some rivers

River	Size of sample	Estimated ichthyomass (kg ha^{-1})	Reference
Apure	1	982	Mago-Leccia (1970)
Atchafalaya	49	245.95–2082.16	Lambou (1984)
Candaba	–	500–700	Delmendo (1969)
Chari	2	701–2166	Loubens (1969)
Kafue	8	444	University of Michigan (1971)
Magdalena	28	122	Kapetsky (1977)
Mekong	1	63	Sidthimunka (1970)
	390		Mekong Fish. Studies (pers. comm.)
Mog i Guassu	1	313	Gomez & Montiero (1955)
Nile Sudd	2	306–433	Mefit-Babtie (1983)
Okavango	2	200–700	Fox (1976)
Oueme	68	1835 ± 825	Welcomme (1971)
Parana			
Temporary lakes	18	959 ± 1512	Bonetto (1976)
Permanent lakes	4	766 ± 348	
Small lagoons		661 ± 517.95	Cordivuola de Yuan (1992)
Medium lagoons		500–600	
Large lagoons		193 ± 132	
Sabaki	1	786	Whitehead (1960)
Senegal	8	110 ± 144	Reizer (1974)
Sokoto	25	661 ± 557	Holden (1963)

migrate for 2000 km from the Caspian Sea to the headwaters of the river, and a coregonid, *Stenodus leuciscus*, moved up to 2100 km. In the Danube and other Ponto-Caspian rivers long-distance migration by large species of salmonid and sturgeon was also common, for example 1500 km by *Huso huso* and 2100 km by *Acipencer nudiventris* (Banarescu, 1965). Shorter migrations of hundreds of kilometres by cyprinids, shads and smaller sturgeons were also general. Even species such as *Rutilus rutilus*, *Abramis brama* and *Cyprinus carpio*, which are not noted for their migratory behaviour in modern regulated systems, were classified as semi-anadromes undertaking shorter (10–100 km) movements. Long-distance migrations are found in some North American species, especially anadromous salmonids, shads and sturgeons. Several of the Australian species undertook long migrations prior to the modification of the Murray–Darling River.

Migration is very common among Latin American species and distances of up to 3000 km have been recorded for migrations of some of the Amazonian catfishes (Barthem *et al*., 1991). Many other species of catfish, characin and prochilodontid undertake long migrations in the main channels of the Magdalena, Orinoco, San Francisco and Parana systems (see Welcomme, 1985, for summary). Migratory patterns can be extremely complex, with shoals of fish, especially characin and prochilodontid species, moving between tributary rivers and the main channel in a phenomenon known as the 'piracema' (e.g. Ribeiro & Petrere 1990). Long-distance migrations by cyprinids have also been recorded from the Tigris–Euphrates rivers and the Mekong and Gangetic systems in Asia where cyprinid and catfish species move upstream to breed in the rocky, well-aerated headwaters of the rivers. Many African species also undertake migrations within the main river channel, although distances recorded so far are low relative to those of other continents.

Longitudinal migrations in some riverine species appear to serve the function of placing eggs on suitable spawning substrates, usually gravels or submerged plants, or in specific conditions of flow and dissolved oxygen which do not exist at the downstream feeding sites. In such species the upstream spawning grounds are well defined by the distribution of these substrates. In large systems, such as the Mississippi, Mekong, Amazon, Orinoco and Parana Rivers, some cyprinid, catastomid, characin and siluroid species have semi-pelagic eggs that form part of the downriver drift during their earlier developmental stages. Such fishes generally move from one area of the river to another apparently similar area upstream. This upstream movement of adults may well be to compensate for the downstream drift of the larval stages. This leads to speculation that the distance travelled upstream by the adults (D) must be approximately the same as the distance travelled downstream in the drift from spawning to the time when fry are sufficiently autonomous as to be able to leave the drift. This may be expressed as:

$$D = \delta v$$

where δ is the time of development from the first appearance of the eggs in the drift to autonomy and v is the current velocity.

This model has several implications. In situations where returning fry are distributed over a limited section of river, discrete populations are likely to arise in different reaches of the river. This may occur in systems with stable thermal and hydrological regimes. In less predictable systems year-to-year variations in temperature (affecting δ) and flow (affecting v), together with turbulence in the river channel, may assure a wider distribution of the progeny from any one group of adults, thereby diminishing the tendency to form subpopulations. There are indications that the latter may be the case in the Parana River (Fuentes, 1998) where very different distributions of fry occur in strong and weak flood years. Little work has been done on the relationship between upstream migration and downstream drift. Such information is urgently needed to resolve questions affecting responses to changes in flow arising from damming and channelisation in rivers. It is also needed to determine the characteristics of artificial flow regimes proposed for the mitigation of spawning failures in this group of species.

Lateral migrations

Lateral migrations take place in rivers between the permanent dry-season habitats in the main river channel or floodplain lakes and the floodplain, and in lakes between the main body of the lake to the shoreline. Migrations of this type are mainly for breeding and to profit from the rich food to be found in temporarily flooded riparian zones. Most riverine species undertake movements onto flooded marginal lands and, where this is not possible, into shallow backwaters and dead arms of the river itself. Many lacustrine species come inshore to breed, depositing their eggs on wave-washed gravels or marginal vegetation. Some have specialised nursery areas, which separate the young from the possibly cannibalistic tendencies of the adults.

Complex patterns of lateral migration are found in some large tropical systems such as the Amazon. Here the 'piracema' involves non-breeding movements of adult fish from large river channels up smaller tributaries and out on to associated forested floodplains. The fish later return to the main channel to move upstream to the next tributary (Ribeiro & Petrere, 1990).

Vertical migrations

Vertical migrations are mainly undertaken by shoaling pelagic species in lakes. They are diurnal and strongly influenced by lunar phase. Such migrations form the basis of light-attracting fisheries for pelagic fishes.

Responses of fish populations to stress

Understanding the way in which fish populations react to externally imposed stresses is important for management as a diagnostic tool, as a basis for developing strategies for management and as a criterion for monitoring.

Responses to fishing and environmental change

Fish populations respond to many types of externally induced stress by undergoing a series of changes in size, species composition and abundance. Fishing, eutrophication, diffuse pollution and environmental degradation all elicit similar responses in the population as a whole (Welcomme, 1999a). Some specific responses may also be observed, such as the elimination of migratory species by the creation of cross-river dams or the local elimination of fish by acute pollution, but these will be described elsewhere. The main change, to which all others are linked, is a decline in the mean size of fish in the population. This is caused by the progressive loss of larger individuals and species and their replacement by smaller ones. The shift to smaller fish also involves a drift from long-lived (K selected) species to short-lived (r selected) ones and is a well-known response to stress by animal communities in general (Selye, 1973; Barrett *et al*., 1976). In inland fisheries it has been amply documented from a large number of climatic zones and continents, e.g. Benin (Welcomme, 1995) (Fig. 11.3),

Poland (Bninska, 1985), Bangladesh (Chu-Fa & Ali, 1987; MRAG, 1994a), Indonesia (Christensen, 1993), Peruvian Amazon (von Eckman, 1983), Shire river, Malawi (FAO, 1993), Lake Malawi (Turner, 1981) and the Niger River (Lae, 1994).

Reduction in size occurs as much in fisheries that are based only on gears that select for a narrow size range of sizes at any one time, as in true multi-gear fisheries where the selectivity range is much wider. Certain factors depend on the fishing strategy. In fisheries based on gears that select for a narrow size range of fish, increasing numbers of species enter the fishery as mesh sizes are reduced. In fisheries with a full range of fishing gears, most species and lengths are targeted and the number of species tends to fall as the fishery selectively eliminates the larger, more vulnerable groups. Some species, especially larger, K-selected ones are unable to accommodate to fishing pressure and disappear from the assemblage. Very often the larger species are piscivorous predators so the process is accompanied by a decrease in the predator/prey ratio. Some species such as the tilapias are able to maintain their place in the assemblage by reducing their breeding size and maturation time.

Levels of catch in multi-species fisheries prove remarkably stable over a range of fishing effort. When catch is plotted against effort there is a long plateau during which the total weight of fish caught varies little. This effect is mainly due to the fact that smaller fish are more productive. Thus, although the standing stock or biomass (B) may decline as fishing or environmental impacts intensify, the production (P) and the P/B ratio both increase (Fig. 4.13). The stability of the plateau is usually apparent in lakes but may not be immediately apparent in rivers and river-driven lakes and reservoirs because the catch varies from year to year. This variation is the result of differences in flood intensity, which influence catches. Examples of plateau effects have been documented in individual rivers (Novoa, 1989), grouped sets of rivers (Welcomme, 1985; Hoggarth & Kirkwood, 1996), lakes (Regier & Loftus, 1972; Turner, 1981) and marine fisheries (Marten & Polovina, 1982). Lae (1997) described the relationship of catch and effort in a multi-species fishery in the coastal lagoons of West Africa by the formula:

$$\log_e(Y+1) = \log_e(Y_{max})(1 - \exp - (af))$$

where y is yield in kg ha^{-1} year^{-1}, f is effort (number of fishermen km^{-2}, Y_{max} is the asymptotic level of yield, and a is a fitted constant. This model describes the form of the relationship based on effort. There is no system available as yet to predict the asymptotic level at which Y_{max} will occur other than the use of generalised predictors such as the Morpho-Edaphic Index in lakes (MRAG, 1995) and yield/area equations in rivers (MRAG, 1994a).

Responses to exploitation under fluctuating water levels

Fish populations in rivers, and river-driven lakes and reservoirs vary widely in abundance from year to year. The relative abundance of the various species in the

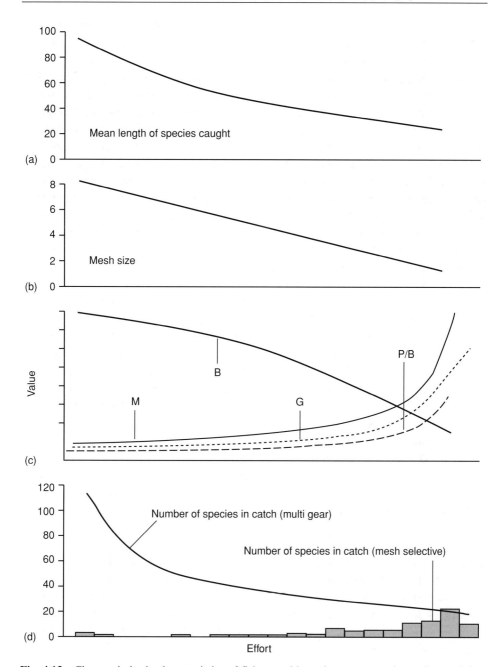

Fig. 4.13 Changes in basic characteristics of fish assemblages in response to increasing exploitation pressure: (a) mean length of fish in catch; (b) mesh size of size selective gear; (c) changes in essential parameters; (d) changes in relative number of species available to broad-spectrum multi-gear fisheries and size-selective fisheries.

population also varies. These variations are in response to fluctuations in precipitation and flood strength. The flood operates on the fish through improved breeding, growth and survival as flood levels and duration increase. Good correlations have been attained between some measure of flood intensity (FI) and catches in subsequent years (see Welcomme, 1995, for a partial review). Here the lag time between the year of flooding (y) and the reaction by the fishery ($y + t$) depends on the time (t) taken for the fish to grow to a size at which they enter the fishery. Many measures of flood intensity have been used. One of the simplest is the area under the flood curve as derived from measurements of water depth, although other measures may be used, such as mean daily discharge. Other measures indicate the amount of water remaining in the system during the dry season. In lightly fished assemblages, floods in a number of years may impact the fishery in any particular year, thus:

$$C_y = a + b(pFI_y + p_1FI_{y-1} + p_2FI_{y-2} \ldots + p_xFI_{y-x})$$

where C_y is the catch in year y, and p_x is the percentage that the flood index of the associated year contributes to the total correlation (see p for alternative form of this relationship).

In such fisheries or in those based on migratory species with long maturation times it may take several years for the fish to enter the fishery and best correlations are with y–3 and y–4. In these cases the mean size of fish in the catch is relatively large. In moderately fished systems the best correlations are obtained with y–1 and y–2. In heavily fished tropical systems lag times are much shorter (y or y–1) and the fish in the catch are correspondingly smaller. For example, 69% of the fish caught in the central Delta of the Niger are less than 1 year old (Lae, 1992). In extremely heavily fished systems the best correlations are obtained with young of the year as they move from the floodplain to the main channel at the end of the floods. For example, in Bangladesh 98% of the catch falls into this category (Halls, 1998).

Where a shift in lag time occurs within a time series because of changes in fishing intensity or practice, analysis of the correlation between flood and catch may not give good correlations unless the time series is split. Two factors in this analysis, 't' and 'the number of years (y; y–1, etc.) influencing the regression' are useful indicators of the overall level of exploitation of the fishery. In highly stressed fisheries t tends to equal zero and the number of years influencing the regression to be confined to one.

In some rivers the amount of water remaining in the system in the dry season appears to influence abundance to a greater degree than the level of flooding, especially in systems where the percentage of residual water is low relative to the volume of the flood.

Summary of changes occurring in response to stress

Table 4.7 lists the major changes that occur in fish populations in response to externally applied stresses such as fishing or environmental degradation.

Table 4.7 Indicators of change occurring in fish populations in response to fishing and environmental stress

Indicator	Trend
Level of catch	Falling levels of total catch levels in single-species fisheries; however, catch levels can be maintained over a wide range of effort and can only be interpreted relative to the catch composition
Mean length	Disappearance of larger fish and falling mean size within species
	Disappearance of larger species and falling mean size of catch as a whole
Number of species	Initial rise in number of species in catch from a few large ones to many small ones
	Later falling numbers until the fishery is confined to few very small species
Type of species	Decline and disappearance of anadromous and long-distance riverine migrants
	Decline and disappearance of native species in favour of exotics where introductions have occurred
	Decline and disappearance of higher trophic levels (predators) and their replacement by lower food-chain species
	Decline and disappearance of species with high oxygen requirements and their replacement by species tolerant of low oxygen (eutrophication)
Response time	In rivers and river-driven lakes and reservoirs, shortened time between flood events and response by population
Other indicators	P/B ratios rise
	Mortality rates (z and f) increase
	Higher incidence of diseased and deformed individuals (extreme eutrophication and pollution)

Notes

1. By Ashley Halls, MRAG Ltd, 47 Princes Gate, London, SW7 2QA, UK, and R.L. Welcomme.
2. By R.L. Welcomme.

Chapter 5
The fisherman and the fishery

The fisherman in society

Any fishery comprises three components: the environment, the fish and the human. The human elements are complex social and economic groupings that extract, process and consume the fish as well as regulate the integration of the fishery into the larger frame of the society as a whole. The fisherman sits within a complex web of social, financial, ecological and administrative influences that condition his life and have to be taken into account in managing the fishery (Fig. 5.1) (see also Townsley, 1997, for a fuller examination of the social issues in fisheries management).

Many of these influences are external to the fishery and define the place of the fishermen in society as a whole. Within the fishery the fisherman, as the producer of fish, is the origin of a flow of resources. This flow is the production chain whereby the fish caught is transferred from the producer (the fisherman), through a sequence of middlemen who transform and transport the product to the consumer. Each step in the chain involves an increase in the price of the commodity.

Many fishermen do not own the fishing sites, nor do they own the gear and boats that they use. The fishermen in these cases are obliged to rent the boats and gear or to work as a paid employee of their owner. Likewise, where the water to be fished is privately owned, the fisherman may have to pay a lease or fee to the landowner for the right to fish. In some cases the owner is the government, in which case the fee usually takes the form of a licence. Fishermen may also be taxed on their income or their landings. This system generally works to their detriment as they are remunerated less for their catch than are other stakeholders for their inputs. Furthermore, fishermen are likely to acquire high levels of indebtedness to boat owners, landowners, gear suppliers and the government, which keeps them in a state of economic bondage. One of the purposes of good management is to break this cycle so as to place the fishermen in charge of their own economic welfare.

The fishermen

Fishermen are far from a homogeneous group and, like any other component of society, have their affluent and powerful as well as their poor and powerless elements (see Figs 5.2 and 5.3). Richer fishermen tend to dominate any fishery by their use of more

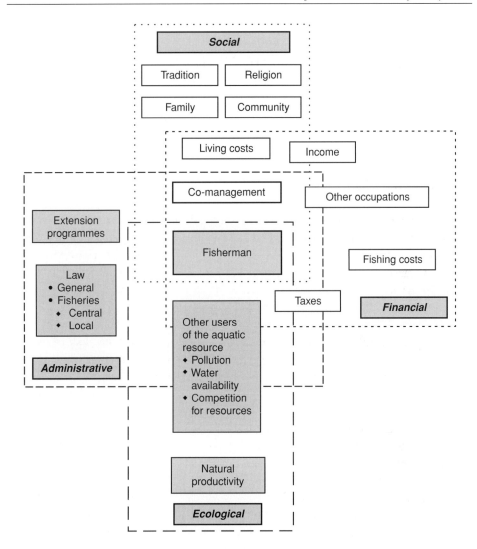

Fig. 5.1 Diagram representing relationships influencing the fishermen.

expensive and efficient gears and by their control of the best fishing grounds. The type and variety of fishermen involved in the fishery will have an important influence on the management structures.

Fishermen can be classified into broad groupings according to their purpose and the way in which they use the resource.

86 *Inland fisheries*

Fig. 5.2 Typical dwelling of lagoon and river fishermen, Ganvie, Republic of Benin.

Food fishermen

Full-time fishermen

Full-time fishermen are one of the smaller stakeholder groups dependent on inland fisheries. They are found in or around major water-bodies where there are sufficient-

Fig. 5.3 Typical temporary camp of migrant fishermen, Niger River, Republic of Benin.

ly large quantities of commercially valuable fish year round. In some systems, full-time fishermen may also migrate between a number of different habitats or river sections according to the season. The livelihoods of full-time fishermen can be very sensitive to the profitability and yield of the fishery, as it is their only source of income and they frequently have a substantial investment in boats and gear. Alternatively, some permanent fishermen are drawn from poorer elements of the community and are employed to fish by owners of boats, gear and fishing rights.

Part-time fishermen

Fishermen in this group usually have other important sources of income, such as agriculture, urban labour or transport, with which they alternate their fishing activities. Such fishermen may also be highly sensitive to the profitability of the fishery, as this is likely to determine the percentage of their time they invest in this pursuit. In other situations, particularly where fisherman–farmers are involved, the fishery is practised mainly during downtime from the alternative activity. Here the fishermen may continue fishing even when catch rates are very low. Even if their opportunity costs are zero, fishing will still be profitable as long as the value of catch exceeds other variable costs (wear and tear on gear, fuel, etc.), which are themselves often very low. Furthermore, frequently no realistic alternative exists for them to occupy their time. Part-time fishermen may have a considerable investment in boats and gear but more often rely on locally produced equipment that is procured at little financial cost. They may also provide a seasonal labour force for boat and gear owners.

Subsistence fishermen

Many people fish to supplement the family diet during slack periods in their daily schedule or seasonal calendar. This form of fishing may occupy members of larger, richer households during periods when they are not needed for agriculture or can be an occupation of last resort for the widowed or landless. Small boys and women are particularly active in gathering small quantities of fish using primitive gear.

Recreational fishermen

Recreational fishermen do not depend directly on the fishery for employment, treating fishing more as a temporary pastime. They are often a relatively wealthy group coming from urban professional backgrounds. They are, therefore, to some degree external to the rural milieu in which they find their sport. Different groups of recreational fishermen have different approaches to the fishery, which also has important consequences for management.

Match

Match fishermen seek to compete individually or in groups according to a set of rules. Rules specify the aim of the match, which is typically to catch the maximum weight or number of fish in a specified period, and the gear and bait to be used. Such fisheries are usually supported by intensive stocking. Maintenance of a pool of large fish in the environment is not particularly desirable and managers will tend to maximise fish density at the expense of size.

Specimen

Specimen fishermen tend to concentrate on particular species and attempt to catch the largest individual of that species. Specimen fish waters need to be managed for a lower density of fish but with adequate environmental quality. Both match and specimen fishermen usually use rod-and-line.

Relaxation

A great percentage of fishermen seeks to relax in pleasant surroundings and are not necessarily bothered as to the quantity of their bag. Management of the environment best supports fishermen of this type. Relaxation fishermen use rod-and-line but may also use more commercial types of gear such as lift nets and gill nets.

Domestic consumption

In poorer countries, and some more prosperous ones such as Finland, recreational fisheries play a great role in supplementing diets and the fish caught is taken home for food. In these quasi-subsistence fisheries the line differentiating between relaxation and subsistence fisheries is sometimes hard to establish. A wide range of gear is used in such fisheries, including rod-and-line, traps, nets and long lines.

Other stakeholders

Other stakeholders derive their livelihoods from either side of the fishery: upstream are those supporting or providing inputs to catching activities; downstream are those involved in the supply chain between the fisherman and the consumer. In some counties, particularly in South and South-east Asia, the same individual can be involved both upstream and downstream.

Upstream

Gear manufacturers

The fishermen or their families make many of their more traditional gears from locally available materials. Specialised gear builders can also be found in rural communities constructing netting from old tyre windings, or traps from chicken wire mesh. More modern gear based on synthetic twine and netting is manufactured in national industrial plants or is imported, and is now readily available in most areas of the world. Because of its cheapness, ease of use and durability, modern netting is supplanting many of the traditional fishing methods, especially among professional fishermen.

Owners of water rights, boats and gear

Most inland fishermen are among the poorest elements of the society. They rarely enjoy financial independence and rely for their livelihood on the owners of the means of production. In some areas inland waters are open to all comers, but there is an increasing tendency to limit access to this previously open resource. In other areas the rights to fish are owned by the government or by private landowners who charge for access through traditionally determined systems of leases and licences. One common system, especially in Asia, is for fisheries to bid for a river reach, arm or lake (lot) which entitles them to fish that water for a specified period. Often entrepreneurs acquire the fishing rights and levy a charge on the fishermen (in cash or as a catch share) or employ them to fish for them. Similarly, poorer fishermen may not own the boats or gear they use but hire them from owners against a fee or a proportion of the catch. More equitable access arrangements also exist, as in some rural Indonesian communities that use a fortnightly lottery system to allocate trap fishing locations to village members.

Tourist industry

The tourist industry supports recreational fisheries by providing transport, accommodation, fish guides, literature, clothing and a range of other services to the recreational fishermen and their families. Sport fishing is used as a major element in tourism and there is considerable competition to attract recreational fishermen from the cities and abroad. A viable tourist industry is a powerful argument in this competition.

Downstream

Once fish are caught they may be reserved by the fisherman for his own consumption, sold fresh or preserved to local consumers, or purchased either at a landing site or directly from the fisherman on the water for onward transmission to distant markets. It is rare for the fisherman to market his catch himself, although this may be done in some local fisheries. It is more usual that the intermediate steps between catching and consumption are carried out by others who, like the fishermen, may undertake these activities on a part- or full-time basis. The number of both activities and actors involved will be determined by both the products of the fishery and the markets that it serves. Where economies of scale allow, each activity may involve a different, specialised actor. Otherwise, the same person may undertake a number of these steps.

Processors

Fish is a highly perishable commodity, especially in the warmer climates. One of the earliest steps in its journey to distant consumers is therefore preservation. The methods used to preserve and market fish are discussed in Chapter 7, p. 125. Preservation by some methods, such as smoking or under ice, may be carried out directly by the fisherman or his family members. In some cases, such as in West Africa, the involvement of the family member is formalised by the sale of the fish by the fisherman to the involved family member. In other areas independent merchants collect the fish and preserve it before onward transmission.

Transporters

Transporters buy fish and transport it for resale. Often the distance travelled by individual transporters is not great as they usually operate from bicycles, small motorised vehicles or canoes. A number of steps may be involved in transferring fish from the catching site, through progressively larger markets where fish from wider areas is concentrated, to an urban market where it is finally bought by retailers.

Retailers

Retailers are responsible for the final sale of fish to the consumer at markets, door-to-door, on roadside food stalls or in fish shops.

Chapter 6
Fishing techniques

Types of fishing regime

Catch and remove

Most fisheries aim at the removal of fish from the water for some defined objective such as food, ornament, destruction of unwanted fish or medicine. In some cases the removed fish are kept alive in aquaria for ornament or for conservation of rare species. Several types of fishery fall into this category.

Natural

Natural fisheries rely entirely on natural reproduction and feeding. Such fisheries are bound by the strict limits of the productivity of the water in which they are practised and by the reproductive potential of the target stock. Reproductive potential may be determined by biotic factors such as abundance of the species, the degree to which it is fished, and competition and predation from other species. It may also be influenced by physical characteristics of the environment such as the abundance of spawning substrates, temperature, the rate of drawdown in reservoirs or the extent of flooding in rivers. The productivity of the water is determined by geomorphic considerations including the nature of the bedrock in the basin, the quantity and type of effluent discharges from human activities, the depth and shoreline development of a lake, the extent of flooding in rivers and the residence time in reservoirs.

Enhanced

When fisheries exploitation exceeds the levels of fishery that can be supported by natural processes, fishery managers seek to intervene in the system to overcome the limiting factors. Most such interventions are aimed at supporting one or a few species that are especially sought after for the objectives of the fishery. Some intervention, fertilisation for example, can operate at the level of the fish community as a whole. Manipulation of the natural cycles of reproduction and productivity can have drastic effects on the species composition of the whole fish community. This may be the intention of the manager in the first place; however, unwanted effects often emerge such as the rise to dominance of a nuisance species.

Stocked

Stocking may be pursued at several levels but, initially at least, is seen as a way of compensating for shortfalls in recruitment by target species that arise through overfishing or through environmental degradation.

Put, grow and take
The most common type of stocked fishery is where juvenile fish are introduced to grow on food organisms arising from natural productivity. The adult fish are later removed from the system by regular fishing for either food or recreation.

Put and take
Put-and-take fisheries reduce the time between stocking and capture to a matter of a few weeks. Fish are often stocked as adults and are removed almost immediately by recreational fishermen. Put-and-take fisheries are practised in some natural waters but are more normally restricted to special ponds often associated with a fish culture station.

Other interventions

A number of other technologies exists to relieve other bottlenecks to production of fish. These include introduction of new species, fertilisation of ponds, engineering of waterways and removal of unwanted species. As the human inputs intensify, natural capture fisheries come to resemble culture systems to an increasing extent. These technologies are described in chapter.

Catch and return

An increasing trend in recreational fisheries is for the fish, once captured, to be returned to the water for the enjoyment of others. Catch-and-return policies are fundamental to the sports fisheries policy of the USA and other temperate countries such as The Netherlands and the UK. They have been questioned by animal rights groups as causing unnecessary distress to the fish. The impact of catch and return on fish behaviour and populations is not well understood. The phenomenon of 'hook shyness' may influence future catching success and fish may sustain damage, which increases their proneness to disease and feeding difficulties. Furthermore, in stocked systems the return of fish to the water may lead to long-term build-up in populations, leading to stunted fish or even mass mortalities.

Fishing technology

Fishing gear[1]

Factors influencing the choice of gear

Many different methods of fishing and types of fishing gear have emerged over the centuries and continue to be used and further developed to meet local conditions. In many areas of the world fishing gears remain rooted in tradition and use simple materials. In others, especially those with recently developed or expanding fisheries, modern gears form the entire basis of the fishery. Traditional gears are often deployed with a great degree of sophistication, which is based on a deep knowledge of the local conditions and the behaviour of the target species.

Many factors enter into the choice of the method and gear used to catch a particular species in a particular area. Principally, the choice will depend on the following variables.

Depth of water
The various types of fishing gear are designed to be operated at particular depths of water. Gear may be set on the bottom, at the surface or in mid-water depending on local conditions and the target species. Depth of water influences the ease with which different gears can be used and exerts considerable influence on the choice of gear and the way in which it is rigged.

Characteristics of lake and river bed
Fishing gears operated in contact with the bottom are as vulnerable to damage from hard, uneven or rough lake or river beds as from submerged vegetation and wood snags. In such areas it is often impossible to use moving gear because of the nature of the bottom. However, some static gear can be successfully placed on the obstructed bed of the lake or river with little problem.

Nature of the vegetation cover
Shallow lakes, swamps and riverine floodplains have characteristically dense vegetation either as a rooted or floating riparian fringe, as in the case of reeds, papyrus, *Vossia*, *Echinochloa* or *Paspalum*, or as floating mats, as with *Eichhornia*, *Salvinia* or *Azolla*. Dense vegetation tends to attract fish but at the same time to hinder the use of many fishing methods. Many of the simple traditional gears operated in rivers and swamps have been designed to overcome these problems.

Species being fished
The various species targeted by fishermen differ in their habits, movements and reactions to stimuli. These differences may also be linked to the age of the organism and to its seasonal behaviour. The fishing gear selected must continually adjust to these habits.

Freshwater shrimps, most freshwater molluscs and numerous species of fish are found at or in the lake or riverbed. Demersal species of this type are caught by bottom-set gear. Pelagic species may be found anywhere between the bed and the surface. These are normally taken by fishing gear that is not in contact with the bottom and is set in mid-water or at the surface.

Some fish congregate during migration, spawning or normal diurnal shoaling. These may be taken in large quantities but their patchy distribution sometimes makes them difficult to locate. Surface water shoaling species can be encouraged to congregate in a suitable area with light. Species, such as freshwater shrimps and many demersal fish, are more loosely distributed, and yet others are found singly or in small, scattered groups.

Season of year and the flow regime
Seasonal migrations are highly significant for fisheries. In rivers the movements of species along river channels and between the river channels and adjacent floodplains provide increased opportunities for capture owing the predictability of the event and the high concentrations of fish sometimes observed. Events in lakes where fish ascend inflowing rivers to spawn and in the sea where anadromous and catadromous fishes move between the saltwater and freshwater environments present similar opportunities.

Excessive flow can wash away gear or choke it with debris and generally make life difficult for fishermen. As a result, fishing during periods of high flow or in permanent rapids is usually avoided unless there is some overriding reason to do so. Migrations of important commercial species provide such an occasion and elaborate gears have been devised to capture such fish in special locations such as the Khone Falls rapids on the Mekong River, the Zaire River rapids and the rapids of the Madeira River in Brazil and Bolivia.

Value
Different species of fish have different characteristics that condition their desirability to society. Factors such as size, taste, firmness of flesh, keeping quality or some other cultural criterion give them very different market values. Furthermore, the value of the same species may vary depending on the way in which it is processed and marketed.

High-value species are normally large and firm fleshed, with a clean taste and good keeping qualities. Small species, which rot rapidly at the high ambient temperatures of the tropics, are generally less valued. The lowest valued fish may be used for reduction into animal feeds, such as pet foods or fishmeal.

The value of the fish influences the choice of gear, as some fishing methods can only be justified for high-value species. For example, fishing with individual hook and line can be economically viable in the case of the large salmon of high individual value or for subsistence fisheries by individual fishermen, but would be inappropriate for the capture of fish for bulk processing into meal.

Financial situation of the fisherman

Certain gears are expensive and others cheap and the choice of gear by any individual is closely related to his economic circumstances. Fishermen have long resolved this problem by hiring gear from a wealthier individual, by working as paid labourers or by grouping into co-operatives.

Legislative, traditional or sporting regulations

The use of gear in any fishery is circumscribed by governmental, traditional and special regulations. Governments usually prohibit damaging gears such as explosives and poisons. They may also rule out the use of individual gears or impose limitations on their use where these are felt to be damaging in the local circumstances. For example, mosquito net seines are banned in many fisheries because they are thought to take an excessive number of juvenile fish. Cross-river barrier traps may be limited to only a proportion of the channel in the interests of sustaining navigation or to ensure some access of migrant fish to their upstream spawning grounds. Cast nets and other active gear may be banned from areas where there are concentrations of static gear. Traditional management systems take into account such factors but also tend to concentrate on fishing methods that have become integrated into local usage. They tend to resist introduction of new gears and also seek to conserve equity among the different groups of gear users. Sporting regulations in recreational fisheries define the type of gear to be used in any particular event or location, as well as such factors as type of bait and weight of line.

Selectivity of gear

Many types of fishing gear select for fish of certain sizes. Such gears are usually constructed in such a way as to have a mesh which allows small fish to pass through, capture fish of intermediate size and are unable to retain fish above a certain length. Mesh selective gears are usually constructed of netting, although woven traps may also show similar characteristics. Typically, the lengths of fish which are susceptible to capture in any particular mesh size are normally distributed. Data from a series of nets of different mesh sizes can be combined to describe the selectivity for a species expressed as percentage of fish of each length retained by each mesh size, Fig. 6.1 gives the example of *Oreochromis leucostictus* in Lake Victoria. The mean length of fish caught can be plotted against mesh size and the resulting regression line can be used in combination with Fig. 4.13 to indicate the numerical composition of catches in different meshes. Each species of fish will have specific selection characteristics because selectivity acts primarily on the girth of the fish. Thus, long, thin species will have a different pattern from short, fat ones, and round-bodied species will differ in their response to nets from flatter ones.

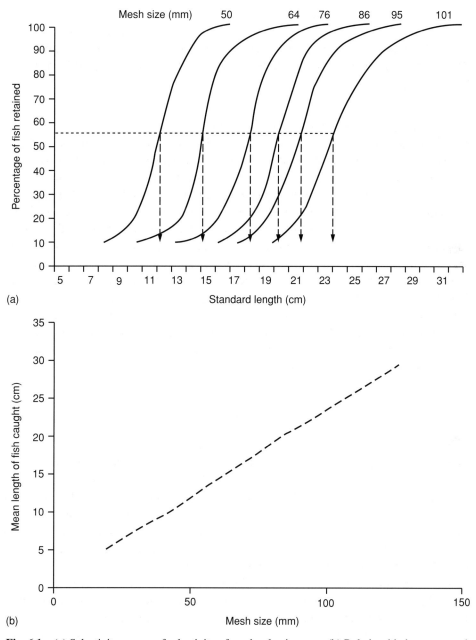

Fig. 6.1 (a) Selectivity curves of selectivity of mesh selective gear. (b) Relationship between mesh size (mm stretched) and mean standard length of *Oreochromis leucostictus* caught.

Principal types of gear

Static gear

Static gears are those which depend on the fish moving to the gear and in some cases taking a bait. For this reason they are also known as set-and-wait gear. They are therefore particularly appropriate for the capture of species during migrations and movements. They are also used in areas where more active methods of fishing are obstructed by underwater obstacles or floating mats of vegetation at the surface. Many traditional methods of constructing and working static gear have been developed around the world to suit particular fisheries. It is possible only to describe some of the more common techniques here. More detailed descriptions of fishing gear are available in Von Brandt (1984) for all types of fishery and in Welcomme (1985) for river fisheries.

Gill nets Gill nets consist of a simple wall of netting maintained by a float line at the top and a lead line at the bottom (Fig. 6.2). The net may consist of one sheet in which the fish are trapped by their gills or several sheets of various mesh sizes in

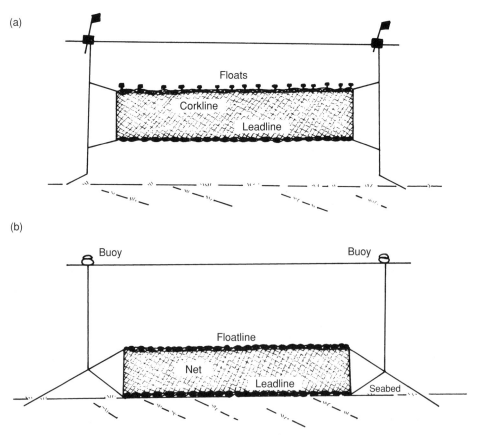

Fig. 6.2 Diagrams of (a) surface-set and (b) bottom-set gill nets.

which they become entangled (tangle or trammel nets). Gill nets may be set either just above the bottom when fishing for demersal species, or anywhere from midwater to the surface for catching pelagic fish, by adjusting the balance between the float line and lead line.

When working inshore in relatively shallow water, the nets are usually set and anchored in position. Alternatively, they may be used as a surface-set drift net in rivers where they and the boat from which they are operated float downstream with the current.

Several gill nets may be set attached end to end in fleets, and rather than being set in a straight line may be placed in complex, trap-like formations. The performance of the net can also be improved by beating the water on either side of the net to scare fish into it, a practice that is illegal in many fisheries.

Gill nets are usually supplied by commercial netmakers, are relatively cheap and, if treated properly, are durable. They are also relatively easy to operate and have formed the basis of many new fisheries that have developed in reservoirs and lakes throughout the world.

Gill nets are highly selective both for upper and lower size ranges of fish and their meshes are frequently regulated by legislation. Their simplicity of construction and operation makes them one of the most basic and widespread methods for fishing inland waters and the preferred method in most lakes and reservoirs. They do, however, require a boat from which the gear may be set and lifted and this imposes an additional cost. In addition, gill nets are vulnerable to poaching and operators frequently spend the night watching over their gear or devise ways to conceal the telltale floats.

Long lines Long lines consist of a long length of line, to which short leaders carrying hooks are attached every 60–180 cm (2–6 feet) (Fig. 6.3). They can be surface or bottom set and adapted for specific target species. Lines may be set from bank to bank across river channels or in lakes anchored between standing lines each with a float and sinker. The line may be unbaited and operate by snagging or foul hooking. Foul hooking is frequently prohibited by tradition because of the dangers to other gears. More commonly the line is baited, the choice of bait depending on the target species. The line is inspected periodically, captured fish are removed, the hooks are re-baited and the line is reset. Long lines may be operated from the shore or by wading in shallow rivers, lakes and swamps, but normally require some form of fishing craft from which they can be set and controlled.

Long lines have an element of selectivity depending on the size of hook used. They are relatively cheap to construct but are time consuming to operate. For this reason their use is frequently entrusted to junior members of the fisher household. Because of their cheapness they are popular with subsistence and occasional fishermen or may form a supplement to the main gear operated by a family.

Traps Traps are devices designed to encourage fish to enter a confined space and to prevent fish from leaving once they have entered. They may be of many sizes and con-

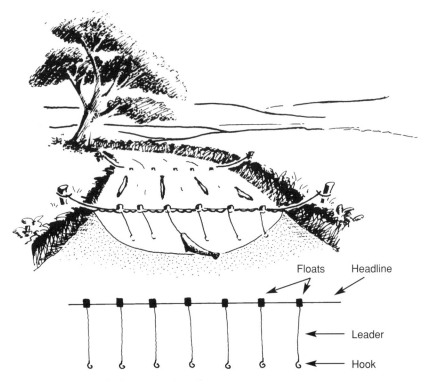

Fig. 6.3 Configuration of a bottom-set long line.

figurations but usually have an entrance, some form of non-return structure and a capture chamber (Fig. 6.4). They may be made out of local materials such as palm fibres or bamboo (Fig. 6.5), or commercially bought wire mesh or netting. Traps may be set unbaited or baited depending on the target species. Baits may include rotting meat, dead fish, palm nuts or corn. Traps are selective for size at the lower size ranges but will capture any size upward that will pass through the entrance.

Usually, traps are set in places through which fish regularly move or congregate, such as river mouths, main river channels and access channels to floodplains. They may also be set in areas not adapted to other gears such as under floating vegetation mats or in relatively shallow water adjacent to the land. Traps may be set as individual engines or associated with barriers that deflect the fish towards the capture chamber. Traps and barriers may extend from the surface to the bottom, at least at the mouth, and may have extremely complex configurations.

Like long lines, traps may be operated from the shore or by wading in shallow rivers, lakes and swamps. More normally they require some form of fishing craft for transport, setting and control. The fish caught remain alive within the trap so they may be set and left for several days. Traps will also continue to catch fish even after they have been abandoned.

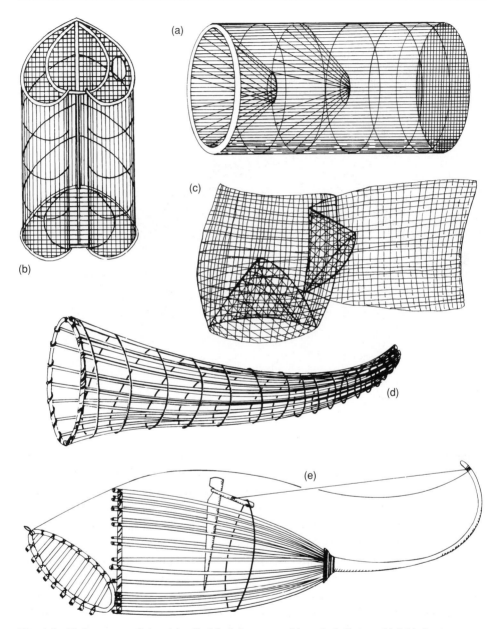

Fig. 6.4 Various types of trap: (a) cylindrical drum trap; (b) vertical slit trap; (c) folded woven trap; (d) funnel trap; (e) spring trap.

To form a viable source of income any fisherman has to set numerous traps. Most types of trap are cheap to construct because they are constructed of locally available materials. Their manufacture is time consuming, however, and often confined to periods of unemployment such as the flood in rivers. Because of their cheapness and ease

Fig. 6.5 Traps used in the Ubolratana Reservoir, Thailand.

of operation traps are among the most common of subsistence and artisanal fishing gears and are found in every traditional fishing community world-wide. Like long lines, the operation of traps is frequently entrusted to junior members of the fisher household and may form a supplement to the main gear operated by a family.

Barriers The efficiency of traps is considerably enhanced by adding wings on either side of them to direct fish towards the opening (Fig. 6.6). These wings are often extended to group several traps. A further modification of this is to extend the structure across the river to form a barrier armed with a series of traps. Barriers in large river and coastal lagoon systems are necessarily big undertakings requiring a considerable labour force to construct and maintain. In such cases they are usually operated by teams working either as a co-operative or as labourers for an owner (Fig. 6.7).

Cross-river traps are sensitive to current and usually have a limited fishing season. Some barrier traps are most effective at the times of fish migrations, especially for longitudinal migrants moving back downstream at the end of the flood. Low water traps are constructed as the falling flood loses its force and are removed or washed away on the rising water. In other cases robust, permanent frameworks are left in place and re-armed with traps when more favourable conditions return.

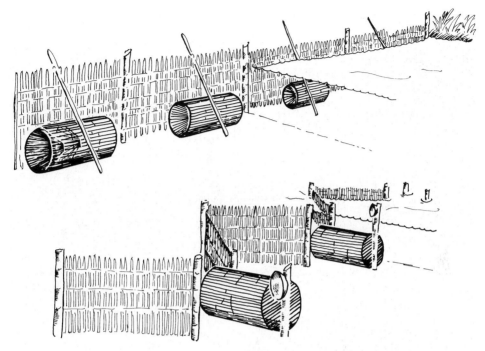

Fig. 6.6 Two types of barrier associated with drum traps.

Fish trapping fences Fish trapping fences are elaborations of barrier traps in which the fence itself is configured into enclosures from which the fish find escape difficult. The fence may be folded into complex patterns of chambers (Fig. 6.8) to capture a wide range of fish and crustacean species. Such structures are necessarily constructed in shallow waters at the mouths of channels draining floodplain lakes, estuarine flats and coastal lagoons. When the flood recedes, or in the case of coastal lagoons and estuaries, the tide falls, fish are collected by the structure as they try to escape the lowering water. Fences may also be used to imprison fish in temporary pools of the floodplain so they are stranded by the falling water and are easily collected. Fences are usually made of locally obtainable branches, bamboo and brushwood. Suitable sites for installing fish fences and barrier traps are relatively rare. There is, therefore, competition for such areas, which are frequently allocated or owned traditionally or leased for a fee (see Fig. 6.9).

Fyke nets and bag nets Fyke nets are traps made of netting (Fig. 6.10). The basic design consists of a cylinder supported by hoops within which cones of netting act as non-return valves. The basic net is supported by systems of wings and leaders, which play a similar role to the barriers associated with other types of traps. These may often be complicated in form. Fyke nets are highly selective for lower size ranges.

Fig. 6.7 Fisherman working in a fish fence. Note the heart-shaped capture chambers that correspond to Fig. 6.8.

Fishing techniques 105

Fig. 6.8 Cross-river fish fence.

Fig. 6.9 Fishing lots in the Tonle Sap of the Mekong system, showing complex fences and trap structures enclosing the lot. (Photo by Niek Van Zalinge.)

106 Inland fisheries

Fig. 6.10 Fyke net.

Bag nets are large fykes usually mounted on rafts in the centre of the stream to catch migrating fish. They are especially common in South-east Asian fisheries such as the Tonle Sap in the Makong, where bag-net sites may be subject to auction and leasing in the same way as fishing 'lots'.

Active gears
Active methods of fishing depend on the gear moving to surround the fish. Gear is therefore pulled either by a boat or by human effort. Active gears are particularly appropriate for the capture of fish when they are relatively static on the bottom, in the water column or in some refuge. Active gears have gained much from the increased power available to fishermen through outboard engines and improved methods for making large sheets of netting.

Trawls Trawls are nets that are towed behind one or two boats to engulf any fish that are travelling more slowly than they are. They consist of a cod end preceded by two wings. The wings are kept open by tension from a pair of boats (pair trawling), a beam (beam trawl) or two flat doors (otter board trawl). The beam or doors are attached to the boat by long warps.

Fishing with trawls is not common in inland waters. The investment needed in boats and gear is rarely matched by the value of the catch. Furthermore most water bodies are not large enough and bottoms are frequently encumbered with mud, logs and rocks. However, in some large lakes the resource is sufficiently abundant and valuable as to justify the use of trawls.

As trawls require a sizeable investment individual fishermen at subsistence or arti-

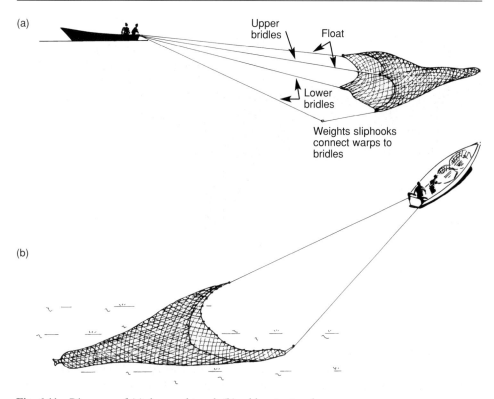

Fig. 6.11 Diagrams of (a) demersal trawl; (b) mid-water trawl.

sanal levels rarely use them. They are essentially a professional, commercial fishing system that may be owned by an individual or a company that employs a number of crew to operate the boat and gear.

Bottom trawls (Fig. 6.11a) are the most common in inland waters and fish for large demersal species such as the Nile perch in Lake Victoria. As bottom trawls operate at or near the lakebed they are very prone to damage and, on occasion, to complete loss from submerged rocks or vegetation.

Bottom trawls have been implicated in damage to the benthic environment by uprooting bottom vegetation, burying benthic organisms and overturning the substrate. They should therefore be used with extreme caution in the often very fragile conditions at the bottom of lakes.

Mid-water trawls (Fig. 6.11b) are much less common in inland waters but are used to capture pelagic species in large lakes such as Lake Malawi. Both single and paired vessels can be used in mid-water and surface trawling. Pair trawlers are particularly suitable in view of the smaller, low-powered vessels that are common in inland waters. A small trawl can be operated from two canoes equipped with an outboard motor, for instance. Two-boat trawling offers the additional advantage in that the towing warps do not pass through and frighten the school of fish before the net reaches them. The

108 *Inland fisheries*

net can be towed from just above the lakebed, to just below the surface the surface. The depth of the net is controlled by its speed through the water and by the length of towing warps. Tow times vary considerably, depending on whether the fish are in shoals or are widely dispersed, and can be as little as half-an-hour where fish are abundant.

Trawls are highly selective for both upper and lower size ranges of fish. The selectivity of the net is determined by the cod-end, which is usually of smaller mesh than the wings. Selectivity towards larger fish declines as these are able to swim quickly enough to escape the trawl. The mesh permitted in any fishery is usually specified by legislation.

Seine nets A seine net consists of long sheets of netting attached to a head line equipped with floats, a weighted foot line and long warps for hauling. The net may have a cod end, although it is more usual for it to be plain. They may be used from the beach or in open waters, bottom set for catching demersal fish or surface set to capture pelagic species.

Beach seines are set from the shore in a semi-circle (Fig. 6.12), usually from a boat. The net is then hauled ashore using long warps. Large nets require a substantial crew to operate and are expensive. They are more of a commercial gear than a subsistence one and are frequently owned by a single fisherman who employs a number of labourers. Alternatively, the net may be owned and operated by a fisher co-operative. Fewer people can operate small nets, by wading or as barrier nets across a river.

The effectiveness of fishing is partly due to the movement of the warps across the bottom, which disturbs and guides the fish within the area being worked. Several species of fish are energetic jumpers and may escape by leaping over the net as it is closed. For this reason fishermen hold the floatline of the net up during the later stages of closing.

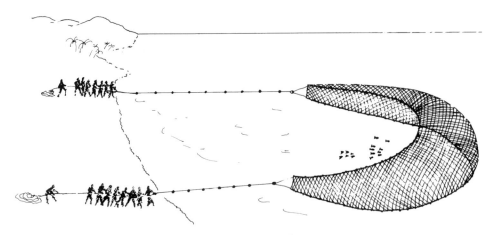

Fig. 6.12 Diagram of a beach seine.

Beach seines are used both in lakes and in rivers. In rivers they are operated from specially prepared sites from which snags have been removed. Such sites are usually jealously guarded by the groups of fishermen who maintain them.

Beach seines for small pelagic species may be set from canoes well offshore and are then hauled in by long towing ropes attached to the wings. The effectiveness of such fisheries is increased when they are associated with lights. For example, in the fishery for *Rastrineobola argentea* in Lake Victoria lamps are attached to rafts, which are set and left to burn offshore for several hours. They are then hauled slowly in until they are grouped together close to the shore. The net is then set around the lamps and hauled in to catch the accumulated fish.

Purse seines are used for pelagic species that form dense shoals at the surface of large lakes, such as the *Limnothrissa* and *Stolothrissa* of Lake Tanganyika. They are also used extensively for large, valuable species such as Nile perch. Purse seines are used in open waters from boats, usually a large mother ship and a smaller boat to lay the net. The basic method of purse seining involves setting a long net to form a wall around the target school. The top of the net is usually at the surface but the net may extend downwards for several tens of metres. When the net has encircled the school its bottom is drawn together to form a purse that is gradually hauled aboard the vessel, restricting the fish to a smaller and smaller area from which they may be removed (see Fig. 6.13). Purse-seine fisheries are often operated in conjunction with lights that may be positioned on rafts or on the fishing boats.

Purse seines are similar to trawls in that they require considerable outlay in the form of boats and nets. As such, they are a commercial gear operated by an individual owner who employs labourers to operate the boat and nets.

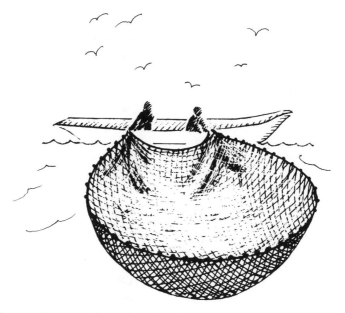

Fig. 6.13 Diagram of a purse seine at the point of closure and hauling.

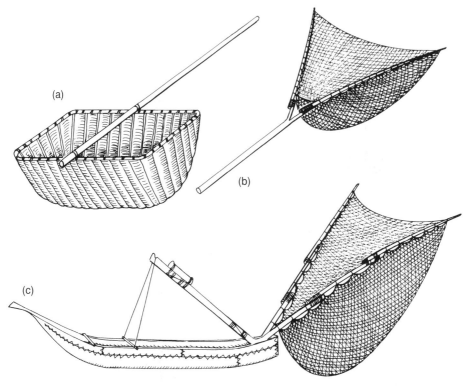

Fig. 6.14 Various types of scoop net and dip net: (a) Basket dip net; (b) V-shaped scoop net; (c) V-shaped scoop net mounted on a canoe.

Seine nets are selective for the lower size ranges of fish but have no effective upper limit to the size captured. The selectivity of the net is determined by the cod-end where one exists or by the inner panels of netting, which are usually of smaller mesh than the wings. The mesh permitted in any fishery is usually specified by legislation.

Scoop nets Scoop nets consist of sheets of netting attached to a frame (Fig. 6.14) that are used in a scooping motion. The netting may be mounted flat or in a pyramidal shape. A wide variety of scoop nets has been developed in various fisheries and smaller versions tend to be popular with artisanal and subsistence fishermen as they are a relatively low-cost gear.

Large pyramidal scoop nets are used from canoes on Lakes Tanganyika, Malawi and Kariba to catch pelagic fish. Fish are usually concentrated by lamps either attached to the boat or on small rafts floating on the water.

In rivers and around the shores of lakes, scoop baskets and nets, often rounded or V-shaped and mounted on a pole, are used for catching small and juvenile fish either on the bottom or at the surface under masses of floating vegetation. The same net may also be used as a type of trawl held upright near the bottom and propelled forwards.

Fishing techniques 111

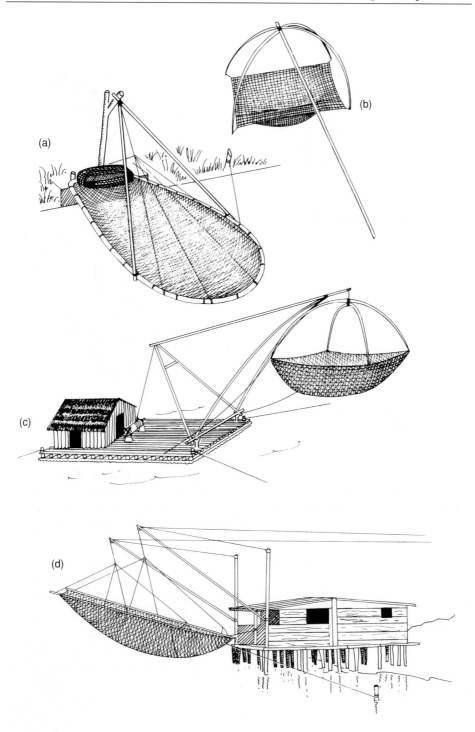

Fig. 6.15 Various types of lift net: (a) bank-mounted lift net (Africa); (b) hand-operated lift net (Asia); (c) raft-mounted lift net (Asia); (d) bank-mounted lift net (Europe).

Although scoop nets normally need a boat, fishermen wading in shallow lakes, swamps and river floodplains can also operate them. In common with other types of gear made from netting, scoop nets are highly selective for both upper and lower size ranges.

Lift nets Lift nets are similar to scoop nets in that they consist of a sheet of netting attached to a frame. They are, however, generally larger and are lifted vertically by some mechanical assistance or leverage. In Lakes Tanganyika and Victoria pyramidal lift nets are used to catch pelagic fishes. The net is mounted below a catamaran and kept open by outriggers. A lamp is placed over the space between the two hulls and may be supplemented by additional lights mounted on rafts. The net is lowered to some depth and after attracting fish is hauled quickly upwards to enclose the catch.

In rivers lift nets are mounted on the bank or on specially equipped fishing craft (Fig. 6.15). The net is lowered into the water and when fish move over the submerged net it is lifted to enclose the fish. These structures are often permanently mounted to capture fish as they migrate along the bank or through restrictions in the channel (Fig. 6.16). In some Asiatic systems fixed engines bearing lift nets may reach considerable sizes.

Smaller lift nets are a popular subsistence and artisanal gear, particularly in Asia and Africa, as they are cheap and easy to operate. In shallow water they can be operated from the bank or on foot (Fig. 6.17). Larger lift nets, mounted in permanent locations may require more capital and be owned and operated by a co-operative. The sites from

Fig. 6.16 Fixed lift net operating in the Tonle Sap, Cambodia. (Photo by Peter Degen.)

which they operate may also be owned or leased from riparian owners. In the Mediterranean lift nets of this type are used recreationally, operated in estuaries from permanent housings along the banks or from bridges to capture mullets.

Fig. 6.17 Small lift nets operated by young boys in rice fields in Thailand.

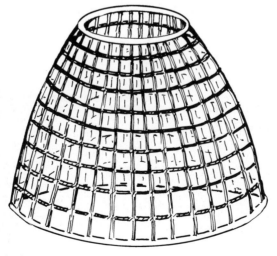

Fig. 6.18 A plunge basket.

Plunge baskets Plunge baskets have wide mouths and open tops (Fig. 6.18). They are plunged into shallow water, after which the operator can feel around inside for any fish captured. Plunge baskets are mainly a subsistence gear attracting the poorest of fishermen. They are also used in ceremonial group-fishing festivals in many of the West African floodplain lakes.

Cast nets A cast net consists of a circle of netting, weighted at the circumference in such a manner that when thrown it opens (Fig. 6.19). It then falls and closes so as to trap any fish within the area covered. The catching efficiency may be improved by the addition of pockets at the circumference. Cast nets can only be used in shallow waters from the shore or from boats. They may be used singly or in groups, in which case the fishermen work together to drive the fish inwards to a point where the group of may cover them with a number of nets thrown simultaneously. They require some expertise to operate, especially when used from a narrow dugout canoe, but their cheapness and transportability make cast nets one of the most common gears in inland water fisheries. They are particularly popular with artisanal and subsistence fishermen. Cast nets are selective for lower size ranges and larger, faster moving fish can escape the falling net but may become entangled in the process.

Clap nets A clap net consists of two matching nets, one operated by each hand so as to bring them together to enclose fish (Fig. 6.20). Each net consists of a wooden

Fig. 6.19 A cast net.

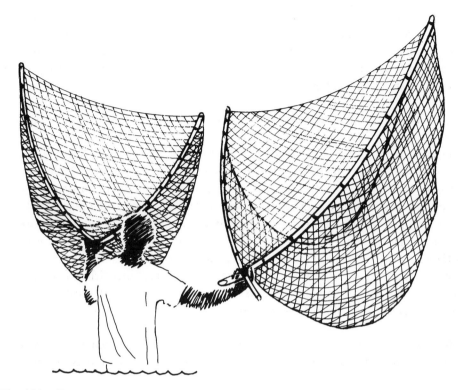

Fig. 6.20 Clap nets.

support armed with netting or basketwork. The clap net is an artisanal and subsistence gear often operated by wading. Clap nets are used in shallow lagoons to capture fish when they are concentrated by drawdown. They may also be used under floating vegetation or as an ancillary gear in fishing fish parks.

Other methods

Trolling Trolling consists of trailing one or more hooks behind a moving boat. The hooks can be baited or equipped with a lure. The method is rarely used in inland waters, being mainly a marine technique particularly applicable to the capture of pelagic species with high individual value, although it has been tried with success for Nile perch in Lake Victoria.

Rod and line Rod-and-line fisheries are used in rivers and lakes to catch demersal and pelagic fish. The line used commercially nowadays is almost exclusively monofilament nylon, and baits may include maggots, meal paste, fish, bivalve molluscs or freshwater shrimp. Fishers may also use flies and lures, particularly in the recreational fishery. Rods and lines operated from the riverbank, from small boats or wading in rivers or shallow lakes are one of the main tools of the subsistence fishery. They are

116 *Inland fisheries*

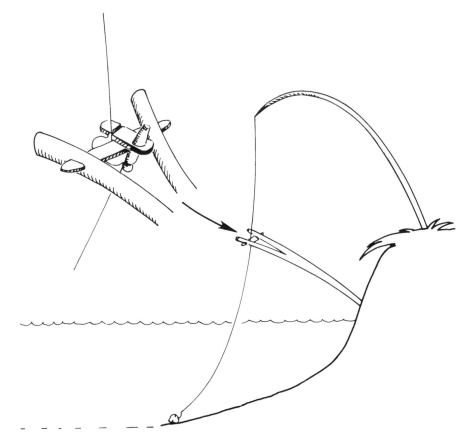

Fig. 6.21 Diagram of a sprung trap line.

also the main basis for recreational fisheries, where the gear can attain great sophistication and cost.

In most cases rod-and-line fisheries are operated manually, although a number of devices can automate the process in the form of sprung trap lines (Fig. 6.21). In some cases the rod is dispensed with and a hooked and weighted line is thrown directly into the water and drawn towards the shore.

Harpoons and bows and arrows Among the most primitive methods for catching fish, projectiles such as harpoons or arrows are still widely used for catching large species in clear waters or near the surface. These methods require great skill and are particularly found among traditional peoples.

Fishing poisons The use of fish poison is an ancient fishing method that is used mainly in enclosed waters such as floodplain lagoons or small residual dams. Fish poisons are currently used in most of the tropics despite being universally condemned and in most cases prohibited. Many species of plant contain the necessary toxins, although rotenone plants are particularly common in Latin America and various types of *Euphorbia* in Africa. In addition, extracts from the bark or roots of a number of local plants (e.g. *Derris* spp.) are prepared and poured into the water. Modern pesticides have now largely supplanted traditional poisons in many parts of the world. The neurotoxic or suffocating effects eventually result in the fish floating belly-up on the surface, where they are collected. Most poisons affect the gills of the fish and the flesh is generally safe to eat, although where pesticides are used residues may accumulated in the fish flesh to toxic levels. Because poisons are indiscriminate, many other benthic organisms may be severely damaged. Often these organisms and small fish, which are not desired, are much more vulnerable to the effect of poisons than the target fish.

Explosives Explosives are widely used for fishing in inland waters and the ready availability of cheap munitions throughout the tropics has contributed to the spread of this practice. In common with poisons, fishing with explosives is prohibited in most countries. Fishing with explosives is extremely dangerous and destructive, indiscriminately killing all species within the radius of action of the explosion.

Electric fishing One of the most modern fishing methods uses electricity to stun fish and draw them to the anode. Fish otherwise unobtainable can be narcotised in this way so that they cannot escape and can thus be easily taken. In some areas of the world such as Thailand massive generators are used to kill and capture fish in rivers. Fishery researchers also favour electric fishing with small-scale apparatus as an efficient method of sampling fish. Fishing with electricity is extremely dangerous and its use as a research tool is governed by strict codes of conduct to ensure the safety of practitioners (Goodchild, 1991). Otherwise, electric fishing is generally illegal.

Fish parks In many areas of the world fish are concentrated and caught by placing vegetation in the water in a more or less structured manner. Brush parks such as the 'samra' in Cambodia, the 'acadja' in West Africa or the 'katha' of Bangladesh cover large areas of the dry-season water-bodies (Fig. 6.22). Fish parks are highly efficient capture methods, initially concentrating fish as refuge traps and giving yields comparable with aquaculture. If properly managed and left in the water for longer periods a fish park can act as a culture system by enhancing food availability and providing a breeding site and refuge (Welcomme, 1972). Individual fishermen can install small fish parks a few metres in diameter, but large installations of up to a hectare in area need co-operative action for their installation and maintenance. All fish parks pose problems of ownership as they occupy areas of lake or river on a long-term basis. The need for large labour forces to install, maintain and harvest large parks is reflected in the social organisation of the communities using this type of structure.

Fish holes Fish holes or fish pits are artificially made drain-in fisheries analogous to those of naturally occurring floodplain pools, in that they concentrate fish during

Fig. 6.22 Various types of fish park: (a) a katha (Bangladesh) being fished; (b) a simple acadja installed in a river (Benin); (c) a field of complex acadjas in a coastal lagoon (Benin).

(c)

Fig. 6.22 (*Continued*).

falling water to retain them until harvesting late in the dry season. Fish holes may be dug by hand in areas where labour is cheap, such as Bangladesh or Africa. Where labour is relatively expensive mechanical diggers may be used. Harvesting is either by concentration and removal of fish with traps and block nets (Africa) or by dewatering with diesel pumps (Asia). Fish holes are found in similar areas of the world to those adopting brush parks where the floodplains of rivers, coastal lagoons and lakes are interspersed with artificial ponds (Fig. 6.23). Normally, fishermen have relied on natural production and capture for the population of the ponds. In recent years the trend to intensify production has led to yields being enhanced through stocking with fish caught from the river and through feeding and fertilisation. Harvests are then comparable to those of more formal extensive aquaculture. In some areas the density of fish holes is such that few natural refuges are left and protection measures are needed.

Associated technology

Fishing technology is far from static and continuous improvement and innovation changes the nature and effectiveness of fisheries. Much of the increase in catch over the last few decades has been achieved through two innovations, the outboard motor and synthetic twine. The outboard motor immediately opened up new fishing grounds

120 *Inland fisheries*

(a)

(b)

Fig. 6.23 Fish holes on the Oueme river floodplain (a) before and (b) after fishing.

by increasing range and thus the accessibility of distant waters. It also reduced the time needed for the fishermen to bring their catch to market and, through the introduction of separate fast transport craft, encouraged the fishermen to stay on the fishing grounds for longer periods. Synthetic twines increased the durability of gear, reduced its cost and above all freed the fishermen from the arduous task of renewing their netting at frequent intervals. As a result, relatively unskilled labour could be drawn into the fishery and the increased labour force, together with the greater catching efficiency, produced a rapid growth in effort. The cost of this has been to fish most stocks beyond their optimum level and to produce a crisis in management that has yet to be resolved.

Further improvements can be anticipated, some of which can be used to conserve the fishery and some of which will increase the pressure on the stocks.

Echo sounding

Few fisheries in inland waters justify the use of echo-sounding equipment, although the drastic reductions in the cost and size of echo sounders in recent years has made them more accessible to fisheries for pelagic shoaling species. Otherwise, echo location is an important research tool for estimating stock size in lakes and rivers and for counting migrating species in rivers.

Mobile telephones

Detection of migrating shoals of fish by the disturbance that they make in the water or by the smell of the shoal is traditional in many Asian and Latin American river fisheries. Fishermen locate such a shoal, fish it, then pack their gear and race up river in trucks to intercept it further upstream. Now, mobile telephones help teams of fishermen to be on station and informed by radio or telephone as to the progress of the shoal upstream.

Fishing craft

Most inland water fishing craft are simple wooden canoes made of local materials. With the increase in fishing pressure and advanced deforestation world-wide, wood for boat building is becoming scarce. Alternative forms of construction that use existing wood more efficiently or substitute fibreglass or plastics are being sought. Larger lakes and reservoirs can have extremely dangerous conditions for navigation. Here larger craft, similar to small marine vessels, are used for trawling or purse seining wherever the value of the fish justifies the expense.

Social and policy implications of fishing technology

Fishing is essentially a social activity. Even the lone fisherman operating with the simplest of gears is constrained by a network of custom, which dictates where and how he is able to fish. Most fishing gear, however, is more complex and implies equally complicated relationships. As a result, fisheries are strongly constrained by traditional and governmental regulation. Some gears are banned outright as being too damaging socially, ecologically or economically. Gear operated from boats requires an understanding between the owner of the boat and the fishermen, which is expressed in a traditional or formal contract. Large gears such as brush parks and cross-river barriers require a considerable degree of co-operation that is usually dictated by traditional usage. In either case there is a tendency for a hierarchy to emerge with the boat or gear owner on one hand and the operatives on the other. There is also a hierarchy among fishermen whereby those owning and operating the larger gears are placed higher than those using more simple equipment. Any multi-gear fishery, therefore, is not homogeneous; rather, it represents a condominium of several fisheries each with its own needs and practices. Many of these are in conflict and such relationships are termed technical interactions, which may be defined as competition between two fishing gears over a unit stock of one or more species of fish. For instance, regulations on mesh size can unduly affect those who use simpler gear, or barrier gears may take an undue proportion of upstream-migrating fish, favouring the fishermen downstream. Unfortunately, where such strong hierarchies exist, the power of those higher in the order to influence policies is generally greater. As a consequence, many of the decisions taken with regard to restricting gears weigh unduly on the poorer fisherman. Care therefore has to be taken in protecting the interests of this group when formulating fishery legislation and management policy.

In inland fisheries these gear-based considerations are complicated by questions of ownership and access rights. Much of the world's inland water is owned either by the state, by collectives such as villages or by private individuals. Furthermore, large fixed gears, brush parks and fish holes require some acknowledgement of long-term rights for the people investing in them. Equally, the labour required to clear seine beaches can only be justified if those doing the clearing enjoy the right to the fishery. Similar arguments apply to systems that are enhanced by stocking or by fertilisation. Ownership patterns are not necessarily static and may vary with season. One group may claim access for their gear during the floods on a river, for instance, whilst another group may claim the same area during the dry season. These factors coupled with modern approaches to management, are tending to restrict fishing to a reserved group who can pursue the twin objectives of conservation and intensification. Nowhere is this trend more obvious than in those countries where recreational interests have confronted traditional food fisheries. Almost always in such cases, the superior influence and economic power of the recreational faction have resulted in the eventual disappearance of the food fishery.

Fishing techniques

Seasonality of fishing

Fishing is a highly seasonal activity in most parts of the world. In Arctic and temperate zone lakes seasonality is determined mainly by the temperature regime. In winter, when the lakes are icebound, fishing tends to be minimal. Special gears such as rod and line operated through holes in the ice, or gill nets attached to special ratchets that enable them to be stretched between two holes, may continue to be operated. Even where there is no ice cover fish are more sluggish in lower temperatures and tend to be caught less than at warmer times of year. In tropical lakes there is less seasonality, although catches may vary between dry and rainy seasons because of differences in fish behaviour.

Regulated rivers and canals tend to behave like lakes. In flood rivers seasonality in the fishery is much more pronounced and is subject to a combination of temperature and hydrological conditions. In Arctic and temperate rivers freezing during the winter is a real possibility and clearly fisheries cease at that time. More normally, however, seasonality is linked to the flood cycle or to behavioural characteristics of the fish. Diadromous fishes, for example, are highly seasonal and migrate into and out of rivers at specific seasons usually associated with spring and autumn/winter flood peaks. Riverine fishes are constrained by the amount of water in the system. Four main seasons are normal:

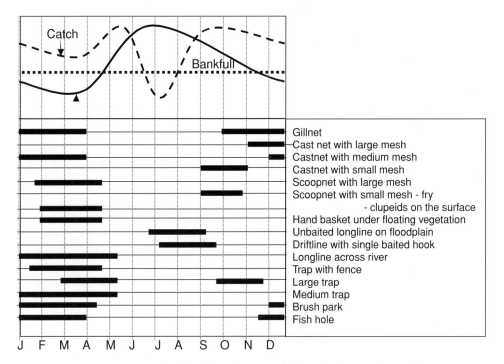

Fig. 6.24 Seasonality in tropical river fisheries: example from the Oueme River, Africa.

- *Rising water*: heavy fishing on fish migrating onto the floodplain from the river or longitudinally in the channel.
- *Floods*: little fishing in most areas of the world as fish are dispersed over the floodplain and fishing conditions are bad. However, in some fisheries, such as in Bangladesh, subsistence fisheries continue at this time. In other areas, such as the Amazon, there are extensive fisheries in flooded forests and in the larger floodplain lakes.
- *Falling water*: heavy fishing for fish returning to the river from the floodplain.
- *Low water*: maximum fishing on stocks trapped within the channel or in floodplain water-bodies.

These various phase in the fishery require different types of gear (Fig. 6.24).

The seasonal nature of the fishery for both volume of catch and gear used has important implications for management, particularly in setting closed seasons, and for regulation of particular types of gear. As gears are usually distributed according to the wealth and status of the fishermen it is possible that, by establishing closed seasons, use of certain types of gear will no longer be possible and the fishermen using them will be penalised. In the Bangladesh fishery, for example, a high-water closed season may benefit the richer fishers working the leased water-body units in the dry season, but may also be disastrous for the poor landless fishers dependent on the flood season. The introduction of such new management regulations must therefore take all of their implications into account, and include adequate supporting measures for any disadvantaged fishers.

Note

1. Co-author: O. Okemwa, KMFRI, Mombasa, Kenya.

Chapter 7
Fish utilisation[1]

Fish as food and nutrition

In most societies a dish is normally not considered complete without some form of meat, poultry, fish or other food of animal origin. Some people eat more fish than others do for natural or cultural reasons. Geographical factors can make fish readily available, as in small island states, in coastal areas or in the riparian regions of rivers and lakes. In some cases the choice of fish is a matter of taste and is selected for its nutritional and health characteristics. In others or it is cultural, based on tradition or religious observance.

Fish is high in protein and minerals such as calcium and selenium. Small fish tend to be eaten whole and in some countries the bones of larger fish are also eaten, contributing, sometimes significantly, to calcium supplies. Recent research on diet has shown that small whole fish tend to contribute far more to dietary balance than do prepared portions of larger fish. Most fatty marine fish and some freshwater fish such as Nile perch are high in healthy unsaturated fatty acids, in particular omega-3 fatty acids. These are associated with health benefits including development of the nervous system and brain in children, development of bones, reduction of blood cholesterol and of cardiovascular diseases, and also aiding against arthritis and asthma.

Dieticians now advise that fish be eaten at least twice a week, a factor that has heightened demand and led to increased prices as well as to greater pressure on fish stocks.

Fish preservation

After capture fish is a highly perishable commodity. Every effort should be made to ensure that delays between catching and sale are minimised. Where this is not possible some form of preservation must be adopted in order to ensure that the product is still acceptable for human consumption when marketed. Many preserved products acquire flavour and quality and thus are considered as delicacies. Some, such as smoked salmon or preserved fish eggs (caviar), have an international reputation for added value. Others, such as the fermented fish sauces of ancient Rome or South-east Asia, become essential ingredients of national or regional cuisine. Yet others, such as the kippers of the UK, the varieties of 'sild' of Scandinavia or the 'stink fish' of West

Africa, have a more local market. However, in all cases these products form important elements of gastronomic culture.

The process of preservation is an essential part of the fishery and forms part of the generalised fishing culture. Preservation of fish may be carried out by the fishermen or by specialised elements within the fishing community. It is frequently a function of the women within the family structure to treat the fish after landing, either by cleaning and filleting it for sale or by smoking or salting the product. In more advanced fisheries, where there is an economically important product, preservation may be assigned to factories where the product is frozen or canned.

Live fish

The best way to preserve the freshness of a fish is to keep it alive. Freshwater fish respond much better to stress (handling, high temperature, low oxygen) than marine fish and it is common to find live fish in restaurants and fishmongers. In the case of high-value products in urban markets fish are transported in specially equipped trucks. Some species are particularly sought after for their ability to stay alive for long periods after capture. These are mostly air-breathing species, which can survive for considerable periods if kept damp. In Europe, eel is sold live in market stalls in this way, while in the tropics a range of snakeheads and catfishes can be transported for long distances over periods of a week or more in basins and other small containers (Fig. 7.1).

Icing and freezing

If the fish cannot be sold live, keeping it at a low temperature can preserve its freshness. Icing immediately after, or even before, death and keeping it in ice will keep it fresh for 2 weeks or more. Tropical freshwater fish keeps very well in ice: a 5 kg Nile perch (whole), for example, can be kept up to 4 weeks. While there are different types of ice, any ice is better then none. If there is a choice then the best is small crushed or flake ice. 'Liquid' ice, a slurry of fine ice particles in light brine, provides quick cooling, while flake and scale ice are also very good. The worst is crudely crushed block ice. A good way is also to cool the fish in a mixture of ice and water and once cool to store it in flake ice. All ice used must be made of potable water.

Iced fish must be kept in insulated or refrigerated rooms, containers or trucks. The cold chain should not be interrupted until the consumer buys – or better – prepares the fish. Depending on the market the fish can be preserved whole, gutted or filleted. In South American and South-east Asian freshwater fisheries many boats are equipped with cold-storage chests and ice is available for suppliers either at the fish landings or at supply stations on the river (Fig. 7.2). In Africa ice supplies are less common, although they have been developed where valuable commercial fisheries, such as the Nile perch, exist.

(a)

(b)

Fig. 7.1 (a) Live fish in the market; (b) basins on a Congo river boat. (Photos courtesy of F. Teutcher.)

Fig. 7.2 Photograph of an ice facility from the back of a truck in Lao PDR.

If the fish cannot be kept live or iced it can be frozen. Freezing can be done in domestic freezers, plate freezers or blast freezers. Larger fishing vessels and cold-storage chains are equipped with freezers, especially where they are associated with formalised marketing networks such as supermarket chains. In other fisheries freezer trucks provide a mobile platform for visiting individual landings. Domestic freezers are not recommended for freezing as freezing proceeds slowly, creating large crystals that damage the tissues and cause drip losses upon thawing. Frozen fish should preferably be covered with ice to prevent it from drying. Alternatively, it can be kept in plastic bags or plastic trays under vacuum. Storage life is limited by temperature and fat content and shows a range from 3 months to 1 year.

The building and upkeep of freezing and ice-producing plants are usually functions of the private sector. Investment in such plants is high and often places pressure on the fish resource in an attempt to recuperate the cost. Experiences in several fisheries, such as that of Lake Victoria, have shown that licensing too many plants can result in increases in fishing effort, leading to a long-term decline in catch.

Smoking

Fish may be smoked to add value, for instance through cold smoking of trout and salmon, and hot smoking of eel. However, the primary purpose of smoking in most

Fig. 7.3 An oven in Africa. (Photo courtesy of F. Teutcher.)

developing countries is to increase the post-harvest shelf life of the fish through a process of smoke drying. The fish is first cooked over a high fire and then dried over a low fire until moist-dry (1–2 days' smoking) or hard-dry (3–4 days' smoking) (Fig. 7.3). After this, the fish can be preserved for some time as a dried product. Shelf life can be increased by periodically re-smoking, although the quality of the product declines with each repetition.

Smoke drying is an unhealthy and dangerous (because of the fire hazard) activity, often carried out by women. The same women may collect firewood over long distances. Each kilogram of fresh fish requires 1–1.5 kg of firewood to be smoke dried. Where wood is abundant little is done to reduce its consumption and a smoking oven may be little more than a raised platform over an open fire; where wood is scarce the ovens tend to be closed in order to reduce the escape of heat. The Chorkor oven (Fig. 7.4) is an example of a fuel-efficient oven useful for large quantities of uniform, small-sized fish.

Smoking has been implicated in massive deforestation, especially of riparian woodlands, because of its high demand for wood. As such, the method is regarded as ecologically undesirable. Establishment of special plantations has been proposed as a method to counter the degradation and at the same time satisfy the needs of the

Fig. 7.4 Chorkor oven. (Courtesy of F. Teutcher.)

fishery. Where no wood is available for smoking, grass or cow dung may be used instead, as in the Sahel countries.

Drying and salting

Small fish are often dried for preservation. If very small they can be dried whole, otherwise they must be gutted and, if large, split. Drying is preferably done on racks in order to speed up the process and also to protect it from dirt, animals, etc. Insects are a problem in tropical countries. Covering the fish during drying is too expensive, as are disinfecting and storing in insect-free places. As a result, dried fish is often heavily infested with insects and losses occur as a result of larvae eating the fish. Sometimes pesticides are overused to combat insects.

Salting and then drying helps to keep insects away but in many areas salting is not a traditional technology. Furthermore, large quantities of salt may not be available and costs may be prohibitive.

Canning

Canning or bottling freshwater fish is uncommon, although specialised products such as fish eggs may be preserved and marketed in this way.

Fermentation

In some countries fish is fermented. In Sahelian countries fish can be left in a container in water without salt for several days before drying. In Sudan *Alestes* is heavily salted and left to ferment until it becomes soft; then it is ground into a paste. In Ghana tilapia is salted and fermented. It is not left until soft but marketed as wet salted fish. In South-east Asia most fish sauces are made from marine clupeids, although freshwater fish can be used for this purpose in areas far from the sea.

Fish meal, fish oil and animal feeds

Inland fish are rarely used for fish meal, although attempts have been made where particularly abundant sources of low-grade fish are present, as in Lake Victoria prior to the changes in fish communities, when an abundant and widespread haplochromine resource existed. More commonly, small fish that are in excess of market requirement may be used to feed chickens or as fertiliser for plant crops. In some areas small shoaling species are caught especially for rendering into fish oil. This is especially common in Latin America, where specialised fisheries for oil-producing species *Lycengraulis olidus* are found in Argentina and Brazil.

Marketing

Once caught and preserved, fish is distributed to the final consumer by a marketing chain (Fig. 7.5), which normally has several steps.

Collection of fish

Fish is collected from the fishermen in a number of ways. In dispersed artisanal fisheries the fish is caught, preserved and stockpiled at the fishing camps until there is sufficient for it to be taken to market. It may also be collected by boats, which circulate around the fishing camps collecting fish for transport to market. In some cases, such as the West African coastal lagoon fisheries, the wives of the fishermen collect the fish as it is caught and take it home for processing. In more coherent fisheries, such as those of lakes, there are distinct landings to which fish is brought for sale by the fishermen to fish traders. In extreme cases, where there is a well-developed export market such as in Lake Victoria, the fish are collected directly by filleting and icing plants at their own landings. Where tourism is well developed, local fisheries supply fish directly to restaurants and hotels. This is particularly common in more developed economies where the sole surviving inland fisheries operate in this way.

132 *Inland fisheries*

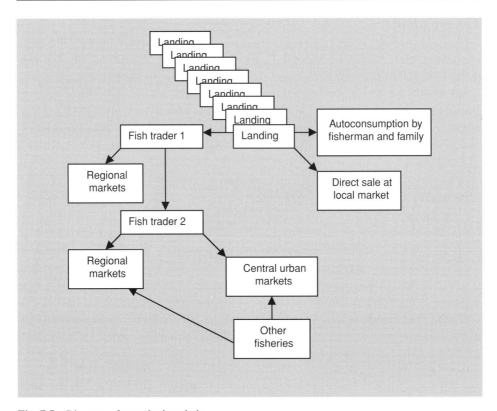

Fig. 7.5 Diagram of a marketing chain.

Autoconsumption

A certain amount of fish is consumed directly by the fisherman, his family and the immediate community. In most cases fish consumed in this way is outside the market structure and usually takes priority over the fish sent to market. Helpers in the fishery are often paid in kind, taking an agreed portion of the catch as their reward. Local consumption, other than by the fisherman, his family and helpers, is often outside the normal economy and is used as part of the barter or favour system of small village life.

Markets

Markets are found throughout most of the world as centres for the sale and exchange of goods. Markets are graded in importance according to the communities they serve and primary, secondary and major markets can be distinguished. Primary markets are those, usually at the fish landing or near to the fishery, which sell fish to local communities and also to small traders who transport it for sale to secondary centres further removed from the fishery. In the case of major landings traders will buy much of the catch for bulk transport to urban centres. Typically, fish are usually only one commodity of the many that are sold in secondary markets, although they usually have separate sections for fresh, live and preserved products. In some areas, especially near major towns, specialised fish markets are also common.

In more developed economies the function of the markets has been usurped by the fish shop and the supermarket. The growth of supermarkets, with their enormous buying power and their demand for predictable supplies of a standardised product, has changed much of fisheries and aquaculture practice. Few fisheries are able to meet the demands of a supermarket chain but where this has happened, for example in Lake Victoria in Africa, all suitable fish was taken by the external market, leaving little for local consumers. By contrast, aquaculture, with its capacity to produce predictable quantities of a standard product, is more adapted to meet the needs of supermarkets.

Traders

Many developing economies are characterised by chains of traders who buy and transport fish from the landing to markets far removed from the fishery. Each step in the chain means an increase in the price of the fish and the traders collectively add value far above the original landed value of the catch. Such chains of small traders, each transporting a small quantity of fish by bicycle or small motor vehicle, may be well adapted to the needs of diffuse rural communities. However, bulk traders more usually serve urban markets, operating directly from fish landing to urban fish shops or supermarkets, using large, refrigerated trucks.

Note

1. Prepared by F. Teutcher, FIII, Department of Fisheries FAO, Via delle Terme di Caracalla, Rome, Italy.

Chapter 8
Resource evaluation

Good estimates of actual and potential catch are essential for a number of purposes, including:

- Management decisions on technical, input and output controls
- Monitoring of the performance of the fishery to establish trends in actual catches in comparison with estimates of potential
- Valuation of the fishery for national planning and resource allocation
- Valuation of the fishery for negotiations with other users of the resource.

Several main approaches are used to evaluate the present performance of the fishery:

- Stock assessment attempts to evaluate the condition of the stock based on a series of biological parameters
- Catch assessment attempts to evaluate the fishery through sampling of the catch and fishermen
- Analysis of markets attempts to evaluate the amount of fish landed by sampling markets
- Analysis of consumption attempts to determine the amount of fish produced from the consumption patterns of the population.

In addition, several methods are used to estimate the magnitude of the stock, including:

- Area-catch methods to evaluate the number of fish present from experimental fishing
- Measures of productivity based on indices derived from experience with other similar fisheries to generate estimates of potential
- Environmental data defining the quality and quantity of habitat.

Most methods for the assessment of stocks and catches have generated a great deal of literature describing and evaluating them. This section does not deal with individual methods in detail but rather aims at providing an overview of the principal current methods, whose relationships one to another are defined in Fig. 8.1.

Guidelines on the selection of methods are given in Guideline 2.

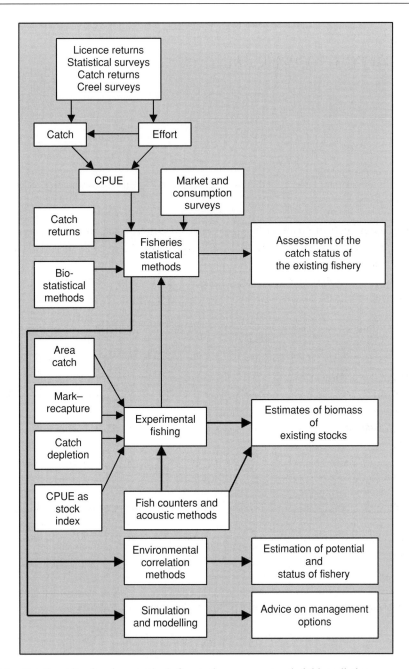

Fig. 8.1 Relationship of various methods for stock assessment and yield prediction.

136 *Inland fisheries*

Guideline 2 Schematic approach to selecting the most appropriate gear and stock abundance methodology (from Cowx, 1996).

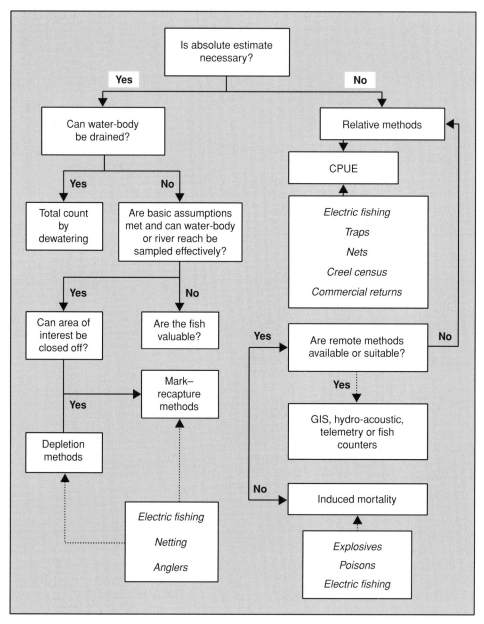

Stock assessment[1]

Fisheries management may be regarded as an attempt by humans to control effectively the exploitation of aquatic biological resources to meet various objectives by intervening in the ecological and social processes which constitute the fishery. Decisions governing such interventions are made on the basis of models and frameworks. These models and frameworks may take a variety of forms including (1) conceptual models of the fishery; (2) empirical models developed on the basis of experience, trial and error or adaptive management (see below); and (3) formal theoretical (technical) models of the fishery (Halls *et al.*, 2000).

This section briefly describes the three main categories of formal models employed in lake and river fisheries management, including their underlying assumptions and strengths and weaknesses. Detailed descriptions and explanations of these and other models, together with the plethora of methods to estimate their parameters, are covered in several excellent textbooks and manuals dealing with fish stock assessment, including Gulland (1983), Sparre & Venema (1992), Hilborn & Walters (1992) and Quinn & Deriso (1999).

Most formal stock assessment models are based on theories of population dynamics and economics. They attempt to generalise the fishery, often in terms of variables that can be controlled by operational rules or external arrangements [e.g. allowable fishing effort, mesh sizes, economic (dis)incentives] and outcomes (e.g. catch or economic rent). In this way they may be regarded as 'theoretical laboratories' for exploring interactions in fisheries systems (Padilla & Charles, 1994).

Models of potential yield and empirical regression models

Models of potential yield

A family of empirical models exists to estimate maximum sustainable yield (MSY) from the generalised formula:

$$\text{MSY} = xMB_0$$

where B_0 is an estimate of the virgin unexploited biomass, M is the instantaneous natural mortality rate, and x takes the value of 0.1–0.5 depending on stock characteristics (Gulland, 1983; Beddington & Cooke, 1983; Sparre & Venema, 1992; Kirkwood *et al.*, 1994; King, 1995; Caddy & Mahon, 1995). The foundations of the model descend from the surplus production model (see below), and that at MSY, the fishing and natural mortality rates will be equal. Although approximate, such estimators provide a useful indication of the magnitude of a potential resource for planning and development purposes. Furthermore, only two parameters, mortality and biomass, need to be estimated.

Empirical regression models

Several linear and non-linear models have been fitted to estimates of potential yield of river systems. A variety of independent explanatory variables may be used to describe their morphological, hydrological and physiochemical attributes, using linear and non-linear regression methods (Toews, 1979; Welcomme, 1985; Halls, 1999). Although these models are easy to construct, the estimates of potential yield, used to fit the models, are usually very approximate and include:

- average annual yield;
- yield corresponding to the mature phase of the fishery according to the generalised fishery development model (Grainger & Garcia, 1996); or
- estimates of MSY corresponding to surplus production models (see later) fitted either by eye or using equilibrium assumptions.

The utility of these models is restricted to offering only very approximate estimates of potential yield for preliminary planning and development purposes. In order to avoid spurious models, independent variables should be selected prior to model fitting on the basis of meaningful ecological hypotheses.

Lakes[2]

In lakes and reservoirs the main factors influencing productivity are:

- The nutrient status of the water, which influences the productivity of the phytoplankton and, through the food web, of all organisms in the water-body. Commonly used measures are the conductivity of the water, total dissolved solids or, more directly, some measure of phytoplankton abundance such as chlorophyll or cell counts.
- The form of the water-body, including shoreline development (the greater the shoreline/area ratio the more productive the water) or depth (the deeper the water the less productive the lake).

These two factors have been combined in numerous ways, of which the most widespread has been the Morpho-Edaphic Index, which related conductivity (or total dissolved solids) to mean depth [MEI = conductivity (μohms)/mean depth (m)]. A typical index derived for African waters by MRAG (1995) is:

$$y = 14.3136 \text{MEI}^{0.0481}$$

Kerr & Ryder (1988) explored the applicability of the MEI and other yield indices in fresh and marine waters. Such relationships are only valid for lakes within similar trophic categories and depend heavily on fisher density. They can only be used where the fisheries are fully exploited, which is generally regarded as being at densities of over 1 fisher km^{-2}. Furthermore, estimates do not seem to be so good where large pelagic fisheries are involved. Part of the variability between lakes using this index is attributable to latitude (temperature), so any attempt to apply such measures must be

Table 8.1 Summary of yields from different types of water-body from different continents

Water-body type	Continent	Range (kg ha^{-1} year^{-1})	Mean (kg ha^{-1} year^{-1})
Coastal lagoon	Africa	5 – 894	231.5
	Asia	29 – 67	42.6
	Latin America	1 – 350	74.8
Lakes	Africa	1 – 2230	130.9
	Asia	1 – 5937	573.1
	Asia (stocked)	56 – 11 625	2635.0
	Latin America	2 – 290	55.2
Reservoirs	Africa	4 – 670	107.9
	Asia	2 – 488	108.4
	Asia (stocked)	1 – 35 583	1573.0
	Latin America	2 – 766	139.2

Table 8.2 Typical relationships calculated for catches from rivers from various continents as functions of basin area and river length

Continent	Relationship	r
Asia	Catch = 7.99 floodplain area$^{0.99}$	0.975
	Catch = 0.0000035 drainage basin area$^{1.792}$	0.92
South America	Catch = 0.069 main channel length$^{1.453}$	0.623
Africa	Catch = 0.032 main channel length$^{1.98}$	0.90
	Catch = 0.03 basin area$^{0.97}$	0.91

limited to sets of lakes of similar geographical area and trophic type. Given these limitations it has been suggested that lake area alone is as reliable an estimate as any, although the great range in yields per unit area (Table 8.1) makes this difficult to apply on a general basis.

Rivers[3]
Similar attempts to apply morpho-edaphic relationships to rivers quickly demonstrated that running waters are less sensitive to edaphic factors. In part, this is due to the much lower variance in conductivity between rivers of similar sizes world-wide. Good relationships can be gained on the basis of form alone, and the following relationships between catch and basin area (or main channel length) have been derived for various continental areas as in Table 8.2.

Regressions relating catch to main channel length can be used to estimate the catch to be anticipated from a specified reach according to the following:

$$C_{kmy-y+x} = a(Ly + Lx)^b - a(Ly)^b$$

where a and b are the regression coefficients, Ly is the distance of the upper limit of the reach from the source and Lx is the length of the reach.

Rivers are far less sensitive to latitude than are lakes and reservoirs, and even cool temperate rivers fall well within the variance of the tropical data sets. They are, however, sensitive to fisher density but support much greater densities of fishers than do lakes and may be considered as approaching full exploitation at densities of about 3 fishers km^{-2}.

One morphological factor influencing river productivity is that the extent of development of floodplains and catches can exceed levels predicted by simple basin area models in rivers with exceptionally large internal or coastal deltas. In such areas simple measures of yield of some 60 kg ha^{-1} of floodplain area can be derived from Africa and greater levels of around 100 kg ha^{-1} of floodplain area are common in Asia.

Surplus production models[4]

Surplus production models are 'holistic models' that encapsulate the net effects of recruitment, growth and mortality in terms of biomass, rather than on the basis of numbers at age (cf. Age-structured models, below). These models are commonly employed to provide estimates of the level of fishing effort that will maximise yield on a sustainable basis. Because of the simplicity of surplus production models, the data requirements to parameterise them [i.e. some measure of abundance, e.g. catch per unit effort (CPUE)[5] and catch data] are also less demanding. Parameter estimation methods that assume equilibrium assumptions (e.g. fitting linear regressions between CPUE and effort) should be avoided. Ideally, non-linear methods should be applied to difference equations fitted to long time series of data exhibiting plenty of contrast. The use of reliable auxiliary information, such as independent estimates of biomass in one or more years, is likely to improve the parameter estimates. Readers are referred to Hilborn & Walters (1992) and Quinn & Deriso (1999) for a fully detailed description and discussion of these models and their extensions, including applications to multispecies fisheries, and Cunningham *et al.* (1985) and Clark (1985) for bioeconomic variants. The computer program CEDA (MRAG, 1992) provides a method for estimating the parameters of three types of surplus production model using non-equilibrium assumptions. Surplus productions models have been used to provide management in advice in both lake (Stanford *et al.*, 1982; Ryan, 1991; Silva *et al.*, 1991) and river systems (Grey, 1986; Montreuil *et al.*, 1990; MRAG, 1994a).

Age-structured (dynamic pool) models[6]

Dynamic pool models represent a population in terms of year-classes or cohorts. They contain three basic components: growth, mortality and recruitment described by submodels. The classic (analytical) dynamic pool model was derived by Beverton & Holt (1957) to predict yield in relation to fishing mortality and age at first capture for given estimates of natural mortality and von Bertalanffy growth parameters. The classic model makes a number of simplifying assumptions, including knife-edge recruitment and vulnerability to fishing gear, isometric growth, and equilibrium conditions. In the absence of information on recruitment, yield is often expressed in terms of yield per recruited fish or yield per recruit (YPR). Modern empirical dynamic pool models make fewer simplifying assumptions, are extremely flexible and can be easily modified to suit almost any situation or fishery (Pitcher & Hart, 1982). When information on recruitment is available, they can also be used to simulate the actual exploitation history of a fishery, or predict the outcomes of management intervention over a number of years. The flexibility of this type of model extends to the submodels describing the processes of growth, mortality and recruitment. For example, fishing mortality may be varied seasonally, or growth modelled according to some density-dependent function (see FPFMODEL, below). Other extensions to the basic model include BEAM4 (Sparre & Willmann, 1991), which can provide bioeconomic management advice in spatially structured multi-species, multi-gear fisheries. Applications of dynamic pool models in lake and river fisheries include: Isaac & Ruffino (1996), Hoggarth & Kirkwood (1996), Madenjian & Ryan (1995) and Bronte *et al.* (1993).

Special considerations of multi-species stocks[7]

Most inland fisheries are based on a large number of species and in many cases are carried out with a wide range of fishing gears. Such multi-species, multi-gear fisheries are not amenable to the more traditional methods of stock evaluation. Whilst individual species in the stock may be subject to classical stock evaluation methods the very complexity of response of numbers of species of different sizes makes it impossible to generate similar models for the assemblage as a whole.

Assemblages of fish exploited by multi-species fisheries react to fishing pressure as described in Chapter 11, p. 207. Increasing effort involves the progressive reduction in the size of the species caught (see Figs 4.13 and 11.3). Reduction in size is associated with changes in mortality rates, growth rate, production and the number of species composing the catch, which all increase. Biomass and CPUE both fall. The combination of falling biomass and rising productivity means that yield remains stable over a large range of effort. The level of stable yield is difficult to determine in advance from the nature of the fish stock alone and managers are forced to rely on indirect criteria such as environmental correlation methods.

Simulation models

Floodplain models

Dynamic models of individual parameters of growth, recruitment, mortality and fishing can be combined into more complex models, which aim to investigate certain characteristics of the fish population and its performance under specified internal biotic and external abiotic environmental and fishery conditions. Models of this type should always be treated with caution as the answers they produce are frequently implicit in the form of the model and the data applied. However, they can elucidate complex situations and can be used to arrive at strategies for management. An example of this is the series of models developed to evaluate the performance of fish populations under the type of pulsed environment found in floodplain rivers. Halls (1998), building on the work of Welcomme & Hagborg (1977), developed a floodplain fisheries simulation model (FPFMODEL) to explore the effects of hydrology and hydrological modification on floodplain fisheries productivity. As in Welcomme and Hagborg's model, the FPFMODEL describes the dynamics of a single species or a group of species having common characteristics in which growth and natural mortality rates and recruitment are assumed to be density dependent. The model is also based on the same weekly iterative interaction of water height (and therefore the area and volume of water on the floodplain) and exploitation, with a fish population (Fig. 8.2). The main difference between the two models concerns the specification of the sub-models describing growth, recruitment and mortality. The model algorithims have also

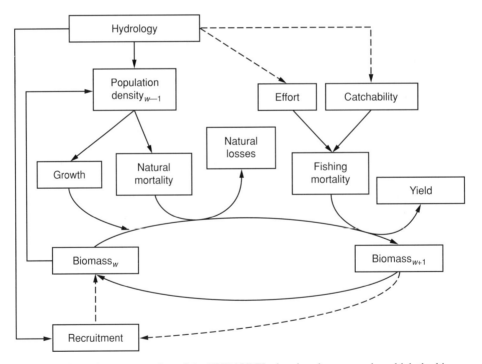

Fig. 8.2 Schematic representation of the FPFMODEL showing the process by which the biomass in week w becomes the biomass in the following week $w+1$. The weekly process is repeated for the 52 weeks of the year, after which recruitment is added in week 52. The process is then repeated iteratively over several years until equilibrium is reached. Solid lines indicate direct influences or operations and broken lines indicate indirect influences or occasional operations. (Adapted from Welcomme & Hagborg, 1977.)

been made explicit instead of being concealed in program code (see Halls, 1998, for details).

The strength of the FPFMODEL lies in its simplicity, generality and flexibility. It may, for example, be easily modified to include other species (without interaction), analogous to the BEAM4 model of Sparre & Willman (1992). No doubt as more tropical floodplain–river systems become modified by hydraulic engineering, this type of simulation model will become increasingly pertinent to multi-sectoral planning and management.

The FPFMODEL simulations indicate that floodplain fish production (and yield) is most strongly dependent on recruitment and therefore on the (density-dependent) survival of the spawning stock. The survival of the spawning stock is most sensitive to the hydrological conditions during the drawdown and dry-season period when fish densities are at their highest, although flood-season hydrological conditions become increasingly important with increasing dry-season water heights (Fig. 8.3). These predictions are consistent with those of Welcomme & Hagborg (1977) and suggest that: (1) yield can be taken under various combinations of high and low water regimes and, therefore, (2) the loss of yield associated with hydrological modifications during the flood season may be compensated by maintaining greater dry-season water heights, and (3) the most important measure for increasing yield is the retention of the maximum possible water height during the dry season, or the creation of more dry-season water-bodies (refuges).

ECOPATH models

ECOPATH models combine estimations of biomass and food consumption at various levels of the food web of an aquatic ecosystem (defined as boxes) with an analysis of the flows between the boxes. A box may represent a life stage of a species or a group of ecologically related species. The relationships between the boxes are defined by a series of linear relationships that have been codified into a computer routine. The outputs of the model give an energetic mass balance of an ecosystem. This enables comparison between changing states, the analysis of the ecological significance of any component of the food web, the identification of 'vacant feeding niches' and the prediction of impacts which change the balance of the system. ECOPATH modelling has been applied to a number of inland systems synthesised by Christensen & Pauly (1993).

Catch assessment

Recording catches

Evaluations of catch may be made directly by recording all catches made. In such cases some type of logbook system is used whereby the skipper of a boat or an individual fisherman records the amount of fish caught during each trip. Recording

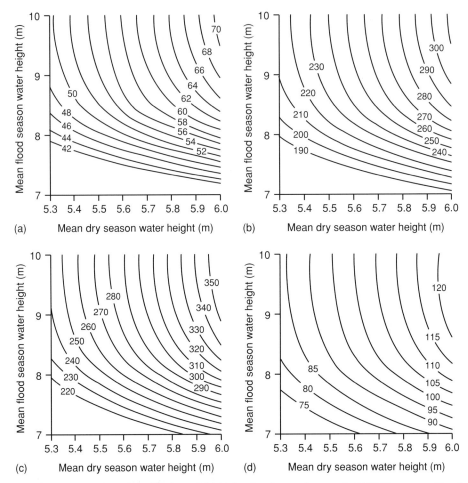

Fig. 8.3 Contour plots of equilibrium yield (*t*) for *Puntius sophore* at the PIRDP generated by the FPFMODEL for different combinations of mean flood and dry-season water height and for a range of fishing mortality, F, rates: (a) F −0.2; (b) F −1; (c) F −2; (d) F −3. (From Halls, 1998.)

catches may be a condition of the fishing licence. This is particularly popular in recreational fisheries where the angler has good control of the fish he or she catches and has an interest in maintaining proper records. It is also practicable in advanced fisheries where the catch is limited to a number of vessels with educated crews. It is less useful where the fishermen have no incentives for proper reporting of the catch. It is also not practicable for larger fisheries involving numbers of widely distributed fishermen, such as most artisanal tropical fisheries.

Catch per unit effort

CPUE methods rely on two main steps:

(1) The determination of the catch per unit of effort (c), and
(2) The determination of the total number of units of effort (E),

after which the simple multiplication of the two parameters gives an estimate of total catch (C) for the fishing gear or fishery concerned:

$$C = cE$$

This apparently simple procedure is very complicated in practice as there are difficulties both with the estimate of catch in any gear and with the definition of effort. CPUE methods are applied to both food and recreational fisheries and have been analysed in Cowx (1991).

Measuring the amount of fish caught is relatively easy in gears such as trawls, seines and fixed lift nets where the gear is of well-defined dimensions and catches are taken at discrete sites where they may easily be sampled. In more diffuse fisheries, where fishing gears such as gill nets, long lines, traps or cast nets are of variable size and used over a widely dispersed area, measurements of catch are more difficult.

Similar problems apply to the definition of effort. Effort may be defined in terms of the gear, fishermen or fishing craft. In gears of stable characteristics where the number of hauls is variable, such as trawls or seines, the measure of effort may be the number of hauls made. In gears such as gill nets or traps that are set overnight the measure adopted may be net nights. In either case estimates of the numbers of each gear in the fishery are fundamental to effort calculations. The fisherman may be used as the measure of effort, and in fisheries largely composed of full-time, professional fishermen this may be satisfactory. However, where the fishermen are only part-time participants in the fishery the definition of the amount of time devoted to fishing becomes more difficult. A third measure of effort may reside in the number and duration of use of the fishing craft.

In multi-gear fisheries the catch and effort for the individual gears have to be estimated separately and the resulting estimates added together. Thus:

$$C = cE_{\text{gear 1}} + cE_{\text{gear 2}} \ldots cE_{\text{gearn}}$$

As the different gears will more often than not have different measures of catch and effort, the combined catch from all gears becomes impossible to calculate with any degree of precision. For this reason it is often assumed that there is some mean performance throughout the fishery which allows the number of fisherman or fishing craft to serve as an approximate measure of effort at least for comparative studies. Because catches and effort are both less variable in lakes, CPUE methods are much more appropriate to lake fisheries than to riverine ones.

Standard CPUE methods are extremely difficult to apply in the complex environment of floodplain rivers, where there may be 10 or 20 different main gear types (not to mention variants of each type), and where the catchability of the gears varies strongly over the seasons. In these fisheries, respondent-based sampling may provide better estimates of catches than attempts to sample by gear type. In such approaches, a sample of fishers is randomly selected from the fishery (perhaps with stratification depending on circumstances) and asked to provide regular information on their catches. Total catches are estimated by multiplying the catch of the sampled fishers by the total numbers of fishers in the fishery. Sampling may also be disaggregated by gear type to provide information on the distribution of catches between gears.

Fish abundance is generally not uniform over the range of fishing operations. This may give rise to imprecise catch and CPUE estimates, which can only be remedied by spatially stratifying the sampling programme to a more local level and/or collecting larger sample sizes. This problem is often acute in the case of pelagic lacustrine species and in floodplains, where gear catchability may vary significantly on a very local spatial scale owing to variations in hydrological or morphological conditions or in fish abundance associated with, for example, local fish-migration routes.

Formalised catch–effort analyses are thus usually applied to commercial catch statistics, which normally present information on both total catch and effort. Other approaches consist of the systematic sampling of a representative grouping of landings, fishermen, gear, etc. Direct reporting by the fishermen may also be used through fishing logs or questionnaires. Here there may be problems with the honesty of fishermen who may sometimes have cause to conceal the real level of their catch and the amount of time they spend fishing. For this reason alternative methods of evaluation are often given preference or used as an independent control on estimates obtained by the CPUE method.

Analysis of markets

Counting fish at markets, customs posts and other localities where fish is concentrated in the course of landing and transport to the final consumer is another way of estimating the absolute and relative magnitude of fish catches. The markets to be surveyed in this way must be carefully defined to avoid the risk of double counting. There is also the risk at higher level markets that fish from other landings and fisheries will be mixed with those from the target fishery. Market data indicate the amount of fish forming part of commercial practice and should be supplemented by consumption surveys to determine the amount of fish eaten by the fishermen, their families and local communities.

Analysis of consumption

Where other methods are not practical because of the diffuse nature of the fishery and market structure, attempts are made at evaluating catches by estimating the amount of fish consumed in the fishing and non-fishing communities. Such studies tend to establish a mean daily ration for various strata of society within the catchment of the fishery. Traditionally, this method tends to overestimate the catch.

Area-catch studies

Direct sampling of the fish community is used to evaluate the standing stock, species composition and main parameters (growth rate, mean length, mortality rate) of the fish in a water-body or river reach. The approaches developed for such studies all attempt to overcome the near impossibility of fishing all fish from the sample site. Complete sampling may be carried out in small lakes or floodplain pools. Normally, however, the water to be studied is too large to permit such an approach and sampling is carried out at specific sites from which the total may be extrapolated. In such cases the area to be studied is usually isolated from the surrounding water by block nets so that the sampling area is defined and the fish have less chance of escape.

Total removal methods

This method aims at the complete removal of fish from a water-body or an isolated area of a larger one. Fish may be removed by narcotising with poisons (rotenone or some substitute is usually the chosen agent) or with electricity. Alternatively, the area may be dewatered or fished to exhaustion with regular gear such as seine nets. Total removal methods provide extremely reliable point estimates but have the disadvantage of being costly and time consuming. Care also has to be taken with the sampling methodology to ensure that the larger, more active species do not escape as the area is being sealed off and to ensure that all fish are recuperated. Where total removal is not practical or acceptable one of the methods below may be used.

Mark–recapture

Mark–recapture methods basically depend on the capture of a sample of fish, which is then marked and returned to the water. A later sample is taken in which a certain proportion will be marked, the rest being unmarked. The population can then be estimated from the ratio of marked to unmarked fish. Variations on the basic method involve different combinations of multiple marking and recapture. Fish can be marked in a number of ways, including injectable dyes, fin clipping or punching, and a variety of tags.

Several assumptions and conditions need to be present for tag–recapture methods to be used successfully.

- There should be no immigration or emigration from the sample site.
- There should be no differential mortality between marked and unmarked fish.
- Marked and unmarked fish should mix into a single population.
- The number of marked fish should form a detectable proportion of the total stock.

Because of these limitations the method is most effective in closed systems such as small to medium-size lakes and isolated floodplain lagoons in the dry season. Mark–recapture methods can provide good estimates of population but they are costly and time consuming and require well-trained personnel. The principles and limitations of mark–recapture methods have been described in Ricker (1975).

Marking is also used to study aspects of fish biology such as migration pathways and growth under natural conditions.

Catch depletion

Catch-depletion methods depend on the reduction in weight of fish caught by successive standard fishings of the same water-body or enclosure. The declining CPUE is plotted against the cumulative weight of fish caught by successive fishings. The total population of fish is estimated from the intercept of the resulting regression line with the x-axis (see Fig. 8.4).

Catch-depletion methods can provide good estimates of population. However, in some cases comparison between estimates obtained by this method and those obtained by fishing out a water-body has shown it to underestimate populations (Halls, 1998). This is mainly because some species appear able to evade capture. Furthermore, certain assumptions conditions need to be present for them to be used successfully:

- There should be no immigration or emigration from the sample site.
- The fish should not develop resistance to capture by the method used.
- The different species should be equally susceptible to capture.

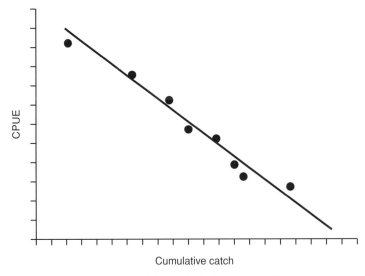

Fig. 8.4 Diagrammatic representation of the catch-depletion method.

The method is costly and time consuming. Because of this and the above limitations it is most effective in closed systems such as small to medium-size lakes and isolated floodplain lagoons in the dry season.

Fish counters and acoustic methods

In recent years the availability of cheap and mobile acoustic equipment, together with the computer software to interpret signals, has introduced a range of methods to detect fish without capturing them.

Acoustic methods

Acoustic methods rely on a transmitter that sends a series of signals through the water. The signals are reflected off the swim bladders of any fish that are in their path and are received by a transducer. The reflected signals are then interpreted by software to determine the number and size of the fish in the path of the signal. To be accurate enough for biomass estimates acoustic equipment has to be carefully calibrated to determine the reflective characteristics of the fish to be sampled, although it can be used in a more general fashion to detect the location of individual fish or fish shoals. Initial successes with acoustic methods were mostly with pelagic stocks in the open waters of lakes. Subsequent experience has shown that the method can be applied to demersal stocks and, if anchored to the shore or on the bottom of a river, can detect the number and biomass of fish moving through river channels.

Despite the relatively low cost of equipment, acoustic methods are still expensive and rely on well-trained technical staff. Their effective use also depends on well-formulated surveys. Johanesson & Mitson (1983) describe the basic techniques for acoustic surveys, but a considerable literature has subsequently grown up around modern equipment and experience (see Cowx, 1997, for examples).

Fish counters

Fish counters are electronic devices that detect the passage of fish through a channel. They are often used in conjunction with fish passes to count highly migratory fish such as salmonids. They may also be used as side-looking or upward-looking arrays in rivers.

Telemetry[8]

In many freshwater environments conventional sampling methodologies have a restricted efficiency because of excessive water velocity or depth, or give a poor temporal resolution, which is hardly compatible with the knowledge required for a proper management policy. Similarly, it is important to determine whether sampling efforts targeting some fish species are carried out over an adequate range, at the correct depth, at the right time of the day and with the appropriate gear.

Underwater telemetry involves tagging fish with an electronic device that sends a signal continuously (transmitter) or upon interrogation by an appropriate reader (transponder). Radio signals (30–170 MHz) can be detected with a portable antenna or with antennae installed on land, car, boat or aircraft. They attenuate rapidly with increasing depth and conductivity, but they are not (or are little) affected by ice cover, thermocline, ambient noise and environmental structure (e.g. aquatic or submerged riparian vegetation). The detection of acoustic signals (30–300 kHz) requires a hydrophone to be immersed in the water. These signals are little affected by depth and conductivity, but they are scattered or reflected by physical structures, such as air bubbles, thermoclines, vegetation or rocks. Hence, radio signals are generally preferred in rivers and streams, whereas acoustic signals are best used in large open environments, such as deep lentic rivers, lakes, lagoons, estuaries or seas. Alternative telemetry systems for fish living in less accessible environments include electromagnetic tags (based on inductive coupling) for site monitoring of highly structured environments (e.g. crevices in rocks) and data storage (archival) tags for fish migrating over considerable distances, exceeding the practical detection range of telemetry systems. The latter tags store the information in an internal memory, which is downloaded after the fish has been recaptured, generally by professional fishermen.

These systems, alone or in combination, provide an efficient set of tools for monitoring the movements of fish that are not readily visible or accessible by conventional means, with a minimum perturbation of the animal tracked, except for the tagging procedure and added weight. With few exceptions, fish are near-neutral buoyant animals and adding weight imposes additional constraints on buoyancy regulation and/or swimming. The smaller the transmitter relative to the fish, the less its potentially negative effect on the animal, but also the shorter its detection range and its operational life. Therefore, it is generally recommended that the weight in water (i.e. weight minus volume) of a transmitter does not exceed 1.0–1.5% of the fish body weight, as a fair compromise between minimum interference and tracking feasibility. Commercially available transmitters can be as small as 0.5 g in air (0.3 g in water), enabling the tracking of fish as small as 20 g. Operational range varies from metres to a few kilometres, depending on the current drain of the transmitter battery and environmental conditions (depth, conductivity, noise). Monitoring the passage of fish as small as 1 g in discrete sites can be achieved with passive integrated transponders (PIT-tags; 11 × 2 mm; 0.1 g in air) that transmit their code upon interrogation by a manual or automatic reader (125–400 kHz), but at short range only (< 1 m).

Transmitters can be attached externally to the fish, inserted into their stomach, or into their body cavity, through an insertion to be closed by stitches (ideally with absorbable material such as catgut), surgical staples (practical for long incisions in large fish with no scale cover) or cyanocrilate compounds. There are risks of increased drag, corporeal damage and lower growth resulting from the use of trans-structural threads. Because of this, external attachment is advised in short-term tracking projects, or when the tag needs to detach from the animal after a certain period (e.g. data storage tags attached with absorbable threads or tags that pop up at the surface of the water after a certain time). Intragastric tags can be easily inserted, with minimum handling time and minimum damage. However, they may interfere with fish appetite and are best when tagging fish during a period of limited feeding (e.g. spawning runs of salmons). The voluntary ingestion of tags wrapped in baits by fish photographed at the time they ingest the bait represents a valuable alternative, as it suppresses capture and handling stress. It also permits the tagging of deep-sea or lake fishes that could not be tagged at the surface because of the variation of ambient pressure. For projects requiring the long-term monitoring of fishes, intraperitoneal implantation has become the most adequate attachment procedure, despite seeming to be the most invasive. The tag is positioned close to the centre of gravity of the fish, causes little or no internal damage, and has almost no effect on growth, physiology and behaviour once the incision has healed, provided the procedure has been tailored to the species of interest, and properly evaluated in the course of feasibility studies. The duration of the healing process is influenced by water temperature, fish age and growth potential, being as short as 4 days in 10-cm African catfishes at 27°C, and requiring as much as 6–8 weeks in adult cyprinids and salmonids at 10–12°C.

Individual transmitters can be identified through their specific carrying frequency, pulse rate or associated electronic code. The latter feature enables the use of several transmitters on the same carrying frequency, it reduces scanning time for automatic monitoring (see below), but it requires more sophisticated receivers with electronic decoding for code identification. Using a directional antenna or hydrophone, the observer can determine the direction of the strongest signal, and either follow this direction to identify the fish's position, e.g. with a global positioning system GPS, or take bearings from several marked places, and locate the signal source by triangulation. While using acoustic signals, automatic positioning can be achieved with an array of omnidirectional hydrophones and time-measuring units, using the inverse principle of hyperbolic navigation to determine the position of the tagged fish (i.e. three and four hydrophones needed for a two- and a three-dimensional positioning, respectively). Fish passage at a certain site can be monitored by fixed automatic listening stations that scan a bandwidth, record all incoming signals on this particular bandwidth, and can be interrogated *in situ* or remotely (radio or telephone). PIT-tag data entry stations rely on a similar principle.

Telemetry systems thus enable the frequent positioning of fish almost independently of environmental conditions, at all times of the day and at all times of the year, provided the transmitter has been manufactured to last that long. Since the first application of telemetry in the late 1950s, there have been about 1000 studies that permitted the collect of key information on the seasonal and daily migrations, home range, foraging behaviour of, and habitat use by Australasian, North American (Gowan *et al.*, 1994) and European (Ingendahl *et al.*, 1996; Marmulla & Ingendahl, 1996) freshwater fishes (chiefly salmonids, cyprinids, anguillids, esocids, percids and percichthyids). Studies of fish swimming trajectories with automatic positioning systems also revealed how fish reacted to environmental stimuli, such as wind, rain or fishing gears. Monitoring the downstream or upstream passage of fish at obstructions provided further information on how these obstructions are perceived by the fish, how much they impact on their migration (delay, mortality after passing through turbines) and which solutions can be implemented to improve the efficiency of fish bypass systems. A similar (and even wider) perspective is in sight for the most diverse fish faunas of tropical regions in Asia, Africa and South America.

Telemetry transmitters can also provide complementary useful information in addition to the fish's position. This can be achieved by coupling the transmitter to an electronic circuit, which is connected to a sensor, and modifies the signal pulse rate depending on the value measured by the sensor. Basically, each and every variable that can be sensed can be converted into an electric, then radio or acoustic signal. This includes environmental variables such as swimming depth, water temperature, salinity, oxygen (short-term studies only) and light intensity, which can provide key information on the habitat preferences of the fish. Combining the information from different sensors, as is the case when these are connected to a data storage tag, may enable the observer to reconstruct the vertical and horizontal tracks of the fish, provided there is enough contrast between environmental conditions within the fish home range. Transmitters can also measure internal variables that give key insight into fish metabolism (heart rate, ventilation rate, internal temperature, swimming speed) and provide a more accurate picture of how fish adapt and react to environmental changes. For example, analyses of electromyographic signals from salmon in fish passes helped in the understanding of how much energy was spent by the fish in different passes. Considering that migratory salmons rely on a definite energy reserve, this enabled the determination of which set of environmental conditions (flood, temperature) could jeopardise their migration, and which bypass system had to be rethought and improved in priority to guarantee their free circulation.

Telemetry systems thus represent a major tool for ecological studies and projects aiming at optimising the management of inland resources and fisheries. They can also provide key information on the behaviour and physiology of cultured fish, which can contribute to a better understanding of how fish adapt to a rearing environment and whether feeding schedules match their activity rhythms. When using transmitters measuring physiological or behavioural variables, telemetry can also be used as a quality-control device for real-time monitoring of the well-being of the fish or of the degradation of the rearing conditions.

The following are recommended for further reading on this topic: Prentice *et al.* (1990), Baras (1991), Baras & Lagardère (1995), Baras & Philippart (1996), Winter (1996), Metcalfe & Arnold (1997), Lagardère *et al.* (1998) and Baras *et al.* (2000).

Environmental data

The evaluation of the condition of rivers and lakes relative to their original or desired state is essential to conservation and rehabilitation. Evaluations are important as a national or regional reference to guide decisions on water allocation, strategies for rehabilitation and potential for fisheries enhancement. Knowledge of the relative state of any water-body is also essential in negotiations with other users of the resource and, in the eventuality of legal action, claims for damages.

Various tools exist for the evaluation of the quality of riverine environments and the possible responses to changes in flow and other features. Many tools have been developed in recent years but most of these have proved overly cumbersome with numerous parameters that call for expensive data gathering. Factors used in developing indices for ecosystem integrity include:

- Combinations of data on river morphology (riparian structure, nature and extent of bottom substrate, depth profiles, slope, etc.)
- Flow (timing and magnitude); water quality (biological oxygen demand, presence of toxic pollutants, turbidity, temperature, etc.)
- Biotic criteria (abundance of phytoplankton, riparian and submersed vegetation, etc.); and fish population structure.

Four approaches have emerged as standards in temperate regions:

- Habitat Quality Indices (HQIs), which measure the general quality of the habitat
- The Instream Flow Methodology (IFIM), a methodological approach to the assessment of future impacts of hydrological change
- Physical Habitat Simulation (PHABSIM), a precise methodology for evaluating instream flow requirements
- The Integrated Biotic Index (IBI), a biotic measure for assessing the relative state of degradation of the system.

Habitat Quality Indices

HQIs set out to estimate the quality of the environment for fish by measuring a series of characteristics of the physical environment. These can vary from system to system and may even depend on the type of river reach. Typical parameters used include the degree of stream modification, bank erosion, bank vegetation stability, riparian vegetation density, pool/riffle ratios, type of bottom substrate, depth, flow, water quality, etc. It is generally supposed that the more factors are included in such models the more they will accurately describe the situation. However, increasing the number of factors to be considered involves escalating costs in money and in time. As a result the more recent trend is to simplify such indices by limiting the number of parameters so as to obtain a timely and cost-effective response. An index is then derived from combinations of the chosen factors, which can be compared with an ideal system, with other similar systems or other reaches of the same river. Such measures give an indication of the state of the river channel as they are derived mainly from instream and riparian conditions.

Where large-scale measures of stream health are to be used, a watershed-based index is often more appropriate. This may include factors such as degree of urbanisation, percentage of land forested, percentage under intensive or extensive agriculture, the degree of industrial development and the degree of modification of the river.

Instream flow methodology

IFIM is a general approach to determining the impacts of changes in aquatic systems and exploring the impacts of proposed changes. The methodology consists of five phases.

(1) *Problem identification*: This involves an analysis of the legal and institutional frame in order to identify key players and an analysis of the physical location and geographical extent of any changes that may occur in the system.
(2) *Planning*: This involves consultations to identify the information needed, the information that already exists and what new information must be obtained. This phase should result in a written work plan.
(3) *Study implementation*: This phase consists of collection of data on a range of parameters, which may include temperature, pH, dissolved oxygen, biological parameters, and measures of flow and morphological parameters such as depth, cover and substrate type. These variables are used to establish the relation between stream flow and stream habitat and should establish a habitat–time series, which estimates how much habitat would be available for each life stage of each species over time. It provides estimates of the relationship between flow and total habitat derived from models such as PHABSIM.
(4) *Alternatives analysis*: Interested groups compare alternative impacts from different flow regimes on the basis of effectiveness, physical feasibility, risk and economics to derive a set of alternative management scenarios.
(5) *Problem resolution*: This involves choosing between the alternatives in the light of the information on their impact. Attempts should be made to reconcile the interests of the various parties, but this is made difficult because the biological and economic values are difficult to interpret, the data and models are never complete and the future is uncertain.

Physical Habitat Simulation System[9]

PHABSIM is currently the most widespread model designed to calculate an index indicative of the amount of microhabitat available for different life stages of fish and invertebrates at different flow levels. It is expressed through a computer program, which has two main analytical components: stream hydraulics and life-stage habitat requirements.

The stream hydraulic component predicts depths and water velocities at specific locations on a cross-section of a stream. Field measurements of depth, velocity, substrate material and cover at specific sampling points on a cross-section are taken at different flows. Hydraulic measurements, such as water-surface levels, are also collected during a field survey. These data are used to calibrate the hydraulic models and then predict depths and velocities at flows different from those measured. The hydraulic models have two major steps. The first is to calculate the water-surface elevation for a specified flow, thus predicting the depth. The second is to simulate the velocities across the cross-section. Each of these two steps can use techniques based on theory or on empirical regressions. The empirical regressions require a lot of supporting data; the theoretical approach requires much less information. Most applications involve a mix of hydraulic submodels to characterise a variety of hydraulic conditions at various simulated flows.

The habitat component weights a series of preselected stream cells according to indices that assign a relative value of between 0 and 1 for each habitat attribute. This index indicates how suitable that attribute is for the life stage under consideration. These indices are usually termed habitat suitability indices and are developed using direct observations of the attributes used most often by a life stage, by expert opinion about what the needs are, or by a combination of the two. The hydraulic estimates of depth and velocity at different flow levels are combined with the suitability values for those attributes to weight the area of each cell at the simulated flows. The weighted values for all cells are summed – thus the term weighted usable area (WUA).

The WUAs for different values of flow are then plotted to obtain a graph which can be used to develop an idea of what life stages are impacted by a loss or gain of available habitat at what time of the year. Time-series analysis plays this role, and also factors in any physical and institutional constraints on water management so that alternatives can be evaluated.

There are many variations on this basic approach tailored to different water-management criteria, or for special habitat needs. However, the fundamentals of hydraulic and habitat modelling remain the same, resulting in a WUA versus discharge function. PHABSIM provides an index to the availability of microhabitat. It is not a measure of the habitat actually used by aquatic organisms. It can only be used if the preferences for depth, velocity, substrate material/cover or other predictable microhabitat attributes under in a specific environment of competition and predation are known. The typical application of PHABSIM assumes relatively steady flow conditions such that depths and velocities are comparably stable for the chosen time step. PHABSIM does not predict the effects of flow on channel change. Finally, the field data and computer analysis requirements can be relatively large.

Index of Biotic Integrity

The IBI is designed to indicate the degree to which a watercourse has been impacted by pollution or morphological degradation through a measure of the health of the fish assemblages. As initially defined by Karr *et al.* (1986), it consists of 12 measures or metrics which fall into three categories: species composition, trophic composition, and

Table 8.3 Metrics used to assess fish communities in the midwestern USA wadable streams

Category	Metric	Scoring criteria		
		5	3	1
Species richness and composition	1. Total number of fish species excluding hybrids and subspecies 2. Number and identity of darter species (habitat specialists in small streams and rivers; sensitive species) 3. Number and identity of sunfish species (species responsive to degraded pool habitat) 4. Number and identity of sucker species (long-lived species with sensitivity to habitat and chemical degradation) 5. Number and identity of intolerant species 6. Proportion of individuals of green sunfish (respond positively to degraded conditions)	Expectations for metrics 1–5 vary with stream size and region <5%	 5–20%	 >20%
Trophic composition	7. Proportion of omnivores 8. Proportion of insectivorous cyprinids (decline in abundance with degraded conditions) 9. Proportion of piscivores (top carnivores)	<20%	20–45%	>45%
Fish abundance and condition	10. Number of individuals in sample 11. Proportion of individuals as hybrids 12. Proportion of individuals with disease, tumours, fin damage and skeletal anomalies	Expectations for metric 10 vary with stream size and other factors 0% 0–2%	 >0–1% >2–5%	 >1% >5%

Modified from Karr (1981) and Karr *et al.* (1986).

fish abundance and condition. Data are obtained for each of the metrics for a specific reach and are compared with values expected for an unimpacted reach. A score is awarded on the basis of the divergence of the observed from the expected value. The sum of the rating then gives a score for the site. The index is able to integrate data from all ecosystem levels into a single comparable score indicative of the quality of the aquatic resource.

The metrics used originally for the assessment of fish communities in streams in the midwestern United States are set out in Table 8.3. The relationship of the scores to integrity is indicated in Table 8.4. This version of the IBI was developed for wadable streams only and other models have been proposed for larger rivers, for example that of Simon & Emery (1995) in the USA and Hugueny *et al.* (1996) for Africa (Table 8.5).

Table 8.4 IBI scores related to the condition of the habitat

Total IBI score (sum of the 12 metric ratings)	Integrity class	Attributes
58–60	Excellent	Comparable to the best situations without human disturbance; all regionally expected species for the habitat and stream size, including the most intolerant forms, are present with a full array of age (size) classes; balanced trophic status
48–52	Good	Species richness somewhat below expectation, especially due to the loss of the most intolerant forms; some species are present at less than optimal abundance or size distribution; trophic structure shows some signs of stress
40–44	Fair	Signs of additional deterioration include loss of intolerant forms, fewer species, highly skewed trophic structure (e.g. increasing frequency of omnivores and green sunfish or other tolerant species); older age classes of top predators may be rare
28–34	Poor	Dominated by omnivores, tolerant forms and habitat generalists; few top carnivores; growth rates and condition factors commonly depressed; hybrids and diseased fish often present
12–22	Very poor	Few fish present, mostly introduced or tolerant forms; hybrids common; disease, parasites, tin damage and other anomalies regular
	No fish	Repeated sampling finds no fish

The steps for carrying out an IBI are described in Fig. 8.5.

The IBI presented here was developed for the rivers of North America. It is an easy model to develop and apply because it is based on relative abundance of fish species, which can be readily determined from market surveys or directed sampling. Criteria for IBI assessments in Europe have also been developed but there is a need to explore the application of these techniques to rivers and lakes elsewhere in the world, especially in the tropics. Alternative IBIs have been developed using invertebrates (B-IBI) as a complement or alternative to indices based on fish (F-IBI). In some cases these have been found to be more accurate predictors of the state of the system compared with other measures such as HQIs and generally reflect the population changes predicted by other models of responses to environmental stress (Table 4.7).

Table 8.5 IBI modified for great rivers of North America

Great river IBI	Definition
Total number of species	Number of fish species including exotics and excluding hybrids and subspecies
Percentage large river faunal group	Percentage composition of Pflieger's large river faunal group. These species are great river specialists and decline with loss of associated habitats
Number of centrarchid species	Number of members of family Centrarchidae, including black basses. Responsive to degraded pool habitat
Percentage round-bodied suckers	Members of genera *Cycleptus*, *Erimyzon*, *Hypentelium*, *Moxostoma* and *Minytrema*. The genera *Catastomus*, *Carpoides* and *Ictiobus* are excluded as they are habitat generalists and omnivores
Number of sensitive species	Number of species of multiple families that are sensitive to a variety of perturbations. Should be limited to 10% of fauna
Percentage tolerant species	Percentage of a number of species from multiple families which increase in relative abundance in degraded habitats
Percentage omnivores	Percentage of animals which take a considerable amount of plant (25%) and animal (25%) material
Percentage insectivores	Cyprinids are not dominant in great rivers, thus this metric reflects all insectivores found in great rivers
Percentage carnivores	Percentage of individuals in which the adults are predominantly piscivores
CPUE	Number of individuals collected with a standard effort and protocols
Percentage simple lithophils	Percentage of individuals which exhibit a simple spawning behaviour
Percentage DELT	Percentage of individuals with poor health, excluding parasitism and protozoan infections

After Simon & Emery (1995).

Notes

1. By Ashley S. Halls, MRAG Ltd, 47 Princes Gate, London SW7 2QA, UK.
2. By R. L. Welcomme.
3. By R. L. Welcomme.
4. By Ashley S. Halls, MRAG Ltd, 47 Princes Gate, London SW7 2QA, UK.

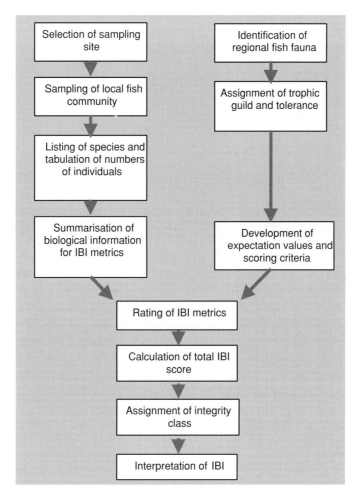

Fig. 8.5 Sequence of activities involved in calculating and interpreting the Index of Biotic Integrity for a stream segment.

5. Although many applications of the model have used CPUE as an index of abundance, it is widely accepted that CPUE may not be proportional to abundance (see Hilborn & Walters, 1992).
6. By Ashley S. Halls, MRAG Ltd, 47 Princes Gate, London SW7 2QA, UK.
7. By R. L. Welcomme.
8. By E. Baras, University of Liège, Aquaculture Research and Education Centre, Liège, Belgium.
9. Information derived from http://webmesc.mesc.nbs.gov/rsm/IFIM_5phases.html

Chapter 9
Social and economic evaluation

Evaluation of the social and institutional context of a fishery[1]

Evaluation of the social and institutional context of a fishery contributes to two important components of management:

- Developing knowledge and understanding of the range of issues to be taken into account when managing fisheries, at its most basic – who are the stakeholders in the fishery and how are the benefits from the fishery distributed among different groups.
- The process of such an evaluation provides the best opportunity to promote stakeholder participation in the development of management strategy, informing them to empower themselves with respect to their role and behaviour in the fishery. The process by which understanding and information about a fishery are gleaned can be very powerful for the resultant management strategy, as motivating appropriate behaviour is fundamental to the sustainability of any fishery.

Several core questions need to be addressed when evaluating the social and institutional context of a fishery. Who are the stakeholders? What is the nature of their interest and relationship with each other? What is stakeholder capacity? What are the different gender roles in the fishery? How can participation be promoted and encouraged? How is conflict best managed? Several methods of enquiry generate both understanding of the issues and an opportunity to engage stakeholders in the management process.

Such considerations are generally more applicable in countries where management regimes are still in a fairly undeveloped state, or not very successfully applied or enforced. In more developed areas ownership and stakeholder issues are frequently better defined, as are the mechanisms for keeping existing institutions under review. However, recently much attention has been paid to evaluation of the social and economic role of recreational fisheries in North America and Europe, and methodologies are being developed for this purpose (see Welcomme & Naeve (in press) for examples of different national approaches).

Stakeholder analysis

Stakeholders are people, groups or institutions that are affected either positively or negatively by management of an inland fishery, or those who can affect the outcome of any management action. Stakeholder analysis is a vital tool in developing an understanding of the social and institutional context of a fishery. It is particularly relevant to issues surrounding management of inland fisheries, which are characterised by a large number of stakeholder groups with a wide range of objectives and interests, multiple sectoral demands on the resource, and complex tenure and access systems. The analysis yields the following insights:

- Who will be affected (beneficially or adversely) by management of the fishery
- Who could influence the management (positively or negatively)
- Which groups of people should be actively involved in management of the fishery, and how
- What is the nature of the interaction between different stakeholder groups, for example, are they co-operative, dependent, competitive or conflict ridden?
- What is the capacity of various players, with particular focus on identifying whose capacity limits their active participation in management and so needs additional support.

As well as an improved understanding of the socio-economic issues, the process of gathering and synthesising the data provides an opportunity to engage stakeholders in a structured assessment of their fishery and to ensure that their participation is integral to the development of a management plan.

Rietbergen-McCracken & Narayan (1998) defined a four-step process to stakeholder analysis:

(1) *Identify key stakeholders*: Assess who potential beneficiaries are; who might be adversely affected; have vulnerable groups been identified; have supporters and opponents been identified; and what are the relationships among stakeholders.
(2) *Assess stakeholder interests and the potential impact of fisheries management on these interests*: Assess: what are the stakeholders expectations of the management plan, what benefits are there likely to be for the stakeholders, what resources might the stakeholder be able and willing to mobilise and what stakeholder interests conflict with the management goals.
(3) *Assess stakeholder influence and importance*: For each stakeholder group, assess its: power and status (political, social and economic); degree of organisation; control of strategic resources; informal influence (e.g. personal connections); power relations with other stakeholders; and importance to the success of the management plan.
(4) *Outline a stakeholder participation strategy*: Plan stakeholder involvement according to: interests, importance, and influence of each stakeholder group; particular efforts needed to involve important stakeholders who lack influence; and appropriate forms of participation throughout the management cycle.

One component of a stakeholder analysis is assessment and development of the capacity of stakeholders. Put simply, capacity is the range of characteristics needed by individuals or groups so that they can participate in the management of the fishery. Four main categories of capacity have been identified in relation to fisheries management (Hoggarth *et al.*, 1999b):

- *Resources*: people and money
- *Skills*: suitably trained people
- *Rights*: recognised roles, responsibilities and the right to manage
- *Motivation*: incentives and disincentives to participation.

Clearly not every stakeholder group will meet all requirements, but it is important to try and match responsibility with capacity and build capacity where skills are poor.

While desk studies and review of reliable secondary data have a role in this analysis, participatory methods – such as stakeholder workshops, local consultations and participatory analysis – should be the principal approach. A series of matrices is an efficient way of organising and synthesising the information gathered as part of a stakeholder analysis.

Rietbergen-McCracken & Narayan (1998) provide a comprehensive guide to carrying out a stakeholder analysis, providing indicative matrices that would assist the structuring of such a process. Grimble (1998) also provides a review of stakeholder analysis as part of the UK Department for International Development (DFID)'s best practice guidelines for socio-economic methodologies. An earlier DFID publication provides guidance on how to conduct a stakeholder analysis of projects (ODA, 1995).

Gender awareness

Gender roles are the socially defined roles of men and women which may be influenced by household structure, ecological conditions, culture, laws, religion, education, access to resources, etc. So gender awareness in the context of managing inland fisheries provides an understanding of the constraints, needs and opportunities of both men and women who have a stake in the fishery.

Long experience from projects and initiatives around the world has indicated that awareness of gender roles is fundamental to success of any resource management plan. The roles played by men and women in a fishery may be quite distinct. For example, in some fisheries women may be directly engaged in fishing activities alongside men, while elsewhere women are only involved in handling, preservation and processing of fish products. The role that women play in a fishery is not static. For example, the recent development of a 'mukene' fishery on Lake Kyoga, Uganda, using small boats seines has attracted women entrepreneurs who control the majority of operations at one landing site. The movement of these women into this enterprise resulted from their need to expand into new sources of livelihoods after the death of their husbands (unpublished DFID report).

Gender analysis sets out to understand and document the differences in gender roles, perspectives, activities, needs and opportunities in a specific context, with the aim of ensuring that an awareness of gender is integral to the evaluation and development of a fisheries management plan. Many international organisations such as the Food and Agriculture Organisation (FAO), World Wide Fund for Nature (IUCN),

United Nations Development Programme (UNDP), the World Bank and DFID are 'mainstreaming' gender into their programmes, moving on from the earlier policy of focusing on 'women in development'. Given the very real differences between men's and women's roles, responsibilities and opportunities in inland fisheries, management plans would be incomplete without acknowledging their importance.

The World Bank Participation Sourcebook (1996) identifies five major categories of information that make up a gender analysis:

- needs assessment
- activity profile
- resources, access and control profile
- benefits and incentives analysis
- institutional constraints and opportunities.

Specifically including someone with training and experience of gender issues in the team undertaking any evaluation of the fishery is probably the most efficient and effective way of ensuring that appropriate care is taken to explore gender issues for management. Training for fisheries managers, whether government employees or nominated members of fishing communities, will help to sensitise stakeholders to gender issues.

Participatory Rural Appraisal

Participatory Rural Appraisal (PRA; also known as Participatory Learning and Action) collectively describes a family of approaches, methods and behaviours that facilitates local people in expressing their experiences and objectives, allows them to identify problems and assists them in the development of solutions. In contrast to the more formal survey techniques and to Rapid Rural Appraisal (RRA) – from which this approach evolved – it is emphasised that this is not a technique for outsiders simply to develop their own understanding. Rather, it is a means to increase the ability of stakeholders to appraise, analyse and plan.

Since the late 1980s PRA approaches have been used successfully to develop policies, programmes and projects across an increasing range of sectors, primarily in agriculture and natural resources, although urban communities and issues have also benefited. The methods used are varied and continually developing. Common tools include:

- semi-structured interviews with key informants
- focus group discussions
- preference ranking
- analysis of temporal variation through daily activity charts, seasonal calendars and time lines (to identify sequencing of events, long term changes or trends)
- mapping of economic and geographical features and institutional relationships
- wealth ranking.

164 *Inland fisheries*

In addition to the central tenet of participation by the local community, some key principles are emphasised. These include:

- flexibility in choice of methods (according to context)
- the need for a team (insiders and outsiders representing the relevant disciplines)
- use of 'optimal ignorance' (ensures that just enough information is gathered to make necessary decisions)
- triangulation to validate qualitative data collected (using partial stratification of respondents and cross-checking of issues by using different techniques).

These methods can be an extremely effective way of identifying viable development options and providing the community with a sense of ownership, but their pitfalls and limitations should be recognised. The methods appear deceptively simple, and as such do not reflect the real need for well-trained facilitators to guide an appraisal and train other participants. The process is very powerful and the potential for a negative outcome from a less than rigorous application should not be underestimated. Great care is needed to ensure that the choice of tools and their sequencing is relevant to the context in which they are being applied. For example, inappropriate timing (usually before trust is established) of questions on wealth can seriously undermine stakeholder willingness to participate. The community can become disillusioned if their recommendations as expressed in the appraisal are not carried through into action. In addition, these methods do not provide quantitative estimates, within statistically definable limits; this can reduce their credibility with some audiences. This last point reinforces the need for experienced facilitation.

There is now a very substantial literature on PRA methods although, since the field is organic and open-ended – in line with its defining principles – there is nothing that presents itself as a standard text. Important works include many by Chambers (1992, 1994) and there has been a steady stream of publications from the International Institute for Environment and Development (IIED) in a series called PLA notes. Pretty *et al.* (1995) provide a guide for trainers that covers a range of techniques and their uses. Rietbergen-McCracken & Narayan (1998) have produced an important compilation of techniques; this may be more widely available. The World Bank (1996) has produced a 'Participation Source Book' which is available on the web (http://www.worldbank.org/wbi/sourcebook/sbhome.htm) or by application to the World Bank. This provides an overview of the bank's experiences, positive and negative, with respect to participation in its projects and programmes around the world.

Participatory Monitoring and Evaluation

Participatory Monitoring and Evaluation (PME) has been defined as: 'a process of collaborative problem solving through the generation and use of knowledge. It is a process that leads to corrective action by involving stakeholders in shared decision-making' (Narayan, 1993). It is closely related to PRA, as the philosophy and many of the tools used are the same. The emphasis of PME, however, is on providing regular re-examination of a situation. So, where PRA encourages understanding of problems and their solutions, PME provides additional insights into the direction and nature of change. PME has building stakeholder capacity for analysis and problem-solving as an explicit principle.

Monitoring and evaluation are common activities in fisheries management: in inland fisheries, data are frequently collected, analysed and used by governments to track resource status. Participatory monitoring and evaluation will primarily provide a tool for stakeholders to track and improve management of resources for which they have responsibility. A monitoring strategy would most effectively use both conventional and participatory methods. For example, an area of floodplain co-managed by local community and government may be set aside as a reserve, providing a refuge for resident fish species. Traditional monitoring and evaluation might be carried out by 'outsiders' such as experts from a project, or a non-governmental organization and focus on the technical performance of the reserve. Community groups might use the tools of PME to monitor the perceptions of a variety of stakeholders to the reserve or the effect of its establishment, and perhaps work towards developing an improvement or an alternative management options for the fishery.

Abbot & Gujit (1998) provide a thorough overview of recent experiences of PME. IIED (1998) provides a useful collection of articles and views on the experiences of practitioners in the field. Rietbergen-McCracken & Narayan (1998) also provide a useful summary of issues and approaches.

Conflict management

Conflict occurs in fisheries at all levels: within a single stakeholder group (e.g. competition for resources among commercial fishers), between groups with different levels of involvement in the fishery (e.g. subsistence versus commercial fishers or water-body lease-holders versus subsistence fishers), between different sectors (e.g. irrigation versus fisheries) and between government departments (e.g. fisheries versus wildlife or fisheries versus agriculture). These conflicts must be actively addressed, as they will undermine the sustainability of any fisheries management plan. Conflicts often arise from the changes associated with management; for example, setting aside areas as reserves or banning a type of gear may affect some stakeholder groups more than others.

The outputs of stakeholder and gender analysis will hopefully identify existing conflict and enable fisheries managers to predict and mitigate conflict arising from changes in management. Similarly, these analyses can yield insight into existing methods of conflict resolution and management, with an assessment of their efficacy. Where possible, conflicts arising from management of a fishery should rely and build upon methods that stakeholders recognise and trust. Mechanisms to address conflict should be fast and low cost and, depending on the context, formal measures such as resolution by the courts may be appropriate, while mediation by local leaders will work in other situations.

The full range of conflict management options should be considered as part of an intervention into a conflict situation, i.e. 'do nothing', force, withdrawal, accommodation, compromise and consensus. In terms of managing fisheries, it is perhaps most important to first acknowledge that conflict will undermine the sustainability and effectiveness of any plan. Capacity building in the area of conflict management would be a proactive component of a fisheries management plan. Initiatives such as clarifying roles and responsibilities among institutions to reduce potential for conflict, re-organising institutions better to match their role in the management plan, and training leaders in consensus building skills to facilitate management of conflict are some options for the better management of conflict associated with fisheries management.

Traditional methods for quantitative data collection

This final section addresses the need for the collection of quantitative socio-economic data through formal surveys. Traditionally, household or market surveys have been the first choice of planners seeking to understand socio-economic issues in fisheries. However, recent awareness of the need for a broader understanding of a range of socio-economic and institutional issues (stakeholders, gender, capacity, participation, monitoring, conflict, encouraging sustainable stakeholder behaviour, etc.) indicates that the use of formal surveys in fisheries needs rethinking. Quantitative socio-economic surveys are most valuable in providing detailed insight into a well-defined problem or issue. The main point is in their phasing. Rather than being the first method used, formal survey techniques are most effectively applied when a fuller understanding of the social, economic and institutional context has been developed and specific problems (that *can* be addressed through the analysis of quantitative data) have been identified. A second important point is that implementation and analysis of formal surveys are too slow, costly and demanding of advanced skills and resources to provide a useful tool for active management. As a result, the more strategic insights gained from these surveys are of most use to policy makers and researchers, setting the wider framework within which management of individual fisheries proceeds.

Household surveys

Formal household surveys are one of the traditional means by which planners gained an understanding of fish catches and their associated income streams. They are a good way of collecting quantitative information and, with a good understanding of the socio-economic and biological characteristics of the fishery and respect for the principles of survey design, they can provide useful estimates of the likely value of the parameters of interest and their statistical confidence limits.

Good surveys are neither easy nor cheap and they are rarely quick. There is a sequence of steps that must be followed. The objective of the survey and the use to which its results are to be put should be clearly defined. Considerable effort should be invested in the design of both the questionnaire and the survey, as it is too easy to ask questions that miss the real issues or provide sample estimates that cannot be extrapolated reliably.

Fish production is difficult to estimate for a number of reasons. Firstly, it is often highly seasonal and, as harvests are both daily and variable, much harder to recall than, for instance, harvests of the main agricultural crops or livestock sales. As a result, a single survey asking for recall of annual catch will often be of questionable reliability. If quantification is the primary issue, then estimates need to be based on monitoring, using a series of interviews repeated at intervals across the season. Secondly, owing to the variability in many areas of households' involvement in fishing and the catching power of the gears they use, there is often a need for both stratification (which is often difficult to achieve clearly) and a large sample size. Both factors substantially increase costs.

Where income and its distribution are the focus of the survey, further issues need to be taken into account. Adjustments must be made to catch quantities to provide economic values and additional information needs to be collected on costs, such as labour, gears, boats and fuel. Data on capital costs are often best collected in a one-off survey, ideally undertaken after the monitoring period, when the full range of fishing activities and gears undertaken by each household is known. If there is a need to quantify the distribution of income, payments for access, interest on credit for gears or other transfer payments need to be calculated. Providing a quantified estimate of such flows can become very complex. Access systems can be highly varied, with different systems co-existing on the same waterbody for different gears or groups. Extrapolation to unstudied water-bodies should only be undertaken cautiously.

Poate & Daplyn (1993) provide a thorough introduction to the strategy, techniques, advantages and pitfalls of formal questionnaire-based investigations in developing countries. Their examples are drawn mainly from rural household surveys and provide specific guidance on estimating income, expenditure and family consumption, price data and household resources. Although directed mainly towards agriculture, its principles are equally applicable to fisheries surveys. Other important, and widely available, sources of data are the books by Casley & Kumar (1987, 1988), to which similar caveats apply.

Market surveys

Counting fish at markets, customs posts and other localities where fish is concentrated in the course of landing and transport to the final consumer is another way in which the absolute and relative magnitude of fish catches can be estimated. This method works best for catches from fisheries where the great majority of fish passes through a single marketing channel. Large and variable subsistence catches, sales by fishermen direct to consumers and multiple markets, or the merging of fish from water-bodies other than that being investigated can all cause complications. Sometimes there is too much 'noise' for the method to be of great value unless supplemented with other (costly) surveys. Floodplains in South Asia and the more densely populated parts of southeast Asia thus present a greater challenge than equivalent habitats in many parts of Africa or South America.

The ease with which these surveys can be carried out and the reliability of the data produced are highly dependent on the way in which the market is operated and used. Both markets and traders within them may differ significantly in the scale and continuity of their operations: larger buyers may operate continuously within a clearly defined time frame, while others may be more ephemeral. Patterns of fish deliveries may also vary, depending on the character of fishing operations on the supplying water-bodies. When the great majority of the catch is delivered first thing in the morning, it is easier to derive reliable estimates than when deliveries are sporadic throughout the day.

In this context, the problem of deriving estimates of total volume or value of catch from a sample is complicated by difficulty in specifying raising factors. It is important therefore that the workings of the market are understood prior to survey design and that this understanding is reflected in the sampling frame and stratification procedures. If traders have to be registered, the sampling frame can be constructed from the list. Otherwise, some form of spatial frame may be needed, with traders interviewed according to the location that they occupy; this must be supplemented by efforts to quantify the number of traders operating.

Conducting interviews while traders are operating is very difficult, as it interrupts business. Arrangements should therefore be made to meet them before or, more probably, after trading has taken place. The most effective means of gathering data will often be to make use of the information that may already be available within the system. Traders often keep records of the weights and values of their transactions. Sometimes these are disaggregated for some individual species or species groups. Where this information is sufficient for the objective of the survey, traders should be encouraged to provide access to their records. Time taken to win trust in these circumstances and scrupulous efforts to make sure that this trust is justified may secure openness from traders on this sensitive information.

Analysis of consumption

Where other methods are not practical owing to the diffuse nature of the fishery or the market structure, attempts are made to evaluate catches by estimating the amount of fish consumed in the fishing and non-fishing communities. Such studies tend to establish a mean daily ration for various strata of society within the catchment of the fishery. The growth in the proportion of fish traded and of the distances across which they are moved means that this approach has its limitations. Cultural differences may influence patterns of consumption, making stratification important. Traditionally, this method tends to overestimate the catch. Conduct of the surveys, at a broad level, should follow the same principles as those required for other forms of household survey (Casley & Kumar, 1987, 1988; Poate & Daplyn, 1993).

Sampling

In all but the smallest systems it is impossible to sample the whole area to be studied for reasons of cost, time and the availability of trained personnel. Some form of sampling protocol has to be devised whereby a limited number of samples can be made to do the work of more comprehensive surveys in a cost-efficient manner. Bazigos (1974) and Caddy & Bazigos (1985) have described such approaches, which fall into two major categories:

- frame surveys, which aim to provide information on the size, structure and spatial distribution of the fishery, markets or environments
- catch assessment surveys, which are aimed at obtaining estimates of catch and effort at selected sample sites,

To which should be added

- social and economic surveys, and
- stakeholder analysis.

FAO has recently developed software to assist fisheries researchers and planners to formulate sampling programmes and to organise and interpret data from such programmes:

- *ARTPLAN*: software to assist users in better planning of surveys and to provide first order of magnitudes of total catch and fishing effort.
- *ARTFISH*: software to organise and process primary data collected by artisanal fishery sample surveys.
- *ARTSER*: software to analyse monthly catch and effort estimates resulting from ARTFISH operations.

The effectiveness of catch assessment surveys depends on the efficient establishment of an initial stratification based on the frame survey, which divides the area to be sampled into areas of broad similarity. Sampling sites usually consist of representative fish landings or markets within each of the strata. Alternative approaches include the selection of entire villages, fishing camps or families (Quensiere, 1994). Similar sampling frames are used in the design of area-catch studies. Ideally, sites within the frame should be selected at random and should be sufficient in number to cover the major variations. However, considerations such as cost, ease of accessibility and co-operation on the part of the fishermen are more likely to influence the selection.

General guidelines for the routine collection of capture fishery data have been issued by FAO (1999c).

Note

1. By V. Cowan and M. Aeron-Thomas, MRAG, 47 Princes Gate, London, UK.

Chapter 10
Integrating information

Once obtained, the information from various sources needs to be drawn together to present a clearer picture of the resource as a whole. Integrated information on the fishery can also be inserted into larger natural resource and agricultural sector planning.

Resource mapping

Information gained from the various types of investigation can be combined into thematic maps, which classify waters according to their physical attributes, types of fish fauna or degree of environmental degradation. These then serve as a basic planning tool at local, regional or national level. Maps classifying waters according to water quality or environmental status are available for many regions of Europe and more are being prepared. However, existing maps are based on different models even within the same country. There is, therefore, a need to harmonise approaches so that maps can be made comparable and extended to wider geographical areas. Current maps, which include basic ecological data, could be extended to include social and economic parameters through GIS systems.

Geographical information systems[1]

The objectives of this section are, firstly, to provide a broad overview of practical considerations for the implementation of geographical information systems (GIS) and, secondly, to highlight recent GIS applications in inland fisheries and to consider their future directions. The perspective is global.

Practical considerations for the implementation of GIS

A GIS comprises a collection of integrated computer hardware, software and trained personnel to compile, input, manipulate, analyse and report spatial (geographical) data. The data may be in any textual, map or numeric form that is capable of being integrated within a single system. A schema of a generalised GIS is shown in Fig. 10.1.

172　*Inland fisheries*

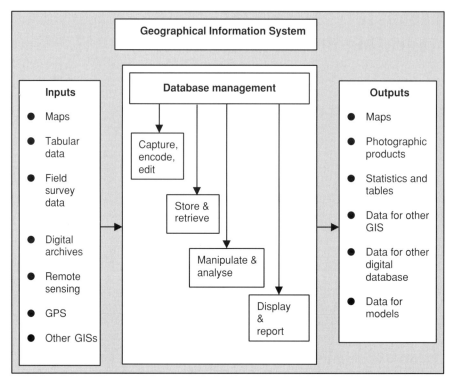

Fig. 10.1　Schema of a Geographical Information System. (After Meaden & Kapetsky, 1991.)

A basic concept is that GIS is a tool that has a broad variety of applications in the fisheries sector. In fact, any fishery resource, fishery or environmental problem that involves spatial location and geographical relationships has a potential solution through the application of GIS. Furthermore, GIS capability extends to multiple objective land (and water) allocation. GIS used in this way can help to solve actual or potential conflicts such as allocation of space between fisheries and aquaculture and definition of essential fish habitat among competing uses for land and water. However, it has to be kept in mind that GIS, like any other tool, requires proficient operators if it is to be effective and cost efficient.

The essential elements for applying GIS to fisheries are personnel trained in GIS technology, fisheries personnel who understand the benefits and constraints of GIS, and the hardware and software appropriate for the level of the application. A single self-taught individual using free GIS software on an ordinary personal computer and peripherals can obtain useful results. Usually, however, formally trained and experienced personnel, relatively sophisticated GIS software, a computer or workstation especially configured for high speed and large memory and peripheral equipment including a digitising board and plotter are required for most current work. For longer-term projects, it is essential that posts dedicated to GIS are established and operating funds are provided to cover not only the costs of the studies but also the periodic upgrading of equipment and training of staff in new methodologies. Ideally, GIS would be established in an 'information-for-fisheries-management unit' that would consist of at least one person specialised in GIS analysis with a natural resources background and several other GIS-capable staff with education and working experience in various fishery disciplines such as resources, management and economics. Other experts would be on call to provide specific advice and guidance as required.

The most time-consuming part of GIS-based studies is the acquisition and preparation of the data for analysis. An important rule of thumb is that it is much more economic to adapt data collected or created for other purposes than to create them for a special fisheries purpose. For example, data originally created for climate change studies were used by Kapetsky (1998) to assess potential for certain kinds of inland fishery enhancements. Digital data used to assess agriculture potential often can be modified for use in inland fisheries and aquaculture. For example, Kapetsky & Nath (1997) interpreted soil characteristics in terms of suitability for fishpond construction and chemical suitability for fish growth. They also used precipitation and evaporation to estimate pond water balance.

A trend important for the implementation of GIS in the inland fisheries sector is that digital data are becoming more widely available. One of the prime sources of digital data is from airborne and satellite remote sensing. Remotely sensed data can provide much of the information usually seen on a topographic map as input for GIS. Because the remotely sensed data can be manipulated through image analysis, and because a variety of specialised sensors is available, information not discernible on aerial photographs can be generated. For example, estimates of the water-surface area covered by floating floodplain vegetation on the Sudd in Sudan have been used by Travaglia *et al.* (1995) to assess annual variations in fishery potential there using thermal infrared satellite data.

Kapetsky (1999) has set out a step-wise procedure for the development of a regional fisheries GIS in the Lower Mekong Basin. The development of GIS capacity in an organisation such as a regional or national fishery administration has to be well planned and the commitment has to be made over a long timespan. If it is not possible to commit for the long term, then an alternative is to contract for the analyses.

Basic to GIS implementation is an appreciation for its strengths and limitations, an awareness that has to be implanted among all staff who could potentially put the technology to use. Thus, GIS awareness raising should be conducted for the whole gamut of inland fishery sectors, including information and statistics, policy and planning, resource evaluations, fishery technologies and marketing. Awareness raising is not only for technical staff, but also for administrators and managers. It is to improve their decision-making capacity that GIS is implemented.

Implementation of GIS can be pursued step-wise. Initial emphasis can be placed on providing a staff training and computer capacity to do mapping (i.e. discerning spatial trends and patterns in fishery data). Those who prove to be particularly adept at mapping can be encouraged to advance towards GIS. In this way, the necessary training can be provided to those most likely to make the best use of it.

As the following review shows, access to the Internet for the implementation of GIS in fisheries is becoming increasingly important. There are several considerations, but basically, all of them relate to keeping up to date with the technology, GIS methodologies, training and applications as well as quick access to the data and technical assistance. The avenues include the websites of government and international agencies implementing GIS, commercial organisations providing GIS hardware and software, and e-mail discussion groups oriented towards specific software or applications.

Basic treatments of the actual and potential applications of GIS in the fisheries sector, including essential theory, have been covered by Meaden & Kapetsky (1991) for inland fisheries and aquaculture, and by Meaden & Do Chi (1996) for marine applications. Simpson (1994) set out a broad framework for remote sensing in marine fisheries linked to GIS. Isaak & Hubert (1997) promoted the use of GIS in inland fisheries science in North America and mentioned some potential applications. Fisher & Toepfer (1998) ran a survey of GIS-related research conducted at US universities with fisheries graduate programmes. Meaden (2000) has completed an assessment of the progress in applying GIS in the fisheries sector.

Overview of recent GIS applications in inland fisheries

Approach

In developing the overview of recent applications three basic considerations shape the format of the presentation: (1) completeness of coverage, (2) timeliness, and (3) accessibility to the original works. As has been made clear by Meaden (2000), a significant number of GIS applications is in the grey literature and many of these are not abstracted internationally. Yet, from a professional viewpoint, it is essential to be familiar with the full depth and breadth of applications. As a compromise between accessibility and completeness of coverage, the following major sources were used to locate and report the material:

- *Aquatic Sciences and Fisheries Abstracts* (ASFA) up to March 1999
- The author's own collection of published papers.
- Papers presented at ESRI Users' Conferences from 1995 to 1998 (http://www.esri.Com/Library/Userconf/)
- Oral presentations and posters found in the book of abstracts from the First International Symposium on GIS In Fishery Sciences, Seattle, Washington, USA, 2–4 March 1999.

The rationale is that using ASFA provides access to literature that usually is easily accessible, at least in abstract form. ESRI User Conference papers are timely and easily accessible and the Internet site is sustained. The symposium presentations are timely and accessible as most of the papers are published in this volume.

In order to convey the maximum amount of information while at the same time conforming to space limitations, the applications are tabulated according to title (Table 10.1). Sources are easily identified as follows: (1) titles with authors' names in parentheses are from the symposium abstracts; (2) titles with authors and year are from the published literature and are found in the References section; and (3) those from the ESRI Conference Library are so identified.

Table 10.1 Titles and authors of recent GIS applications in inland fisheries

Fishery resources

Direct assessments

- GIS supported assessment of fish stocks in Austrian lakes (Wanzenböck)

Fish movements and migrations

- Integration of juvenile chinook salmon location estimates and water velocities to assess migration behavior in the forebay of Bonneville Dam, Columbia River (Hardiman)
- Using GIS and GPS to map the seasonal distribution of Independence Valley speckled dace and tui chub (Rissler *et al.*)
- The animal movement programme: integrating GIS with statistical analysis and modeling of animal movements (Hooge *et al.*)

Habitats

Classification and inventory
- Use of the Geographic Information System in aquatic habitat management (Starostka, 1994)
- Using address matching to derive in-stream fish habitat unit locations (Martischang & Carlson, ESRI User Conference, 1995)
- Display and analysis methods for aquatic habitat data (Wing)
- Building foundation spatial data layers for aquatic analyses (Clarke)
- Spawning abilities and quantity of northern pike (*Esox lucius* L.) in Lake Constance (Wittkugel & Fischer)
- Assessing spatial patterning of in-stream habitats using GIS (Young *et al.*)
- Raster-based thermal habitat suitability analysis for selected freshwater fish (Doka *et al.*)
- Arrangement of habitat inventory information on GIS platform to identify optimum and degraded areas of endangered fish tor putitora habitat (Srivastava *et al.*)
- Evaluating fish habitat in a South Carolina watershed using GIS (O'Brien-White & Thomason, 1995)
- Where the wild fish are: documenting salmonid presence and use for effective management (Hudson *et al.*)
- The North Pacific Rim salmon ecosystem project (biogeographic zones of salmon in NA and Asia) (Rodgers *et al.*)
- Spatial patterns of aquatic habitat in Oregon (Jones & Flitcroft)
- The distribution of historical salmon habitat in an upper reach of John Day Reservoir of the Columbia River (Sheer)
- Interpretation of reconnaissance fish habitat inventory data For BC (Cheong *et al.*)

(*continued*)

176 *Inland fisheries*

Table 10.1 (*Continued*)

Fishery resources

- An initial analysis of the numbers, distribution and size of Zimbabwe's small dams (Chimowa & Nugent, 1995)

Habitat quality and quantity linked to fish abundance and distribution
- The role of GIS in area fish habitat management plans in the Great Lakes basin (Minns *et al.*)
- Defensible methods for pre- and post-development assessment of fish habitat in the Great Lakes (Minns *et al.*, 1995)
- A desktop GIS of northwest salmonid resources in the Columbia river basin (Genovese & Emmett, 1997)
- Estimating the abundance of stream fish populations using geographic information systems technology (Fisher *et al.*)
- Potential for use of geostatistics in combination with biological data to assess the freshwater production of Atlantic salmon in Norway (Erikstad *et al.*)
- Predicting the spatial distribution of *Dreissena polymorpha* (zebra mussel) among inland lakes of Wisconsin: modelling with a GIS (Koutnik & Padilla, 1994)
- Applications of GIS to map and manage fish habitat: The highly migratory species example (Perle *et al.*)
- Population and conservation biology of the threatened leopard darter (Toepfer, 1998)
- Application of geographic information systems in fisheries: habitat use by northern pike and largemouth bass (Rogers & Bergersen, 1996)
- Crayfish distribution as a function of hydrology in Everglades National Park (Hendix)
- A volume-based habitat model using GIS, bathymetric maps and field measurements: Huntsville Lakes case study (St. Onge, 1995)
- Effects of reservoir drawdown on littoral habitat: assessment with on-site measures and geographic information systems (Irwin & Noble, 1996)
- Geographic mapping of spawning areas of fish in the littoral zone of Lake Constance – long term changes (Wittkugel & Fischer)
- Macrophytes, hydrology, and aquatic ecotones: a GIS-supported ecological survey (Janauer, 1997)

Habitat approach to aquatic biodiversity
- Freshwater biodiversity: a preliminary global assessment (World Conservation Monitoring Centre, 1998)
- National GIS fills 'Gaps' in biological diversity (Corbley, 1996)
- Aquatic gap analysis: demonstration of a geographical approach to aquatic biodiversity conservation (Bain *et al.*)
- The Missouri aquatic gap pilot project: a multi-agency GIS tool for assessing Missouri's aquatic resources (Sowa & Heverland)

Essential fish habitat
- Identifying habitats essential for pike, *Esox lucius* L. in the Long Point region of Lake Erie: a suitable supply approach (Minns *et al.*, 1999)
- Identifying and assessing essential fish habitat using GIS (Brown *et al.*)
- Spatial methods being developed in Florida to determine essential fish habitat (Rubec *et al.*, 1998)

Modified habitats
Enhancements:
- Geography and constraints on inland fishery enhancements (Kapetsky, 1998)
- Biotic analysis of lower granite reservoir drawdown using GIS (Pinney, 1994)

(*continued*)

Table 10.1   (*Continued*)

Fishery resources

- Etude expérimentale de la qualité des alevins de saumon, *Salmo salar* L. Destines au repeuplement. Contribution a l'analyse hydraulique de l'habitat (Quality of hatchery salmon, *Salmo salar* juveniles for restocking. Contribution to the habitat hydraulics study) (Charles, 1996)
- The basinwide prioritization of stream habitat enhancement sites in Western Oregon (Thom *et al.*)
- Reservoirs, GIS, and bass habitat (Cross, 1991)

Artificial habitats:
- A role for comprehensive planning, geographical information system (GIS) technologies and program evaluation in aquatic habitat development (Gordon, 1994)

Restoration and rehabilitation:
- Hydrologic and hydrographic database design in support of anadromous fisheries restoration planning (Beachler, ESRI Users' Conference, 1998)
- A landscape approach to wetlands restoration research along Saginaw Bay, Michigan: baseline data collection and project description (Burton & Prince, 1995)
- Measuring the effects of reservoir operations using GIS (Cunnigham *et al.*)
- A stream classification system for identifying reintroduction sites of formosan landlocked salmon (*Oncorhynchus masou formosanus*, Jordan and Oshima) (Tsao *et al.*, 1996)
- The role of GIS in selecting sites for riparian restoration based on hydrology and land use (Russell *et al.*, 1997)

Fisheries

Planning and potential

- Global distribution of inland capture (Section 1.8 in: FAO Inland Water Resources and Aquaculture Service, Fishery Resources Division, 1999)
- Recreational boating on Delaware's inland bays: implications for social and environmental carrying capacity (Falk *et al.*, 1992)

Inland aquaculture for culture-based fisheries

- A strategic assessment of the potential for inland fish farming in Latin America (Kapetsky & Nath, 1997)
- A strategic reassessment of fish farming potential in Africa (Aguilar-Manjarrez & Nath, 1998)
- A strategic assessment of the potential for freshwater fish farming in the Caribbean island states (Kapetsky & Chakalall, 1999)

Holistic inland fisheries information system

- GIS 'fishery resources of russia' (Babayan *et al.*)

Management of fisheries together with aquaculture)

- Lease site considerations for hard clam aquaculture in Florida (Arnold *et al.*, 1996)
- Integrated resource management using GIS: shellfish aquaculture in Florida (Arnold & Norris, 1998)

(*continued*)

Table 10.1 (*Continued*)

Environment

Terrestrial effects on habitats and fishes

- Evaluating the biotic integrity of watersheds in the Sierra Nevada, California (Moyle & Randall, 1998)
- Shoreline alterations on the Muskoka Lakes, Ontario – mapping techniques, methods development, preliminary results and interpretive framework (Hutchinson *et al.*, 1995)
- The use of a geographic information system to evaluate terrain resource information management (TRIM) maps and to measure land use patterns for Black Creek, Vancouver Island (Brown *et al.*, 1996)
- Using GIS to find the relationship between fish diversity and stream bank cover in Illinois (Kompare, ESRI User Conference, 1998)
- Impacts of land use on stream habitat for the threatened leopard darter (Toepfer *et al.*)
- The influence of catchment land use on stream integrity across multiple spatial scales (Allan *et al.*, 1997)
- GIS in the coastal landscape analysis and modelling study (clams): a prototype application to predict juvenile salmonid summer use and habitat from landform and landuse data (Burnett *et al.*)
- Spatial pattern of forest clearing in watersheds containing essential fish habitat for Pacific salmon (Kelly *et al.*)

Risk evaluation and environmental health

- Geographic information systems for risk evaluation: perspectives on applications to environmental health (Nyerges *et al.*, 1997)

Water quality and water quantity

- GIS modelling of hydroperiod, vegetation and soil nutrient relationships in the Lake Okeechobee marsh ecosystem (Richardson & Hamouda, 1995)
- Combining field sampling, a geographic information system, and numerical modelling to analyze sediment distribution in a backwater lake of the upper Mississippi River (Gaugush, 1994)
- Development and primary application of a GIS database for Miyun Reservoir water quality protection (Jia *et al.*, ESRI Users Conference, 1998)

Climate change

- Using GIS and individual-based modelling to predict global climate change effects on trout in the southern Appalachians (Levine *et al.*, ESRI Users Conference, 1998)
- Thermal limits to salmonid distributions in the Rocky Mountain Region and potential habitat loss due to global warming: a geographic information system (GIS) approach (Keleher & Rahel, 1996)
- Potential habitat loss and population fragmentation for cold water fish in the North Platte River drainage of the Rocky Mountains: response to climate warming (Rahel *et al.*, 1996)
- Predicting the impact of climate change on the spatial pattern of freshwater fish yield capability in eastern Canadian lakes (Minns & Moore, 1992)

Multi-sectoral planning and management including fisheries

Watersheds and catchments

- The digital watershed atlas – creation and use (Hawthorne)

(*continued*)

Table 10.1 (*Continued*)

Multi-sectoral planning and management including fisheries

- Environmental mapping and modelling of a catchment using GIS (Viers *et al.*, ESRI Users' Conference, 1998)
- California rivers assessment: assembling environmental data to characterize California's watersheds (Viers *et al.*, ESRI Users' Conference, 1998)
- Natural resource mapping using GIS: coastal and watershed applications (Halls *et al.*, ESRI Users' Conference, 1996)
- Aquatic conservation strategies and the Navarro River watershed: applied GIS for total maximum daily load decision support (Viers *et al.*)
- Combining tree growth, fish and wildlife habitat, mass wasting, sedimentation, and hydrologic models in decision analysis and long-term forest land planning (Olson & Orr, 1999)
- Integrating social and ecological suitability in a community-based, spatially explicit, adaptive watershed conservation programme (Habron)

Floodplains

- Hoh River floodplain inventory: integrating GIS and GPS to redefine floodplain management (Taylor & Silver, ESRI Users' Conference, 1998)
- GIS and the floodplain management planning process in Washington State (Henderson *et al.*, ESRI Users' Conference, 1996)
- Using GIS for managing flood plain ecosystems (Miyamoto *et al.*, ESRI Users' Conference, 1997)

Characterisation of recent GIS applications

A meaningful framework is required in order to categorise the applications. Ideally, the framework would not only characterise recent applications but at the same time be indicative of needs for new applications. Thus, applications were considered in four broad categories pertaining to:

- Fishery Resources
- Fisheries
- Environment
- General Planning and Management, Including Fisheries.

A schematic breakdown of the major and subcategories is shown in Fig. 10.2. The subcategories are intended to show not only where applications are, but also where there are opportunities and needs as evidenced by relatively few entries.

Among the four major categories there is an uneven distribution of applications. Sixty per cent of the total of 85 are in fishery resources; however, nearly all of them address habitats. Of the habitat applications most are directed towards classification and inventory studies, and habitat quality and quantity in relation to distribution and abundance of fishes. Thirteen per cent of all applications concerned modified habitats.

Fisheries, in four subcategories, accounted for only 9% of the total. Although the overall classification scheme is somewhat subjective in terms of what are fisheries

180　*Inland fisheries*

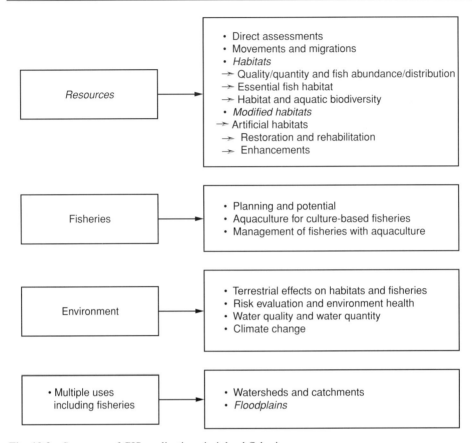

Fig. 10.2 Summary of GIS applications in inland fisheries.

applications and what are fishery resources applications, the results still indicate a paucity of applications aimed directly towards the management and development of fisheries.

Environment accounts for 19%. Estimating terrestrial effects on the aquatic environment is the most frequent kind of application.

Multi-sectoral management that specifically includes fisheries amounts to 12%, of which most applications are from a watershed or catchment perspective and others concern floodplains. It is encouraging to see fisheries integrated with other uses and included in overall management plans.

Geographically, there is an overwhelming representation of applications in North America.

Selected applications of GIS in inland fisheries

The papers presented in this section have been selected with two criteria in mind: (1)

breadth of applications, and (2) innovative use of GIS. In order to facilitate comparisons with the spectrum of applications as illustrated by the titles in Table 10.1, the same outline for the commentary has been followed as in Fig. 10.2.

Fishery resources

Direct assessments
In marine systems GIS provides a natural framework for the acquisition and analysis of geo-referenced biological, oceanographic and environmental data (Kieser *et al.*, 1995). Fisheries acoustic survey data are generally collected along transects. Transects and individual acoustic measurements can be accurately geo-referenced and a GIS can thus be used to map individual transects and the qualitative and quantitative data recorded along them. Use of GIS allows for the integration of ancillary data such as coastline, bathymetry, trawl sets and oceanographic measurements.

Direct estimates of fishery resources over large areas, usually using hydroacoustics, are not nearly as common in inland fisheries as in marine fisheries. There is one GIS application in inland waters (Table 10.1).

Movements and migrations
Applications of GIS to movements and migrations of inland fishes are scarce; however, attention is called to the 'Animal Movement Programme: Integrating GIS with Statistical Analysis and Modelling of Animal Movements', which is a collection of spatial statistical tools in GIS running in ARC/VIEW. An important point here is that this toolkit has applications not only to fish movements as intended by the authors; it also has potential applications to fishing, for example, in relation to variations in the spatial distribution of fishing, particularly on seasonally varying systems such as floodplains and reservoirs.

Habitats
Inland fishery resources, resource potential (and loss of potential) and fish catch are usually estimated by using the distribution and expanses of aquatic habitats as the baseline for interpolation and extrapolation of sample results. Therefore, GIS applications are relatively numerous with regard to habitats.

Inventory and classification
Many of the applications in this section involve inventory and classification over broad geographical areas and they are aimed at management as an end. For example, O'Brien-White & Thomason (1995) made a catchment-wide evaluation of the fish habitat of the Edisto River Basin in South Carolina. A committee of fisheries biologists was formed to evaluate fish habitat within the basin, and the GIS was used to compile and analyse data and generate maps to represent quality of fish habitat. It was concluded that GIS enabled the analysis of fish habitat over a larger area and with more comprehensive data sets than fisheries biologists traditionally use. Benefits from

GIS in this study included the prioritisation of sampling efforts and locating mitigation and restoration sites.

The channel type, an inventory and mapping tool for stream classification based on stream reaches, was incorporated into a GIS to facilitate manipulation and storage of stream inventory data by Starostka (1994). The basic component of the channel type is the fluvial process group that describes the interrelationship between runoff, landform relief, geology, and glacial or tidal influences on erosion and depositional processes. Channel-type inventories provide key information on fish habitat utilisation, habitat capability and enhancement options.

Habitat quality and quantity linked to fish abundance and distribution

Measures of habitat quality and quantity are among the most important in arriving at an understanding of the abundance and distribution of fishes. Therefore, applications are relatively numerous.

On a broad spatial scale a desktop GIS of north-west salmonid resources in the Columbia River Basin was developed by Genovese & Emmett (1997) based on a series of ARCVIEW projects. The system includes data on five species and is categorised into salmonid spawning escapement, hatchery releases and presence/absence tabular data. Salmonid spawning escapement data sets are further divided by observation type (i.e. dam counts, fish per mile) and linked to spatial data sets representing streams and rivers within the Columbia River Basin. Additional coverage of dams, hatchery locations, state boundaries and land ownership has been incorporated into each project to provide geographical reference and to permit data analysis. This desktop GIS allows scientists and managers to view salmonid population characteristics and geographical features simultaneously over varying spatial scales. The addition of detailed habitat information (e.g. land use, roads per mile and water temperatures) will enable scientists to identify how past and future watershed activities will affect salmonid populations in the north-west.

Janauer (1997) made a broad perspective study of macrophytes, hydrology and aquatic ecotones in the backwater system of the River Danube in Austria. The GIS overlays revealed spatial coincidence between the centres of sports fishing activities and survey stretches with high densities of macrophyte vegetation. Interesting features of this study were that it incorporated the macrophyte survey, a dynamic hydrological model and a socio-economic study of sports fishing activities. Conclusions were drawn from this database by GIS thematic map interpretation and overlay techniques.

Irwin & Noble (1996), using on-site measures, quantified the effects of reservoir drawdown on littoral habitats in two reservoir embayments and remotely sensed data entered into a GIS. The GIS allowed large-scale assessment of changes in size and configuration of habitat patches, whereas on-site measures provided finer resolution of habitat characteristics. Because some littoral habitats support higher densities of fish, differences in habitat on embayment-wide scales should be considered when managing water levels.

Abundance of leopard darters was determined in a creek and river in Oklahoma using a mark–recapture (Schnabel) technique and GIS (Toepfer, 1998). Differences in abundance among streams and among years were related to stream geomorphology and hydrology. Swimming performance was measured in a laboratory flow-through apparatus, and the results were compared with velocities measured in culverts under road crossings. A non-point source model and GIS were used to estimate sediment yields associated with silviculture and to relate yields to quality of leopard darter habitat.

Habitat approaches to biodiversity

The World Conservation Monitoring Centre (1998) used river basins as habitat frameworks in which to make a GIS-assisted assessment of freshwater biodiversity at a global scale. World-wide, among 151 river basins, 30 have been identified that support high aquatic biodiversity on the basis of high fish family richness and high vulnerability to future pressures. Vulnerability was based on a low score for wilderness and a high water resource vulnerability index. Basically, these are high-level measures of the state of catchments and the pressures exerted on them. Of the 30 river basins, 39% (by total area) are in Africa, 35% in Asia and 26% in Latin America. Using these biodiversity-vulnerability measures, the most stressed catchments are to be found in South Asia, the Middle East and western and north-central Europe. The least stressed are those in the north-western part of North America.

Habitat approaches to biodiversity of fishes in the context of GAP analysis, as practised in the USA, are very comprehensive and are highly structured so that results are comparable from location to location. The US National Biological Survey programme for GAP analysis (Corbley, 1996) has a centralised GIS that provides an overview of vegetation and animal species distribution across the country in private and public conservation lands. Although not dealing exclusively with the biodiversity of fishes, this approach represents a vast body of experience from among the 200 organisations involved that could be put to work for fisheries purposes.

Essential fish habitat

Efforts to conserve and protect fish habitats are frustrated by the key unanswered questions as to which habitat types and how much must be protected to ensure natural self-sustaining fish stocks (Minns *et al*., 1999). The need to define essential fish habitat, mandated by the Magnuson-Stevens Fishery Conservation and Management Act in the USA, has provided a huge stimulus to the use of GIS for habitat analyses, but much of the work so far has been accomplished in marine and estuarine waters. Nevertheless, the general approaches and techniques are the same as those that would be applied in inland waters.

Minns *et al*. (1995) developed a methodology for use in the Great Lakes for the pre- and post-development assessment of inshore fish habitat. The prototype methodology focuses on shoreline development projects. The purpose is to provide a quantitative, defensible assessment protocol which proponents and fish habitat managers can use to assess compliance of shoreline development projects with the federal habitat policy of 'no net loss' of fish habitat in development. The protocol combines published habitat requirements of three thermal guilds of Great Lakes fishes with GIS-based areal estimation of lost or modified inshore habitat due to placement of headlands or offshore structures in pre- and post-development assessments of adult and spawning habitat of fish, and habitat for community production.

A volume-based habitat model using GIS, bathymetric maps and field measurements was carried out on four lakes as a case study (St. Onge, 1995). Applications include accurate calculation of the habitat volume available for a given fish species at a given time by using temperature and dissolved oxygen profiles to determine the upper and lower depth boundaries of fish habitat. The application should be readily extendable to reservoirs and to the inclusion of other water-quality measurements that affect fish distribution, such as turbidity.

A suitable supply approach to identify habitats essential for pike *Esox lucius* L. in the Long Point region of Lake Erie has been put forward by Minns *et al.* (1999). A habitat supply method is elaborated in three ways:

(1) The basic physical habitat assessment is derived from a remote-sensing inventory database.
(2) Methods of quantifying the thermal regime and integrating it with other habitat elements are examined.
(3) Habitat supply estimates are used in a pike population model, and pike biomass and production are simulated and compared with available records.

The roles of error and uncertainty are examined for all elements in the estimation and for the application of suitable habitat supply values. It is concluded that there is potential for supply measurement and analysis to guide fish habitat management.

Florida is developing a database and conducting modelling to identify and spatially delineate fish habitats. These methods may be useful to fishery management councils seeking to determine essential fish habitat under the US Magnuson-Stevens Fishery Conservation and Management Act. The Florida Estuarine Living Marine Resources System consists of a relational database that summarises bibliographic information on the habitat requirements of fish and invertebrates important to fish. Rubec *et al.* (1998) use Habitat Suitability Index (HSI) modelling to estimate the geographical distributions of species by life stages. They use water-column and benthic habitat data from each estuary to create gridded environmental maps. Suitability indices of relative abundance across environmental gradients are derived from fisheries independent monitoring data. GIS is used to run the HSI model to produce predicted distribution maps.

Modified habitats
Modifications to habitats can be considered in three ways: enhancements, rehabilitation and restoration of degraded habitats, and creation of new or artificial habitats.

Enhancements Enhancements are being increasingly practised to expand benefits from inland fisheries, both in developed and developing countries; however, there are limits, some of which can be assessed geographically. Kapetsky (1998) describes a GIS-assisted approach to address the following questions from global and continental perspectives:

- What quantities of water resources are available to be enhanced?
- What are the prospects for an increase in enhanceable surface area from new reservoirs?
- How are enhancement prospects affected by climate and availability of inputs?

One fundamental idea that is reinforced by this comprehensive GIS approach is that, although the possibilities for increasing inland fishery production through enhancements seem nearly infinite at first glance, in fact, they are quite finite. For example, one subinvestigation asked the question: 'Where are there prospects for the fertilisation of small water bodies using crop by-products and manures as inputs (i.e. natural fertilisers) and also where is continuous growth and reproduction of Nile tilapia possible?' The results showed that only about 4% of the Earth's land surface encompasses conditions combining sufficient to ample water to maintain small water bodies and moderately high to high quantities of natural fertilisers along with continuous growth and reproduction of the Nile tilapia.

GIS was used in a reservoir study by Pinney (1994) to evaluate proposed enhancement efforts aimed at increasing the survival of Snake River salmonids, specifically migratory smolts and adults of salmon protected under the Endangered Species Act. A relational database was used to evaluate different flow option effects for proposed drawdowns of the reservoir on the salmon stocks. GIS allowed the assessment of how parameter changes in habitat quality and quantity across trophic levels affect salmonid survival. Benthic invertebrate and fishery data constituted biological layers, transects and blocks that could be used analytically in relation to a three-dimensional bathymetry layer. A graphical interface was used to model spatially the effects of changing water surface elevations and flow rate on critical elements affecting anadromous salmonid survival. This included measures of flow rate, critical water velocities, fish travel time, fall chinook rearing habitat distribution and availability, trophic availability, spawning habitat, predator–prey interactions and distributional behaviour in a dynamic environment. It was found that the visualisation of critical habitat and physical processes through GIS is an innovative approach that provides a much-needed bridge between the fields of planning and empirical science. It provides a visually based, easily understandable and communicable medium, allowing science to be used to improve model performance for more biologically realistic planning activities.

A stream classification system was developed by Tsao *et al.* (1996) to identify potential reintroduction sites for a self-sustaining population of Formosan landlocked salmon. To understand why the landlocked salmon had disappeared from most of the streams in its historic range, intensive habitat surveys were performed in six river tributaries. Habitat information was collected at three spatial scales: that of the stream reach, the mesohabitat and the microhabitat. Data were converted into a GIS model that facilitated analysis of temporal and spatial relationships between the salmon and its environment. Similarly, Charles (1996) dealt with the causes for the failures of stocking in the natural environment within the framework of an enhancement programme of migratory fishes in the Aquitaine Region of France. This experimental approach, using GIS, was an analysis of parts of the local hydraulic surroundings exploited by fry. Information on the physical environment in relation to its biological use should lead to a better definition of fry preferred habitats.

Artificial habitats The use of GIS in artificial aquatic habitat planning is taken up by Gordon (1994). Although the paper deals with marine enhancements, it is relevant to inland fisheries in several ways concerning a comprehensive or systematic approach to planning, infrastructural support and offshore user and non-user considerations.

The inland homologue of marine artificial reefs is brush park fisheries that have a long tradition in running and standing waters in Cambodia, Bangladesh and among several countries of West Africa in lakes and coastal lagoons (Welcomme & Kapetsky, 1981). Although the brush parks functioned well to concentrate fishes, there have been social problems associated with allocation. In this regard, GIS provides a range of analytical techniques including multiple criteria for decision making and multi-objective land (and water) allocation. It can be used for the selection of the sites and for the allocation of space for these gears among competing uses for open-water fisheries and for navigation, or for the better allocation of other moving and fixed gears among habitats.

Rehabilitation and restoration of habitats The significance of GIS in rehabilitation and restoration for inland fisheries is that basic studies of habitat expanse and quality are needed in order to make decisions about where to site ameliorative activities. Two papers place emphasis on the use of GIS to acquire the baseline data. For example, in 1993 and 1994, Burton & Prince (1995) collected baseline data on five wetland complexes along Saginaw Bay as part of planning for a wetland restoration project. This project may restore hundreds of hectares of wetlands on hydric soils presently drained for agriculture. Wetland complexes are a mosaic of different types of wetland and upland. For example, one study site was mapped using GIS into 13 cover classes relating to plant community composition, invertebrates and fish. Russell *et al.* (1997) note that successful long-term wetland restoration efforts require consideration of hydrology and surrounding land use during the site-selection process. They describe an approach to initial site selection that uses watershed-level information on basin topography and land cover to rank the potential suitability of all sites within a watershed for either preservation or restoration.

Inland fisheries At first glance, there appears to be a paucity of applications of GIS to fisheries. This is, indeed, the actual case; however, it is somewhat tempered by the fact that many of the applications allocated to the various subject areas of habitats as well as could have been placed in fisheries under a subheading of applied fisheries resources research.

Planning and potential Estimates of fisheries potential must be based on knowledge of the present performance of fisheries. From a global perspective, the FAO (1999a) assessed the performance of inland fisheries by comparing the geographical distribution of inland capture on a continental basis with various measures of surface water resources available for exploitation. Estimates of surface areas of various kinds of water bodies were obtained using GIS data originally generated for global climate change studies, and Kapetsky (1998) describes the data sources and GIS methods used. GIS was then used to estimate the amounts of land areas and surface waters within FAO's continental capture fishery statistical regions. This study showed that the continental shares of inland capture do not relate closely to the relative amounts of land and water available. Taking Asia as an example, that continent produces nearly 64% of inland capture, but has only a 20% share of the total continental area, a 23% share of swamps, marshes and other wetlands, a 7% share of lake area and an intermediate index of river density. However, it does have a relatively large number of reservoirs. A variety of factors is involved. Heavy exploitation of virtually the entire available water surface and the widespread use of fishery enhancements, mainly stocking, to increase food fishery yields are the most important ones. In this way the Asian achievements provide an indication of the extent to which inland fisheries can be intensified through supplementation. In addition, many fisheries probably produce relatively high yields because of heavy nutrient inputs from human settlements around lakes and from soil erosion. The limited data showing that eutrophication of lakes and reservoirs in China is proceeding at a rapid rate, with more than one-half those surveyed classed as eutrophic, would support this view.

Inland aquaculture for culture-based fisheries Applications of GIS in aquaculture are of increasing importance to inland fisheries as more and more inland fishery benefits have their source in culture-based fisheries and other fishery enhancements. GIS is employed to estimate two basic criteria:

- Suitability of the site for the culture method, usually production of fish in ponds, raceways or floating cages.
- Suitability of the site for the organism to be cultured.

Continent-wide estimates of the potential for inland fish farming provide an illustration of site-selection fundamentals that could be easily extended to culture-based fisheries. Studies recently have been completed for Latin America (Kapetsky & Nath, 1997) and Africa (Aguilar-Manjarrez & Nath, 1998), as well as for the Caribbean Island states (Kapetsky & Chakalall, 1999). Four criteria were used to estimate potential for small-scale fish farming in ponds: water loss, potential for farm gate sales, soil and terrain suitability for ponds, and availability of agriculture byproducts as feed or fertiliser inputs. A fifth criterion was added in order to estimate potential for commercial fish farming: urban market potential. These criteria were weighted in different ways to make small-scale and commercial fish-farming models on the basis of expert advice. Numbers of crops per year of five widely cultured species were predicted based on monthly climatic variables. By varying feeding levels and harvest sizes small-scale and commercial level outputs were simulated. Combining the small-scale and commercial models with the simulations of fish production provided overall suitability ratings for each five arc minute grid cell (approximately 9 × 9 km) in Latin America and the Caribbean and three arc minute grid cell (5 × 5 km) in Africa. Potential was reported as the absolute and relative amounts of surface area in each country meeting the criteria with various levels of suitability.

These studies were carried out to estimate inland fish-farming potential. With some modification the same approach could be used for the siting of large-scale and village-scale fish hatcheries, for the selection of appropriate species and sites for stocking and to predict the growth of fingerlings up to stocking size. The proximity of large and small water-bodies into which fish could be stocked would constitute an additional criterion and data layer in the GIS. Finally, this approach lends itself to any level of resolution for which data are available, for example, a county (Kapetsky *et al.*, 1988), a state (Kapetsky *et al.*, 1990) or a country (Kapetsky *et al.*, 1991).

Management of fisheries together with aquaculture
GIS would seem to provide opportunities for determining the allocation of land and water resources between fisheries and aquaculture. However, no studies of this type have been undertaken to date in freshwaters and it has been necessary to extend this review into brackish waters to find a good example.

Hard clams of the genus *Mercenaria* support an important commercial fishery in the Indian River lagoon on the eastern coast of central Florida (Arnold *et al.*, 1996). In response to the extreme variability in landings (and resultant income) that characterises the hard clam (*Mercenaria mercenaria*) fishery in Florida, many local residents have become involved in the aquaculture production of hard clams. As the industry continues to develop and expand conflicts have arisen. Original conflicts focused on the allocation of submerged bottomland between site-specific aquaculture operations and fishing for naturally occurring clam populations. However, those conflicts have expanded to involve upland property owners, environmentalists and natural resource managers. Further, the geographical scope of conflict has expanded concomitantly with the expansion of hard clam aquaculture to other regions of Florida. Arnold & Norris (1998) have developed, and continue to expand and refine GIS technology for conflict resolution in the hard clam aquaculture industry. Site-specific information on water depth, water quality, upland land use, site accessibility, transportation corridors, the distribution and abundance of clams and submerged aquatic vegetation and other parameters are used to determine the suitability of a site for hard clam culture. GIS has been employed to allocate hard clam leases at several sites using a proactive approach, to maximise the probability that the lease-site applicant will be successful in obtaining their lease and in realising an economically successful operation upon that lease, while contemporaneously resolving complaints from competing user-groups.

Environment

Degradation of the environment, including loss of habitat, is widely recognised as the most serious threat to the sustainability of inland fisheries (FAO, 1999a). GIS can be used in several ways to link inland fisheries with the environment. Applications have been allocated among terrestrial effects on aquatic habitats and fishes, risk evaluation and environmental health, water quality and quantity, and climate change.

Terrestrial effects on aquatic habitats and fishes
Knowledge of the state of the terrestrial environment, watershed by watershed, is essential for forecasting the state of the aquatic environment both locally and regionally. It also provides the basis for integrated fisheries management throughout river and lake basins.

There is an increasing amount of effort being put into using GIS to relate landscape and land uses to the quality of the aquatic environment and to water quantity. For example, O'Neill *et al.* (1997) called attention to monitoring of the environmental quality at the landscape scale using a combination of remote sensing, GIS and principles of landscape ecology. Allan & Johnson (1997) have reviewed catchment-scale analysis using the same tools to characterise aquatic ecosystems according to landscape.

A GIS-assisted measure of the state of the aquatic environment for fisheries is provided by the analysis of 145 large watersheds around the world that account for 55% of the land surface, not including Antarctica. The analysis focuses primarily on watersheds as ecological units and on the risks of degradation from human activities that may undermine their ability to provide ecological services and maintain intact the biodiversity within them. The analysis showed that watersheds ranking highest in biological value are also generally the most degraded. Biological value was based on the number of fish species and fish endemics, and the number of areas with endemic birds. Stresses were found to be especially severe in watersheds already substantially modified or degraded. In particular, India, China and South-east Asia stood out as places where pressures on watersheds are intensifying. This is of concern because these correspond to the most important areas of inland fish production globally. Other major watersheds that are less degraded, such as the Amazon and the Congo, nonetheless are beginning to experience rapid change. GIS was used in several ways in this study. One was to establish the 'content' of the watersheds in terms of the indicators of stress. Remote sensing was used to estimate land cover.

At a higher level of resolution, Moyle & Randall (1998) developed a watershed index of biotic integrity in order to assess the biological health of 100 watersheds in California. It used various measures of abundance and distribution of native fish and frogs for metrics. The index was based on whole watershed variables. The broad-scale variables indicative of human disturbance included dams, reservoirs and diversions, roads alone and in proximity to streams, and introductions of trout into fishless areas. The most important factors contributing to low indices of health were large dams and introduced fishes, although some measures of the intensity of use of terrestrial habitats also were important. Evaluations of the biotic integrity of watersheds over wide regions can help managers to set priorities for watershed-oriented systems of aquatic conservation and provide starting places for more intensive studies. The role played by GIS in this study was to standardise the variables in terms of occurrence in one-hectare units.

Another practical example of the catchment approach was the use of a distributed parameter model linked to a GIS by Allan *et al*. (1997) to predict that an increase in forested land cover would result in dramatic declines in runoff and sediment and nutrient yields. Habitat quality and biotic integrity varied widely among individual stream sites in accord with patterns in land use and land cover. In this study the extent of agricultural land at the subcatchment scale was the best single predictor of local stream conditions.

Land-use patterns measured in 1994 were compared with historic land use in relation to increasing pressures by encroaching urban development, agriculture, and forestry practices on Coho salmon (*Oncorhynchus kisutch*) streams on the east coast of Vancouver Island by Brown *et al*. (1996). GIS was used to evaluate terrain resource information management (TRIM) maps and to measure land-use patterns for Black Creek, Vancouver Island. A unique feature of this study was that not only the methodology, but also the costs in using the GIS for Black Creek are presented.

Risk evaluation and environmental health
According to Nyerges *et al*. (1997) GIS applications for risk evaluation concerning environmental and ecological health are appearing with greater frequency. Their paper is mentioned in this review to call attention to the breadth of applications available within the conceptual framework for risk evaluation. Risk evaluation encompasses and synthesises several component frameworks [including risk scoping, risk communication, risk assessment (risk analysis), risk management and risk monitoring] concerning environmental health, and hence human and ecological impacts. An examination of the research done in the use of GIS for risk evaluation concerned applications in risk assessment rather than risk scoping, management and monitoring.

Water quality and water quantity
Lake Okeechobee, the third largest lake in the USA, provides an example of an application in this category. Here, Richardson & Hamouda (1995) related vegetation changes occurring in the marsh between 1989 and 1992 to various soil and hydrological parameters. GIS modelling of hydroperiod, vegetation and soil nutrient relationships in the Lake Okeechobee marsh ecosystem was carried out. In order to develop the environmental parameter databases, a hydrological model was used to portray spatially the movement of water over the marsh and a grid sampling survey of soil nutri-

ents was conducted to map soil parameters. Changes in vegetation communities were related to short hydroperiods, low soil organic matter and high soil bulk density. With these results, together with information on the ecology of the fishes of the lake, it should be possible to infer linkages among hydroperiod, vegetation, and the distribution and abundance of fishes.

Climate change

Limitations surrounding studies of climate change derive from three sources: climate observation and prediction, fish distribution mapping, and prediction of climate impacts on fishery indicators (Minns & Moore, 1992). A continuing problem with predicting climate change on fishes is that the capability to extrapolate and interpolate climate variables at useful resolutions, say 1 km^2, is not yet good (Kapetsky, 2000). In addition, data sets of extremes on relevant time scales (e.g. weeks, months) that would allow modelling to estimate best and worst cases are not yet readily available.

There are few references to GIS and climate change relating to fisheries, and much underlying work on habitat suitability and on prediction related to the factors that determine habitat suitability for fishes has to be carried out in order to estimate the effects of climate change. Therefore, there is a close relationship between climate change and essential fish habitat studies.

Thermal limits to salmonid distributions in the Rocky Mountain region and potential habitat loss due to global warming were examined by Keleher & Rahel (1996). As warming proceeds, salmonid populations would be forced into increasingly higher elevations and would become fragmented as suitable habitat for coldwater fish became separated from main river channels and restricted to headwater streams.

Multi-sectoral planning and management including fisheries

It is well perceived that the inland fisheries sector is in competition with other users of land and water. Nevertheless, planning and management often tend to be single sector. The content in this section is heartening in that it shows that fisheries interests and needs are specifically being taken into account in multi-sectoral planning. This is being done from both a watershed, catchment perspective directed at the 'environmental health' of aquatic systems, and a floodplain perspective as a biologically productive, sometimes vital, habitat for fishes.

Particularly noteworthy in this section is the description of an approach by Olson & Orr (1999) in which the emphasis is on preventive multi-sectoral management. This involves matching land use activities that include forestry, fisheries and wildlife to the capabilities of the landscape in an ecosystem management planning approach. GIS was used to target a desired future condition in the planning process. The approach has been successfully implemented for landscape-scale sustained yield plans and is currently being used for multi-species habitat conservation plans.

Summary and conclusions

Sources of information on applications

The distribution of sources of information about GIS in inland fisheries clearly indicates that interested fishery workers have to go further afield than mainstream journal literature if they are to keep up with applications. Just under one-half of the titles came from published sources. At the same time the importance of periodic symposia (39% of titles herein from the Seattle symposium) and user conferences (13% herein from ESRI conferences) to bring out recent and innovative applications is underlined. Finally, the Internet, via web pages of individual projects, educational institutions, applications-oriented discussion groups and commercial GIS organisations, is an important source of comprehensive and up-to-date GIS activities that should not be ignored.

Geographical distribution of applications

It is remarkable that there are so few GIS applications in inland fisheries outside North America. One reason is that, although there are inland fisheries, GIS activities in some countries (e.g. China: Miyun Reservoir, Water Quantity and Quantity, Table 10.1), much of the material does not find its way into English-language publications or abstracting services. Another fundamental reason is the lack of higher educational opportunities for GIS in fisheries outside North America. For example, there are only two universities in the UK, Stirling and Canterbury Christ Church University, which offer GIS in the fisheries sector. From an FAO Fisheries perspective, the promotion of the Code of Conduct for Responsible Fisheries among developing countries relies heavily on the dissemination of information and techniques of management. Clearly, more emphasis should be placed on awareness raising along with training on the applications of GIS.

Given this situation, applications that could help developing country inland fishery administrations to assess food fishery potential, particularly benefits and associated risks from enhancements, to judge the environmental health of river and lake basins, and to allocate resources between management of fisheries and development of aquaculture are, as yet, scarce. There are several problems:

- Lack of digital data with which to inventory and characterise inland water bodies
- Low level of awareness of the benefits of applying GIS to inland fishery problems
- Poor quality of fishery statistical data as independent variables in GIS.

The first of these will be remedied by the more widespread use of available remote sensing technologies (e.g. Travaglia *et al.*, 1995), cost reductions for access to the data and further technological improvements in them. The second is in progress. For example, the Fisheries Unit, Mekong River Commission Secretariat, has recognised the need for incorporating GIS into fisheries and is now preparing for the implementation of GIS in fisheries and aquaculture at a regional level. The third, incomplete and poor quality fisheries data, will require much effort over a long period for improvement. A fourth reason for the lack of application of GIS in developing countries is the absence of suitable institutions for their application. Local planning agencies use administrative boundaries, not watershed boundaries. Such countries do not therefore seem to have the appropriate environmental management agencies with a catchment-based remit that would be required for useful GIS studies in inland fisheries. This may take some time to rectify.

Present and future directions of applications

Inland fishery workers are finding a wide variety of applications for GIS that broadly include fishery resources, fisheries, the environment and multi-sectoral management that includes fisheries (Table 10.1). The applications are heavily weighted towards applied research and the acquisition of information for management, as indicated by the distribution of titles among categories. GIS as a tool for assisting fishery administrators in decision making in management and for the allocation of land and water resources between fisheries and aquaculture. One explanation for the few applications is that GIS has not sufficiently matured as a technology in order to be effectively used in inland fisheries management. In fact, just the opposite is the case: the technology is sufficiently mature and its analytical capabilities are more than adequate. There are two problems: (1) the lack of availability of reliable inland fisheries data for analysis, particularly in developing countries; and (2) that fishery managers and administrators, unlike fishery researchers, have not yet had sufficient awareness raising and training in order to ensure that GIS is effectively deployed for decision making.

Another explanation is that, in North America, where GIS applications are by far the most prevalent, nearly all inland fisheries are recreational. Fisheries are highly regulated there, enforcement is very effective and management is synoptic. Thus, the main concern is not directly with management of the resources, but rather with the sustaining the aquatic environment for fisheries. It follows then that the main applications of GIS are aimed at managing the environment with most efforts focused on habitats as the environmental units of concern.

Integration of applications

Ultimately, inland fisheries will be sustained only by improvement in two directions: (1) better identification and spatial quantification of water quality and water quantity needs; and (2) increasing co-operation and co-ordination with other sectors sharing the same or similar concerns for the quality and quantity of water. GIS provides a tool to help to achieve both of these in terms of estimating the importance of each sector's share in ecological, economic and, eventually, political terms, and more importantly, to identify areas of overlapping concern and mutual benefit.

Key elements are in place to increase the momentum already realised in this direction. The results show that there are relatively strong bodies of GIS experience in several critical areas of interest:

- Identification and classification of aquatic habitats in relation to fish distribution and abundance
- Definition and quantification of essential fish habitat and estimating the effects of terrestrial land uses on aquatic habitats and fishes.

The significance is that there already exists a large body of information and experience that can be brought to bear to solve practical problems. Taken together and integrated, advances in these three directions mean that fisheries interests will be much more equitably represented in multi-sectoral planning and management, including fisheries, a direction that is itself advancing.

Another important area for integration is the landscape approach to estimating terrestrial effects on aquatic systems with quantification of essential fish habitat. If the former can be widely institutionalised and made dynamic by periodic synoptic assessments along the lines of the EPA Index of Watershed Indicators (United States Environmental Protection Agency, 1997), then the state of fish habitat can be interpreted in terms of cause–effect relationships.

Note

1. By James McDaid Kapetsky, 5410 Marina Club Drive, Wilmington, NC 28409, USA.

Chapter 11
Fishery management

Management has been defined as the 'integrated process of information gathering, analysis, planning, consultation, decision-making, allocation of resources and formulation and implementation, with enforcement as necessary, of regulations or rules which govern fisheries activities in order to ensure the continued productivity of the resources and accomplishment of other fisheries objectives' (FAO, 1997a). In a narrower sense the term is applied to decisions and actions affecting the magnitude and composition of fishery resources and the distribution of benefits from its products. There are two main approaches to management, that based on the resources, mainly advocated by natural scientists, and that based on society, mainly advocated by socio-economists. Modern management seeks to reconcile these two views using them as tools for reaching balanced decisions on the resource with the participation of all stakeholders in the fishery. Inland fisheries managers are also increasingly called on to reach compromises with other users of the aquatic resource. This and the following sections deal primarily with those interventions that aim at:

- ensuring the sustainability of fisheries and the equitable distribution of its benefits (management of the fishery)
- maintaining the environment on which they depend (management of the environment), and
- increasing production over those attained by natural processes alone (management of the fish).

Related issues of management of aquatic biodiversity and of legislation are also examined.

Social and policy consideration[1]

Inland living aquatic resources are often of major economic significance. The nature of the derived economic benefits depends on the both the type of assemblages present and the forms of exploitation that take place. These may vary from subsistence fishing, where fishermen fish with the primary purpose of feeding themselves and their families, through commercial exploitation of fish for sale both for consumption by others and as ornamentals, to recreational fisheries, where local economic benefit is derived from the sale of services to fishermen who gain pleasure from the act or prospect of capture.

The challenge to management is to ensure that the flow of benefits, however derived, is sustained at a level that is at or near its potential. This flow may be suppressed for two principal reasons. Firstly, actors outside the fishing sector can compromise the functioning of the ecosystem supporting the fishery by moderating hydrology, reducing water quality (through nutrient loading or pollution) or altering habitats critical to fish reproduction or growth. Secondly, actors within the fishery can alter the characteristics of the species assemblage, most often when excessive fishing pressure causes recruitment failure of the larger and/or more valuable species. The task of fisheries management, therefore, involves the encouragement of non-destructive patterns of behaviour by either actors outside the fishing sector or by the fishermen themselves, or both.

Success in this task is dependent on two different, but not necessarily separable, activities. The first of these is the identification of changes needed in patterns of behaviour that are actually or potentially destructive to the fish resource or the ecosystem supporting it, i.e. knowing what is needed. The second is the establishment of a set of institutional arrangements and incentives for resource users that are capable of inducing such changes, i.e. making it happen. The challenge of inland fisheries management is that both aspects are usually difficult but that success in both is essential. Knowing what changes would be desirable is of little benefit if you cannot make them happen; and making changes happen can be positively destructive is you are unsure or confused about what changes are needed.

This section considers how best this may be achieved in different circumstances, with a focus on the management of activities within the fisheries sector. It argues that stock status for many lake and river species is dependent on ecological features and levels of fishing effort that are essentially local in character, and that local people should therefore play an important role in the determination and enforcement of rules relating to their fisheries. The conditions facilitating local management are then considered. As there are many problems that cannot be solved simply by local action, higher levels of co-ordination and decision making are then considered.

Identifying needs

Identifying whether changes in fisher behaviour or environmental modifications are needed to improve economic outcomes (i.e. greater sustainability, increased catch, improved equity) and what these might be depends on recognition of changes in the status of fish resources and of the factors influencing any trends observed.

Fish are both mobile and, until the moment of capture, largely invisible. As a result, assessing the status of the resource must rely on inference from catch characteristics rather than on direct observation. Identification of trends in stock status is further complicated by the fact that fish biomass is naturally variable, both from season to season and from year to year, and that fish assemblages are often made up of species that differ significantly in their response to environmental change.

Evaluating the effects of factors influencing fish resources can also be complex. While more directly visible than the fish stock, fishing effort is also difficult to quantify systematically. In tropical floodplain fisheries, which tend to be highly seasonally variable and spatially heterogeneous, the interaction of multiple gears of varying selectivity and the large numbers of economically important species quickly becomes a highly demanding exercise in assessment.

Because of the effective impossibility of achieving numerical control of individual species, Wilson *et al.* (1994) suggest that management needs to focus on the stable parameters of fisheries systems, namely habitat and biological processes. Changes required in the use of particular gears – embodied by rules on the type, timing and level of effort – can, however, be useful. Stopping indiscriminate and environmentally damaging practices, such as dynamiting or fishing with poisons, can yield benefits widely. Other rules of gear use would be determined in relation to the geographical range of the stocks that they affect. Blackfish species that move little during the course of their life cycle can be well managed at a local level. Species that are more strongly migratory will be better managed on a wider spatial scale. Whatever the scale of management, the response of both fishermen and fish stocks to rules will be uncertain; therefore, it is important that rules are responsive to experience and changing circumstance: rule making should be adaptive.

Environmental changes affecting the fishery can be clear and comparatively easy to rectify, such as siltation of a connecting channel that blocks fish migration. They can also be more subtle, difficult to identify and harder to reverse.

Encouraging change

How changes can be brought about will depend on the relationship to the resource of those responsible for the activities that damage it. The approach to actors whose livelihoods do not depend on fishing will clearly have to differ from that to fishermen who often stand to gain directly from improvements in resource management.

When dealing with fishermen, it is essential that initiatives recognise the nature of the resource. With the exception of privately owned water-bodies such as ponds, where an owner can exclude others and claim all of the benefits available, the majority of inland fisheries in developing countries are common-pool resources. This means that the resource units are subtractable (i.e. what is taken by the user is no longer available for another) and users are difficult and costly to exclude (Ostrom, 1992; Bromley, 1992). Bringing about the aggregate change in fisher behaviour needed to achieve improved outcomes from the fishery thus depends on fishermen being persuaded that it is in their best interests to modify their own individual activities in line with the rules proposed, even when this is contrary to their short-term interests.

The type and variety of stakeholder dependence on the fishery will influence how easily this can be accomplished, owing to differences for each group in the potential costs and benefits arising from adherence to the rules. Catches in the short term can be sacrificed with far less impact on household well-being in a fishery that is largely recreational than in those that are primarily commercial and/or an important source of animal protein. Variety in stakeholder involvement can also create problems due to the differential impact of the changes in behaviour required. For instance, on many tropical floodplains, some households fish full time, making use of the river channels and the residual water-bodies during the dry season, while others fish briefly on the floodplain when the opportunities arise. These differences between groups in the livelihood importance and seasonality of fishing are frequently reflected in the gears they use and the species they capture, as well as the water-bodies on which they operate. Rule changes restricting particular gears or areas of fishing may thus entail high costs and little benefit for some groups.

Even when the potential benefits of moderating behaviour exceed the potential costs, fishermen may not follow proposed rules. This may be due to lack of assurance, if they doubt that others will modify their fishing practices sufficiently for the potential gains to be realised, even if they follow the rules themselves. Alternatively, they may wish to free ride, by seeking to derive a share of the collective benefit from others changing their behaviour without changing their own. Success is therefore intimately connected to the degree to which fishermen are committed to following the rules and, if they are not, the probability of detection and effective sanction.

In all river systems, and particularly those where floodplain catches are important, detecting when fishermen are not complying with the rules can be difficult. While some fishing gears, such as the elaborate fixed structures sometimes used on the rivers of Africa and South and South-east Asia, are highly visible, much fishing can take place largely unobserved in areas that are remote or disguised by vegetation. This means that monitoring at a level that results in a high probability of detection is expensive, unless undertaken mutually by the resource users themselves. Enforcement can also prove difficult or costly, even when violations have been identified, if the appropriate institutional structures are not in place.

Responsibility: the possible actors

Since the early 1950s many governments have taken responsibility for inland fisheries management, a trend that accelerated with the formation of many new post-colonial nations in the 1960s. It is a task that many have found difficult. In part this is due to the cumulative stress placed on aquatic ecosystems by the process of development and the increased pressure of fishing from rapidly growing populations. It also reflects fundamentally flawed assumptions about the ease with which appropriate fisheries management plans can be developed and enforced. For fisheries where a substantial proportion of the fish assemblage is made up of species that reproduce and live locally, management plans need to reflect local stock status, local conditions and the needs of local stakeholders. Moreover, given the very limited resources of fisheries departments and the alternative priorities of collaborating agencies, such as the police, success in changing patterns of behaviour through detection and enforcement is rare (Satia, 1990).

The growing awareness of the limitations of government capacity in developing countries – either to formulate locally appropriate plans or to motivate compliance with them – has seen increasing advocacy for and reliance upon the community in the management of fisheries (Pomeroy & Williams, 1994). In many countries, this involves a return to or increased recognition and support for traditional local management arrangements.

In the past two to three decades advances in the management of common-pool resources other than fisheries (such as forests, water for irrigation, land used as shared grazing pasture) has often been based on the involvement of local communities. Experience from managing these non-fishery resources, most notably community forests and irrigation schemes, has generated many insights into the collective management of renewable resources. Managers of inland fisheries, therefore, have an opportunity to learn from their methods and approaches as well as from the now rapidly growing experience within the fisheries sector.

Communities have a detailed knowledge of the local area, of the characteristics of the fish stock, of the features of local water-bodies, of the types, timing and selectivity of fishing operations, and of some of the non-fishing activities that may affect fish production (Berkes, 1995). They are equally aware of the strengths and weaknesses of the individual stakeholders at the local level. If their collective knowledge can be effectively orchestrated and potential conflicts mediated, they are thus in a position to contribute very significantly to and, in some circumstances, even determine what needs to be done.

Communities can also motivate changes in behaviour. If resource users have participated in identifying what changes of behaviour are desirable, the rules agreed can acquire a legitimacy that reduces the chance that they will be broken. In addition, fishermen operating on the same water-body may be in a good position to monitor each other and detect rule breakers. The use or threat of sanctions from the community may thus reduce infractions.

The need for co-management

Achieving effective community involvement in fisheries is, however, neither easy nor a universal panacea. Communities exploiting inland fisheries rarely comprise a single group of fishermen all with the same interests. Differences in their dependence on the fishery, as described above, which can be complicated further by differences in their social, cultural and financial status, mean that it is not easy to create 'community' organisations able to generate the level of trust and commitment required to manage a fishery, even when the fish assemblage is made up by 'local' fish.

The additional problems arising when a significant proportion of the catch comes from species whose stocks are either shared with other villages or undertake substantial longitudinal migrations, require the creation of overarching structures to mediate between and co-ordinate communities (Hoggarth et al., 1999b). Similarly, while some of the effects of local non-fishing activities on the ecosystem can be dealt with best by mediation within a community, the impact of more distant activities requires action at a higher administrative level.

The ideal structure is thus one involving the government, the resource users and the community in a co-management arrangement. In some countries, non-governmental organizations (NGOs) have been shown to have a clear comparative advantage over government in the critical tasks or developing community management groups and co-ordinating them when this is needed.

Developing a co-management structure

Development of a co-management structure involves three distinct aspects. Firstly, the responsibilities, rights and relationships of different stakeholders with respect to the different management roles have to be made explicit and agreed. Secondly, a local capacity for fisheries management needs to be legitimated and developed, where it exists, or created, where it does not. Thirdly, overarching institutional structures, involving both representatives of local communities/user groups and government, need to be created to mediate on wider issues, such as ecosystem impacts and the fishing regulations applicable to important shared stocks. These steps are outlined below.

Responsibilities, rights and relationships for co-management of inland fisheries

Successful management requires that stakeholders take responsibility for a range of roles. In their discussion of Asian floodplain fisheries management, Hoggarth et al. (1999b) identified more than 15 individual roles (Fig. 11.1). While noting that the list is not exhaustive, it does serve to highlight some roles (such as the need for effective mechanisms for communication) that resource management plans often neglect to address explicitly.

Given this list of roles, the next question is: who should take responsibility for each role? To answer this question, it is important to assess who *can* take responsibility, and also who has the incentive to take responsibility. As indicated above, for riverine fisheries some stakeholder groups have a clear advantage over others with respect to certain functions. However, it is important to clarify all of the roles, agree on the division of responsibilities and identify areas where skills need to be developed.

Devolving responsibilities to communities and user groups

Devolution of responsibility to communities is not a simple task. The following list

Fishery management 201

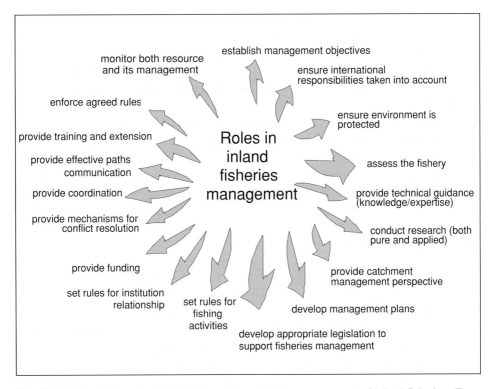

Fig. 11.1 Roles that need to be filled to achieve effective management of inland fisheries. (From Hoggarth *et al.*, 1999b.)

draws together the lessons and ideas from the literature (Pinkerton, 1989; Ostrom, 1990, 1992; Pomeroy & Williams, 1994) to provide an initial set of co-management guidelines. Points with common themes are grouped together under a single heading.

Participation must be rewarded by positive returns
(1) Individuals are more likely to participate when it is clear to them that the benefits exceed the cost of their involvement, i.e. that they gain a positive return.

Unit of management
(2) The boundaries of the management unit must be clear and of a manageable size. This includes both the physical boundary of the resource, e.g. the agreed area covered by a management plan, and who is part of the management unit, e.g. a list of legitimate/licensed fishermen and the membership of management committees and their structure. Fish, of course, do not recognise administrative boundaries and their movement may be on a scale that erodes the incentives for unilateral action by a single village. This is one of the main reasons why overarching structures are needed (see below).

Operational management: resource side-rules and regulations
 (3) Rules that specify who can use the fishing resources, how, where and when must reflect local conditions (i.e. the fishery, other resource use in the catchment, etc.).
 (4) Rules are best made by the individuals affected by them. This includes detailed fishing or collection rules as well as rules governing who can make and change the arrangements guiding management of the resource.
 (5) Communities should set up a system of penalties to deal with people who break rules. The system should include a mixture of light through to more serious penalties to make allowances for different levels of rule breaking and individual circumstances. The penalties are important to ensure that everyone keeps the rules, and seeing that all others are obeying is a key incentive for individuals actively to support and maintain the management system.
 (6) Communities should establish ways of resolving conflict. Mechanisms should be fast and low cost, relying on both formal (e.g. law courts) and informal (e.g. committee meeting) methods.
 (7) The 'community' should live near the fishery and have a common approach to collective problems. A community with previous experience of solving problems facing many of its members and that has a shared understanding of key objectives, will have more chance of successfully meeting the challenges of managing a fishery than a community with lots of internal conflict.

Institutional framework: capacity to manage
 (8) Previous experience of managing natural resources within a community provides a good foundation for stakeholders to take on responsibility for fisheries management. Existing institutions may be used, if suitable.
 (9) Communities must have external recognition of their right to manage. For example, government legislation may allocate tenure over a specific resource or government policy may delegate responsibility of a well-defined management unit to an appropriate user group.
 (10) Management should be supported by a nested arrangement of organisations. This means that there is an appropriate forum for people involved at all levels in the planning process, and that these forums are linked together.
 (11) There should be a core group within the community that takes leadership responsibility for the management process. Individual resource users should have incentives and be willing to commit time, money and effort into fisheries management.
 (12) Communication between government and the community requires a joint body to be established. Membership should include representatives from both stakeholder groups (community and government) and should have a remit to monitor progress, resolve conflict and reinforce local decisions.
 (13) Institutions governing common property resources should be democratic and representative. This means that procedures must be established to ensure that all

stakeholders have a voice on decision-making committees and that representatives should be elected or nominated in an agreed way.

Monitoring
(14) Management needs monitoring of both the fishery and the activities of stakeholders relying on the resources. Monitors must be either the resource users themselves or, at least, accountable to them.

This range of resource and community characteristics provides guidance to where the chances of success will be intrinsically higher and of what additional support by government, or other organisations, may be needed.

Building local skills

Enabling local institutions to take the initiative in the management of the fisheries in their local waters is not easy, quick or to be achieved without cost in time, effort or resources. However, it is critical that this stage is not cut short, as such institutions are the essential foundation of an effective co-management structure.

While there are villages with a tradition of managing their own fisheries through a process of consultation and consensus building and which achieve fisheries outcomes that are both sustainable and equitable, these are the exception rather than the norm. Replicating such institutions is not simply a matter of requesting a meeting at which a series of institutional rules and regulations is announced; there is no predefined package that can be disseminated.

Although there are similarities to the process of forming a group to receive or adapt a technical extension message, group formation in this context is much more demanding. This is because the resource is shared, so success depends on a very high proportion of local fishermen becoming committed to the process. The first step is therefore to ensure that they are informed and consulted at all key stages and that all stakeholder groups are fully represented in more detailed discussions. Initial commitment, once gained, must be maintained by ensuring that management decision-making remains open and transparent.

Within the wider user community, institutional rules must be agreed on: the size and composition of the management committee; the means by which members of the committee might be appointed or removed; and the procedures through which different types of decision (operational fishing rules, penalties, etc.) can be determined. Unless this process is transparent and seen to be fair, the operational rules that result will lack legitimacy among stakeholders, decreasing the chances of rule adherence and of rule breaking being reported.

Setting up a group in this way is complex and time consuming and demands specialist skills – reaching the poor and helping them build community-wide institutions in which they can participate – which lie beyond the experience of many fisheries departments. Additional support from other agencies is therefore required. In many countries, NGOs are well equipped for this role.

Defining an overarching management structure

Levels of management

The overarching management structure needed to support local management of fisheries must be designed to orchestrate and co-ordinate the different institutions that both support activities at a local level and perform functions with a wider geographical scope. This structure can be envisaged as a series of horizontal and vertical linkages within and between government departments, supporting a hierarchy of spatial management units. Where NGOs are to play an important role, these too may be included.

At the base of this structure will be local management units. Dividing the fishery into smaller, local units within the lake or river catchment gives the flexibility needed for effective management of complex inland fisheries. The most basic unit is simple village management areas (VMA), where a single village/community/user group is given control of a defined water-body or set of water-bodies. This can work well where a large proportion of its blackfish stocks is local. However, as the proportion shared with other communities increases, the reward to unilateral action by a single community diminishes. This creates the need for intermediate management areas (IMAs) in which the actions of different communities are co-ordinated.

The highest spatial unit is the catchment management area (CMA), where conflicts relating to the exploitation of highly migratory fish stocks and major intersectoral interactions can be resolved. The size of CMAs and the number of boundaries (national and international) that they cross is highly variable. Some alternative typologies are illustrated in Fig. 11.2. A fuller discussion of management units is to be found in Hoggarth *et al.* (1999b).

At the apex of the fisheries management structure, although not a management unit in the same sense, is the national government.

Role of government

The role of government clearly changes significantly from level to level within this hierarchy. For local management units, it is largely a facilitator and provider of technical advice. It would not be responsible for most decisions, although it could provide additional technical information upon which decisions could, in part, be based. For intermediate management areas, the government would again principally act as facilitator and adviser, bringing representatives of community fisheries groups together to discuss management of shared stocks, encouraging exchange of experiences, and helping to identify common lessons and alternative approaches through which local management regimes could be improved by adaptation.

At the catchment level, government would be responsible for organising research into whitefish stocks and the development of initiatives that might help to sustain their productivity. It would also play a key role in mediating between different sectoral interests to ensure that fisheries concerns were properly reflected in overall policy and in the design of initiatives that might modify hydrology or water quality. It should also

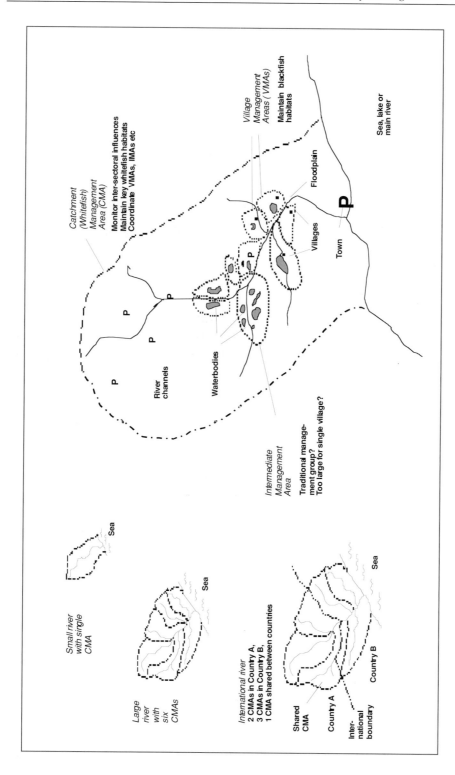

Fig. 11.2 Alternative types and configurations of spatial management unit as applied to a floodplain–river system.

Table 11.1 Possible division of roles at various hierarchical levels of the organisation of the fishery

Hierarchical level	Institution	Function
International	Agencies	Formulation of global policy Regulation of trade Depository for international agreements
	International basin organisations	Formulation of basin policy Centre for agreements on discharge, migratory fish passage, etc.
	Non-governmental organisations (NGOs)	Representation of international sectoral interests
National	Central government	Select national objectives Develop and enforce national policy framework Promote development strategies Mediate negotiations on water and environmental use Carry out, co-ordinate and disseminate results of research Create an enabling legal, educational and administrative framework
	NGOs	Represent sectoral interests at national level Participate in regional and unit management where appropriate
Catchment management	Regional governmental authorities	Select catchment objectives Develop and enforce regional policy framework Manage sectoral interactions Identify management units Allocate unit management responsibilities Co-ordinate management units Enforce rights of individual management units Resolve conflicts between individual units Conduct local research and information collection to assess resources
Fishery unit	Village/landing committees	Assess local needs for resource management Select unit objectives Design management plan Implement management plan Enforce decisions and resolve conflicts at local level

play a role in defending individual structures when these are too weak to enforce their own policies and to prevent incursions by outside interests.

Table 11.1 presents a simplified allocation of responsibilities between the various levels of administrative hierarchy and different stakeholders.

Strategies for regulation of fisheries

Fisheries for a limited number of species are common in the temperate zones where biodiversity is not high. Here commercial fisheries have been managed on the basis of maximum sustainable yield using the predictions of surplus yield and other population dynamic models. The utility of the maximum sustainable yield (MSY) concept, the fundamental reference level (Caddy & Mahon, 1995) of these models, has largely been discredited at the scientific level. However, it has long captured the minds of policy makers and legislators at global level, and no alternative concept with equal appeal has yet been found. MSY is particular suited to management regimes based on strong central control. Consequently, it has never formed a major component of inland fisheries policy except for some stocks in large lakes because it is poorly adapted to the unstable systems and fragmented artisanal fisheries found in most freshwater environments. However, there is now an increasing trend to maintain production by enhancement in such waters. By contrast, most inland fisheries of the subtropical to equatorial zones are based on a large number of species and in many cases are carried out with a wide range of fishing gears. Such multi-species, multi-gear fisheries are not amenable to the more traditional methods of stock evaluation because of the complexity of response of numbers of species of different sizes.

Multi-species fisheries react to fishing pressure. Increasing effort involves the progressive reduction in the size of the species caught (see Fig. 11.3). Reduction in size is associated with changes in mortality rates, growth rate, production and the number of species composing the catch, which all increase. Biomass and catch per unit effort (CPUE) both fall. The combination of falling biomass and rising productivity means that yield remains stable over a large range of effort. This close association of effort and the length of the fish caught implies that the fishery can be managed entirely on the basis of control of length both in terms of the assessment of the status of the fishery and through promotion of mesh or fish size limitations.

Goals for regulation

The behaviour of multi-species fisheries illustrated by Fig. 11.4 suggests that there is:

- an initiation phase of growth;
- a prolonged phase of sustained production (B); and
- a collapse when excessive effort is applied (C).

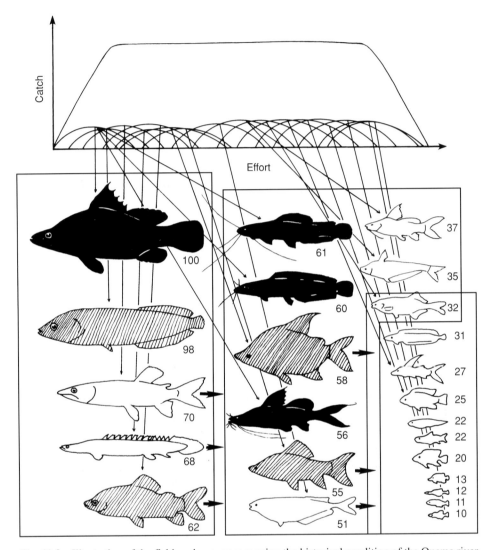

Fig. 11.3 Illustration of the fishing-down process using the historical condition of the Oueme river fishery (West Africa) illustrated with representative species. Black = species that disappeared from the fishery before 1965; hatched = species whose numbers were seriously reduced; Black arrows = species that reduced their breeding size. Peak yearly yields in the 1950s–1960s were around 10 000 t.

Collapse of fisheries due to overfishing has been well documented in lakes [e.g. the North American Great Lakes (Regier & Loftus, 1972) or Lake Victoria before the advent of the Nile perch (Fryer, 1972, Fryer & Iles, 1972)]. By contrast, the resilience of flood rivers is such that the author only knows of one river, the Oueme, where a collapse in the actual catch together with the elimination of many of the largest species from the fishery was induced by fishing alone. However, many cases are documented

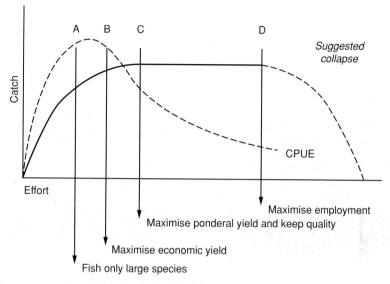

Fig. 11.4 Diagram of the stages in exploitation of multi-species fisheries.

where fishing and environmental pressure has together produced such a collapse (see Welcomme, 1995).

Such a model makes conventional concepts of overfishing difficult to apply until the collapse phase is underway. Until that occurs individual species may be overfished and disappear from the fishery, but the assemblage as a whole continues to produce at a high level, albeit of fish which may not have the same quality as those that have disappeared. Consequently, overfishing can only be judged with reference to some defined value such as a particular group of species, quality or size. Fisheries showing the plateau effect can absorb greater amounts of effort, either as labour or as improved technology, than would a fishery concentrating on only the larger species in the assemblage. Management is complicated further when several fisheries, each of which may react differently to the changes in size and species composition, exploit the same assemblage.

The following four basic strategies can be adopted for management in inland fisheries (Welcomme, 1985) (Fig. 11.2).

A. Manage the fishery only for large species of high commercial value

This strategy is particularly relevant to commercial fisheries aimed at providing fish for export or for urban markets. It means that yield will be less than maximal in terms of absolute tonnage, although the value of the catch may offset this. Regulation of the fishery is principally by strict control on the mesh size of the gill or seine nets forming the main gear of such fisheries. Data such as those in Fig. 4.13, combined with data on gill-net selectivity, can be used to calculate optimum mesh sizes for a particular fishery In fisheries where enforcement is difficult, mesh restriction on its own is insufficient. There is always a tendency for fishermen to reduce their mesh size to compensate for the inevitable reduction in the size of fish caught, which negates this strategy. Consequently, control of effort combined with severe restrictions on fisheries for smaller fish is needed to contain any risk in the collapse of the stocks of the major target species. On the whole this strategy is undesirable on the grounds of equity as it usually penalises poorer fishermen. It has also proved unsuccessful on a sustained basis when demand for fish is high. For example, the central delta of the Orinoco system maintained a fishery for large pimelodid catfishes for a considerable period when the fishery was only lightly exploited. At the same time catches in the heavily fished lower Orinoco were moving to progressively smaller species (Novoa, 1989). More recently, rising demand for fish in Colombia stimulated a more active fishery, leading to a rapid decline in the size of fish caught. In many cases this decline in abundance of larger species, and the fall in CPUE to levels which make the commercial fishery unprofitable, leads to its displacement to new areas further from the main landings (e.g. the Amazon at Manaus; see Petrere, 1983).

B. Maximise economic yield

This strategy relies on maximising value from the fishery by striking a balance between the size (quality) of fish and their number. In some fisheries for urban supply consumer preferences are for larger fish, although this may be culturally dependent. This strategy usually fails because the consumer adjusts to smaller fish and prices for these rise as larger specimens become depleted. Similar social responses on the part of fishermen to those to strategy C also apply.

C. Maximise yields but conserve the fish assemblage as far as possible

This strategy is optimal in that it allows for the taking of a maximum yield without allowing the fishing-down effect to proceed to species of overly small size. It is, however, extremely difficult to implement and implies rigid enforcement of regulations on both mesh size and access. It also requires banning of gears and other fishing practices that impact species of smaller size or, alternatively, by the establishment of closed seasons and/or closed areas. This degree of control is liable to introduce social inequalities at the artisanal level such that the poorer fishermen who usually concentrate their efforts on species of small size using inexpensive gears are actively selected against. Regulations to maintain the fishery at this level in the past have usually been poorly formulated with little understanding of the processes involved in multispecies fisheries. They are also extremely difficult to enforce by central authorities given the diffuse nature of many inland fisheries. As a consequence there is a high degree of failure in applying this type of management and the fishing-down process usually continues to stage D. Failures can usually be traced to fishermen objecting to centrally formulated and enforced policies. To correct this, co-management may be adopted whereby responsibility for the fishery is transferred to the fishermen. Co-management assumes that the fishery communities are sufficiently educated to assume the task of management and given support in their enforcement of locally oriented fishing strategies.

D. Allow the fishery to become fished-down

Decisions to allow the fishery to become fished-down are usually made for social reasons rather than ones based on the fishery resource. Major among these is the need to provide for large numbers of landless peoples who have no other form of sustenance. In this respect the fishery represents a social and political choice of last resort. In other instances the sheer difficulties of enforcing regulations in a milieu basically hostile to control leads to lethargy on the part of government. This results in *laissez-faire* policies that allow the fishing-down process to continue to the levels seen, for example, in the Oueme river in Africa or in Bangladesh in Asia (MRAG, 1994a). In lakes, reservoirs and regulated rivers this policy has proved disastrous leaving damaged faunas that constantly deliver less than their former potential. The decrease in length of catch and the diminished CPUE at higher intensities of exploitation tend to force out commercially oriented fisheries in favour of artisanal and subsistence activities. In rivers that retain much of their original form and capacity to flood, this process may not be disastrous in that the resilience inherent in flood regulated systems allows for recuperation even in the face of extremely heavy exploitation.

When fisheries become fished-down to the extent that favoured species are severely reduced in abundance there is a growing trend to compensate either by stocking to maintain the target species or by introducing other species that may resist exploitation pressure better.

The changes induced by the fishery do not act in isolation, as external effects arising from the impact of non-fishery uses mimic the changes brought about by fishing. Such external effects reduce the capacity of the fishery by accelerating the fishing-down process. Decisions on the management of the fishery, therefore, have to be taken in full knowledge of any non-fishery activities that may impact the fish assemblages and in consultation with the other users of the river.

Measures for regulation

Central and traditional regulatory authorities use a number of mechanisms to manage fisheries. Which ones are successful are conditioned by the behaviour of the exploited fish stock, and the policy, social and economic characteristics of the exploiting society. The role of fisheries management in responsible fisheries has been considered globally by FAO (1997a). This section describes the more common interventions to regulate inland fisheries so that they can achieve societal objectives.

Technical measures

These attempt to limit the fishery by placing restrictions on the types, characteristics and mode of operation of the gear used.

Mesh limitations

Limitations on minimum mesh size are common in centrally regulated inland fisheries. They tend to be popular with administrators as they provide a single figure that can be incorporated into legislation and subsequently enforced. Mesh size limitations are imposed to protect breeding stocks of specified species and are normally based on scientific assessment of selectivity compared with size of maturity. They can also be used, theoretically, to fix a multi-species fishery at a certain point in the fishing-down process, when an inflexible mesh limitation will limit exploitation to a few species within the selected range of capture sizes and will exclude many smaller species from the catch. This type of regulation can work well in fisheries based on few species of similar size that are based on a few landings or employ a number of large fishing vessels. They tend to fail when applied to multi-species fisheries in diffuse artisanal tropical river and lake fisheries. Here, single mesh limitations fail to take into account the differing performance of the various gears, many of which may not be mesh selective. Furthermore, social and economic pressures force fishermen to maintain catch levels by reducing mesh size in response to falling catches of fish in the target range. Mesh limitations imposed on a national scale which fail to take into account the differing needs of individual fisheries within the territory are equally inappropriate.

Gear limitations

Both centrally and traditionally regulated fisheries generally prohibit certain types of gear. Genuinely damaging gears such as poisons, explosives and electric fishing apparatus are widely condemned but are still used where controls are slack. Ultra-fine mesh gear such as mosquito net seines may also be prohibited as they are felt to take too great a proportion of the juvenile fish. Perceived social injustice may also result in some gears being banned locally, particularly by traditional regulation. In such cases users of the gear are typically seen to be taking more than a fair share of the catch, as in the case of brush parks and cross-river barrier traps, or to be compromising users of other gear such as in certain types of long line. In such cases the limitation of the gear may be spatial or temporal. Cross-river traps, for instance, may only be permitted over a certain proportion of the river width to allow for passage of fish and boats. Long lines may be permitted during certain periods or in locations where active gears such as cast nets, drift nets and seines are not generally in use.

Recreational fisheries tend to impose internal restrictions to improve the quality of the sport. These can include the type of gear and type of bait that can be used for any species or in any locality.

Closed seasons

The total seasonal closure of the fishery or seasonal prohibition of certain gears is common in both centrally and traditionally regulated fisheries. The closed season is usually to protect fish during a vulnerable part of their life cycle such as spawning or migration. The fishery may also be seasonally closed to certain gears to avoid conflicts. Closed seasons are relatively easy to enforce, especially if accompanied by education programmes that reinforce traditional regulatory mechanisms.

Closed areas

Fishing may be prohibited in areas of lake, river or floodplain to protect the reproduction and growth of certain species. Such prohibitions are often a feature of traditional management systems where the set-aside is supported on religious grounds. Areas may be closed seasonally in the case of certain types of spawning ground or passage for migration. More usually exclusion is total so as to form a permanent reserve for particular groups of fish. Two main types of reserve or protected area exist, conservation reserves and harvest reserves. Of the two the harvest reserve corresponds most closely to the type of closed area used to promote the fishery.

Restrictions on area may also be imposed to avoid conflict between different groups within the fishery or between the fishery and other users. Zoning, whereby the fishery is restricted to certain areas, has become common in countries where the use of any body of water is intensive for a number of interests, including wildlife conservation, boating and angling.

Input controls: control of access and effort

Limitation on access to the fishery is one of the most widespread methods of management. In many societies tradition limits access to the fishery to certain ethnic or social groupings and even within these there are controls on who actually fishes. In Africa and South America many governments opened access to the fishery in the interest of social equity, especially during the more socialist oriented eras of the 1960s and 1970s. The opening of the fishery resulted in rapid increases in fishing pressure as numbers of landless and emarginated persons turned to the fishery as a source of employment. This in turn led to the breakdown of traditional management systems. Open access policies have since been reversed in many areas and traditional approaches to limiting access have been resurrected. Access limitations lead to social inequalities by excluding potential fishermen from the privileged group, but such exclusion is now seen as a necessity in all natural resource management. Limitation of access and the consequent creation of a privileged group with guaranteed access is advantageous in that it creates a climate within which better management practices can be fostered.

The following mechanisms exist for regulating access.

Licences

Access to the fishery may be confined only to those fishermen or fishing craft holding a license issued by central government or its delegate. Ideally the number of licenses issued should be limited by scientific calculation of the potential of the waterbody related to effort and CPUE. Such calculations are difficult. Licences are appealing to governments as they represent a source of revenue as well as a direct control over the fishery. They are relatively easy to enforce in recreational fisheries and in commercial fisheries on large lakes that are based at a few landing sites, use large fishing craft and have a permanent cadre of professional fishermen. They are almost impossible to enforce in artisanal fisheries, especially in those of river–floodplain systems where a large population of part-time fishermen is distributed over a wide geographical area.

State-regulated access

In state-regulated systems the state retains ownership of the fishing rights but assigns them explicitly or tacitly to a particular user group. In most countries open waters such as large lakes and rivers belong to the state, although the rights to fish such waters may be designated to a particular group. In other categories of water, ponds, reservoirs and small streams ownership is more complex. In some countries artificial waters and the fishing rights therein belong wholly to those constructing them. In others, while the water may belong to the constructor, the fish stock may be in the public domain. The status of small rivers and streams is often equally unclear as the stream bed, the water and the fish may belong entirely or in part to the state, to the riparian owner or to some third party.

The state may also lease out part or the entire fishery for a fee. This pattern is especially common in Asiatic 'lot' fisheries, such as those of the Tonle Sap of the Mekong river, where the fishing rights to a reach of river, an area of lake or a specific fishing site are auctioned periodically. The successful bidder then manages the fishery for a specified period. Frequently the successful entrepreneur subleases parts of the fishery and extremely complex client–patron relationships can result from multiple subleases. At one time the period of the lease was for one year, leading to unrestricted fishing in which the lessees tried to recoup the maximum profit from the concession. It is now thought better to extend the lease period to 5 or even 10 years, giving an incentive to the lessees to apply more sustainable methods of management or enhancement. In addition, they may be subject to technical and output controls of one type or another.

Ownership

Ownership may be communal or individual. Communal ownership is often conferred by the state and individual ownership purchased. There is a growing tendency to define ownership of what were previously open-access resources, and much of the current trend to community management or co-management is linked to this process. Ownership, by either a group or an individual, has the advantage that this group then manages the resource, granting access according to their own criteria. It also improves confidence to invest in the fishery by using enhancement and rehabilitation techniques to improve yields and sustainability. However, management practices are usually constrained within guidelines set by government.

Output controls

Controls on the quantity and quality of fish landed are one of the major approaches to the management of marine fisheries. Similar types of control are also applied in large lake fisheries where there is a substantial commercial component. They are generally more difficult to apply in inland fisheries because of the high percentage of artisanal fishermen and the diffuse nature of the fishery and its markets. Two main approaches to output limitations are used in inland waters.

Quotas

These set limits to the total amount of fish or the amount of a specified species that can be landed from any water body in any one year [total allowable catch (TAC)]. This amount is then divided among the participants in the fishery. Alternatively the amount can be formalised as individual quotas which can then be inherited or traded (ITQs). This system is practicable in lakes where yields are relatively constant and there is a definable core of permanent fishermen among whom the quotas can be allocated. In river fisheries the concept is harder to apply because of extensive year-to-year variation in catches and the large number of part-time participants. The quota system works well in recreational fisheries where creel limits are often specified in the licence, which define the number, weight and species of fish that are allowed to be removed from the water, usually on a daily basis.

Size limits on fish landed

Some of the most ancient controls on fisheries appear to be based on minimum sizes at capture. For example, there is a plaque of a sturgeon in the Capitoline Museum in Rome which is reputed to have been mounted at the fish landing of that city to specify the minimum size of fish remitted to be landed and sold. Limits are usually specified by central government on the basis of scientific criteria similar to those used to determine mesh limitations. While size limits are difficult to enforce on the water, they can be controlled in markets where traders are prohibited from buying and selling fish smaller than the specified minimum. Such a system was used, for instance, to control the Loire River fishery in France in the seventeenth century by Colbert,

Minister to Louis XIV. Market controls of this type work where there are few large markets or commercial purchasers such as filleting factories. It is almost impossible to control in the situation of most rivers where there are numerous and ill-defined points of sale.

Note

1. By M. Aeron-Thomas and V. Cowan, MRAG, 47 Princes Gate, London, UK.

Chapter 12
Environmental management

Other users of the inland water resource

Fisheries in the context of multi-purpose use

Inland aquatic ecosystems and the organisms that they support are among the most vulnerable natural systems on the planet. Almost all natural and human activities taking place within a basin are reflected in the quality of the water, its quantity and timing and in the form of the system. This effect is so marked that it may be said of inland water fisheries that fishery managers are not in control of the resource they manage. Knowledge of the effect of these other users is important in drawing up plans for management. It is of little use, for instance, to project for a growth in a river fishery when a dam is to be built upstream, or to plan a reservoir fishery when drawdown is going to be excessive for the fish to spawn. Knowledge of the impacts of other users is also vital in quantifying the damage they do to the fishery, in negotiating with other users for provisions that will allow the fishery to continue and for planning mitigation or restoration.

The basis for planning inland fisheries rests on a clear understanding of the place of the fishery in national or local interest. Only rarely do inland fisheries have value sufficient to place them in the higher ranks of economic and social priorities, and inland waters are themselves subject to demand from many other users. Nevertheless, locally such fisheries may provide significant food and employment and may add value to the total aquatic resource. In highly urbanised countries the use of inland fisheries for recreation can add even further value to the sector. Some countries, especially in the temperate zones, assign increasing importance to the conservation of healthy and natural fish populations, which may preclude further development of the fisheries sector.

Properly managed inland fisheries are normally compatible with and encourage the maintenance of natural waters at a satisfactory quality for fisheries and other users. In this sense fisheries under natural regimes add value which is obtained at little cost to the society. However, some portion of it must be reserved for the fish and some portion of the basin must also devoted to providing suitable habitat. When demands on water are intense this allocation of water and habitat to the fishery may be contested.

Inland fisheries involve costs and benefits that can be balanced against the costs and benefits from other uses. These are often economically and socially more powerful

Table 12.1 Impact of other users of river and lake basins on fisheries (see also Table 12.2)

Use	Mechanism	Effect
Power generation	Dams	Interrupt longitudinal connectivity
		Stop water flooding the floodplain
		Change water discharge patterns
		Sedimentation changes
	Uptake of cooling water	Entrainment of juvenile fish
	Discharge of cooling water	Entrainment of fish
		Changes to thermal regime
Flood control	Dams	As above
	Levees	Interruptions to lateral connectivity
Navigation	Dams	As above
	Channel straightening and deepening	Loss of habitat
		Changes in basin morphology
		Wave creation and turbidity
Domestic use	Dams	As above
	Water transfers	As above
	Domestic sewage	Eutrophication or pollution
Agriculture	Dams	As for water transfers
	Water extraction	Altered flow regimes
	Diffuse fertilisers and animal waste discharges	Eutrophication
	Pesticide discharges	Pollution
Forestry	Removal of vegetation cover	Altered runoff, increased sedimentation
	Monoculture of pines	Acidification
Industry	Waste discharge	Pollution
Mining	Waste discharge	Pollution
	Discharge of tailings	Increase in sedimentation
Water transfers	Movement of water from one river to another	Changes in hydrology in donor and recipient basins
		Risk of transfer of organisms
Wildlife conservation	Set aside of areas as national parks and conservation areas	Usually positive reinforcing fisheries needs but may conflict

than fisheries and the conflicts and tensions resulting from the choices between objectives and users have to be resolved politically.

The main other users of the basin and the mechanisms by which they impact on fisheries are shown in Tables 12.1 and 12.2.

Power generation

Power generation is normally carried out by the state or by special power-generation companies. Such companies usually own the generating station and, in the case of hydroelectric plants, the dam or other associated infrastructure and sometimes the reservoir behind the dam. Power-generating companies often have a considerable amount of political and economic power and are sometimes considered concerns of national and military interest. They may therefore be prone to secrecy about their operations and are often inflexible to the demands of other sectors.

Flood control

Flood control is a function of national or regional governments. Flood-control mechanisms usually involve dams, overflow channels, levees and other structures designed to evacuate water from the land to the sea as quickly as possible and to confine water in the main channel. Many flood-prone areas were originally floodplains that have been drained for human occupation. This process is self-sustaining, as once occupied, these areas demand increasing and expansive interventions to protect them from further flooding. Effective flood-control schemes often transfer the problem further downstream as flows there are increased by the accelerated runoff upstream. A cascade situation is then set up whereby flood risk increases because the storage of the natural floodplain is removed. Where such areas are agricultural, modern approaches would suggest that the occupants be compensated and removed so that the area can be restored to floodplain. Where the flood-protected area is urban, the level of investment is such that all attempts will be made to improve the flood-control system to cope with any level of future flooding.

Navigation

Navigation in a river calls for the deepening and straightening of the navigable channel. This is associated with bank training structures that remove diversity and often employ levees to contain the water in the channel. While private companies usually own the boats and barges on the river, the responsibility for channel maintenance usually lies with government. Operations of this sort are very expensive and are often heavily subsidised. Navigation interacts with fisheries through the high degree of environmental modification associated with the creation and maintenance of a navigable channel as well as through environmental disturbance by the wash of passing boats. Some solutions to this exist and can be negotiated with the appropriate national authority. Most national and international river basin authorities are primarily concerned with navigation and usually do not incorporate fisheries interests.

Recreational boating is an intensive use of navigable waters in more prosperous regions. This can cause considerable nuisance to wildlife and fish through the constant disturbances to the natural system. Recreational boating interests are largely concentrated in the hands of individuals but are strongly regulated in some areas by national and local governments. Regulations include licensing and restriction of recreational navigation to designated areas.

Domestic use

Domestic use is one of the highest priorities in much of the world. The trend towards increasing urbanisation has intensified the problem of potable water supplies. In response, networks of reservoirs and water-transfer schemes have been developed. These are generally associated with water-treatment plants to ensure the quality of the water used for human consumption. Because urban supply is closely associated with public health there is usually strong government intervention even where private water companies assure the actual supply. The major impacts from urban use are through the discharge of urban wastes as either treated or untreated sewage. Other impacts arise where large-scale reservoir construction or water-transfer schemes are involved.

Agriculture

Agriculture is the biggest consumer of water through spray or ditch irrigation systems. The need for water, particularly in arid areas, requires impoundment in reservoirs, transfers from distant sources or the extraction of subterranean water. Severe ecological problems have followed many major water-transfer schemes and, in some cases, such as the Nile in Egypt or the Colorado in the USA, rivers no longer discharge to the sea. Allocation of water for agriculture is usually a national or regional responsibility and functions through systems of permits for water withdrawal. Such systems may be extremely ancient and based on traditional management. The allocation process for agricultural water has rarely taken into account the needs of fisheries. However, with the development of instream flow criteria, fisheries interests can open a debate with those allocating water rights to assure that at least minimum flows are provided for the fish and fisheries.

Agriculture also employs enormous amounts of chemicals that appear in the waters through diffusion over large areas. These include fertilisers, which eutrophicate the waters, and pesticides, which are powerful and toxic pollutants. In some societies fertiliser and pesticide use is regulated by central government.

Forestry

Forests and woodlands occupy much of the landscape. A forestry department at national level usually controls exploitation and reforestation projects, although private forested land is also common. Intact forest conserves soil through the binding effects of the root masses. It also retains precipitation, releasing it over a period of time into the rivers. It thus serves to even out flows. The principal effects of forestry on fisheries are through deforestation, for either timber or land clearance for agriculture. The newly denuded soils are washed into rivers by rainfall producing a flashier hydrograph and increased sedimentation. Local effects of forestry are habitat destruction by log floating and eutrophication under compacted log rafts. There is also evidence that reforestation with the wrong species, such as conifers, can acidify waters to a point where fish are endangered.

Animal husbandry

The luxuriant grasses found on riverine and lacustrine floodplains and water meadows have always been highly prized for grazing and support high densities of cattle. In many arid areas, traditional transhumance migrations have developed whereby cattle move between the floodplain in the dry season to adjacent rain-fed pastures during the flood.

Free-range cattle can break down bank structure and increase sedimentation at watering points. Animal husbandry usually falls within the jurisdiction of individual farmers, although in some countries access by cattle to public waterways is subject to central or local government legislation.

All forms of intensive animal husbandry, including aquaculture, can produce eutrophicating wastes that can locally deoxygenate the water, resulting in fish kills. In such circumstances major impacts can be avoided by requiring animal and fish farmers to install settling tanks downstream of the rearing facility so that the worst of the organic load can be removed.

Industry

Industry is a major user of water in developed economies. Water is usually extracted for cooling, washing or as a solvent as part of the production process. Major impacts are from pollution by a range of substances including heavy metals and toxic organic wastes or by substances having a heavy biochemical oxygen demand (BOD) such as paper waste. Industries usually fall within the private sector and should be prohibited by legislation from discharging toxic material.

Mining

Mining wastes and tailings cause a number of problems including pollution of the water either with material being extracted or with solvents used as part of the extraction process. Examples of this are the contamination of water with toxic copper compounds in copper mining and with mercury used in extraction of gold. Tailings can form an ongoing source of pollution where toxic matter enters the aquatic system through leaching long after mining has ceased. Water can be used as part of the mining process either as high-pressure jets at the face or for washing extracted material. These processes can increase siltation in the river.

Gravel extraction from riverbeds and margins is cause for concern in some areas. This activity not only changes the form of the river locally but also increases siltation, which may cause problems for a considerable distance downstream.

Permits to mine or to extract gravel are usually issued to private companies by local government authorities. Such permits should include specifications that limit the damage during extraction through limits to extraction rates, measures to prevent escape of toxic material and incorporation of settling ponds to remove toxic material and sediment from any discharge waters. They should also provide for rehabilitation after extraction has ceased.

Wildlife conservation

Aquatic ecosystems are not only inhabited by fish, they are also in demand for a wide variety of wildlife. Wetlands are crucial for waterfowl and as resting sites on long-distance migrations. They also support characteristic vegetational complexes. Rivers and lakes support a rich mammalian, reptilian and amphibian fauna. As such, these environments are targeted by numerous conservation organisations. Internationally, particular sites may be scheduled under the Convention on Wetlands of International Importance Especially as Waterfowl Habitat (Ramsar) or through the UNESCO programme for Biosphere reserves. Nationally countries may designate areas as reserves or Sites of Special Scientific Interest. Many countries support energetic non-governmental organisations (NGOs) that lobby for particular conservation interests. Mostly these interests parallel those of fisheries and their proponents can be used to reinforce pressures brought by fisheries. Sometimes conflicts may occur in that conservation interests may wish to limit or prohibit fishing for faunal preservation or for animal rights.

Impacts of other users

It has been estimated that over 75% of temperate waters are severely impacted by dams and by water regulation resulting from reservoir operation, interbasin diversion and irrigation (Dynesius & Nilsson, 1994).

Most tropical rivers have cascades of dams and large-scale water abstractions for agriculture and in emergent countries, such as China, nearly all of the waterways are severely degraded by pollution and environmental modification.

Water quality

Four major groups of substances influence water quality:

- *Nutrients* which contribute to eutrophication and when excessive may themselves become toxic through their BOD, which extracts the oxygen from water

- *Toxic pollutants*, usually originating from industrial and mining activities as well as from agriculture. These may be directly poisonous or may change the pH of the water and include:
   metals (zinc, mercury, nickel, cadmium)
   pesticides and herbicides
   cyanides
   nitrites
   chlorine
   ammonia
   surfactants
   acids
   drilling fluids
   pulp and paper effluents
- *Fine particulate matter* which contributes to siltation
- *Thermal waters* which contribute to temperature changes.

Eutrophication

Eutrophication is the enrichment of water through inputs of nutrients, principally of nitrogen and phosphorus. Eutrophicating inputs come mainly from agro-chemicals and from sewage effluents. They are particularly difficult to control because most agricultural sources are diffuse and, although discharges from major urban areas may be point sources, the increasing disposal of sludge on agricultural land is making this diffuse as well. Eutrophication in lakes gradually shifts the trophic status of the water and produces changes in their fauna and flora. At its most extreme severe oxygen deficits can occur as the BOD of the water rises. This can accelerate the deoxygenation of the hypolimnion and eventually render the lake unfit for most forms of life. In rivers natural eutrophication occurs as the river flows towards the sea and accrues organic loads. In floodplain rivers this can lead to natural anoxia in the main channel and in the floodplain water-bodies, especially during the hot, dry season. Because rivers are turbulent, mixing and re-oxygenation occur naturally, giving the river a capacity for auto-purification. This capacity can be exceeded, and river channels that would not normally become anoxic can do so. On floodplains the rapidly growing vegetation takes up most of the nutrients. However, extra nutrient loads deposited in floodplain lakes can accelerate and aggravate the onset of dry-season anoxia.

Pollution

Pollution occurs when toxic substances are discharged into water from industry, mining and agriculture. At their most extreme, toxic discharges can produce mass mortalities and prolonged contamination can render systems fishless. Even where discharges have no immediately apparent impact, long-term effects through the accumulation of pesticides and heavy metals in the food chain can produce grave consequences for fish, wildlife and humans dependent on the system. Pollution reaches the aquatic system in two forms.

Point-source pollution is discharged into the system through a single source such as a pipe or outlet, usually from an industrial or sewage treatment plant. Most toxic industrial and urban wastes arrive in a river or lake in this way. As such effluents are relatively easy to control and trace, responsibility for any pollution incident is correspondingly easy to assign and prove.

Diffuse-source pollution is mainly the result of agricultural and forestry activities, although it can also occur where there is small-scale mining over a large area. Pollutants are usually pesticides, herbicides and agricultural antibiotics, which are transmitted to the aquatic system with seepage of rain and groundwater. This type of pollution is extremely difficult to attribute and control, as it is often a function of large-scale cultural practices over very wide areas. It tends to be chronic rather than acute, causing dysfunction rather than massive mortalities. However, unfortunate combinations of rainfall, temperature and biocide applications can result in large fish kills. One example of the effects of such pollution is where heavy spraying for Soya culture around the Pantanal of Brazil coincided with heavy rains. These washed the active material into a river where the polluted discharge met an up-river migration of *Prochilodus*, which all died.

Sediment

Sediment erosion, transfer and deposition are natural processes responsible for much of the morphology of rivers and their floodplains. The amount of sediment in water rises over natural levels when the terrain in the basin is disturbed. Exceptional discharges of sediment can arise naturally through landslides and storm events, although forest cover normally minimises such effects. With deforestation, poor agricultural practice, mining and gravel extraction additional soil is washed into rivers and transferred to lakes, reservoirs and floodplains lower in the system.

As flowing water is able to carry a load of sediment proportional to its flow, any excess is deposited in riverbeds, lakes and reservoirs and on floodplains. This shortens the life of the individual water-bodies. Sediment deposits can also raise river channels above the surrounding plains, increasing the risk of flooding and eventually of lateral displacement of the channel. The deposition of fine sediment suffocates bottom-living organisms, chokes nesting sites and provides anchor points for invasive vegetation.

Conversely, where sediment has been deposited in reservoirs and other areas of low flow, and where flows are accelerated through channelisation, the water downstream has the capacity to pick up sediment. As a result the bed of the river is excavated, often to a point where it can no longer overbank and flood the surrounding floodplain.

Acidification

Recently, water in thousands of lakes and many rivers in the northern hemisphere has become acidic. This process has been traced to rain which dissolves the sulphur dioxide, nitrogen dioxide and nitrogen oxide released into the air when fossil fuels are burned and transforms it into sulphuric and nitric acids. Increased acidity in the water also frees large amounts of aluminium from the soil. Acidity of water combined with high concentrations of aluminium kills fish eggs and fry, and also kills adult fish. It also eliminates most aquatic plant and animal life, depriving fish of food. Countries affected, such as Scandinavia and Canada, have to spend considerable sums on liming programmes to combat this process.

Interruptions to connectivity

Dam building

Most river systems throughout the world have been dammed at some point (see Fig. 12.1). Dams have several effects.

Firstly, by impounding water they create a reservoir, which replaces the flowing river with a more static lacustrine water-body. In extreme cases, such as the Parana River in Brazil or the Volga river in Europe, the system has been converted into a cascade of reservoirs with very little of the original free-flowing channel left.

Fig. 12.1 Nam Ngum dam, Lao PDR. An illustration of the blocking of a river channel by a dam wall.

Secondly, by barring the river they create an insurmountable obstacle to the passage of migratory species. These then cannot reach their spawning grounds and may become locally extinct. Attempts have been made to limit the impact of dams on migrating fish through the construction of fish passes. Fish ladders may be successful for some energetic migrants such as the salmon but other types of structure need to be developed for other species.

Thirdly, dams and their operation for power generation and irrigation change discharge and silt load patterns in rivers.

Levees

Artificial levees may be regarded as linear dams whose function is to keep water off the floodplain. They may be extended to enclose complete areas of wetland as polders for the same purpose. In so doing they isolate floodplain water-bodies which follow their natural succession and disappear through sedimentation and vegetation growth. At the same time the lack of flooding prevents the formation of new floodplain features. Levees also prevent fish from migrating laterally to the richest breeding, feeding and nursery sites in the system unless some fish can penetrate sluice gates or other water-control structures. Great reductions in overall productivity and the local disappearance of some species usually follow levee construction. The enclosure of the channel, which is a natural extension of the channelisation process, also alters the discharge pattern in the river.

Channelisation

Channelisation is the constraining of a natural channel into an artificial feature designed for flood control, navigation or land reclamation. This process involves straightening, narrowing and deepening the previous meandering channel. The banks are usually leveed to prevent flooding and are lined with riprap, gabions or concrete to prevent erosion. The result is the conversion of a natural, diverse and functional aquatic ecosystem into a sterile, uniform one.

This process usually results in a loss of river length; for example, the Missouri river lost 120 km between 1890 and 1947 and the surface area was reduced from 49 250 ha in 1879 to 24 645 ha in 1972. Channelisation also results in loss of species, reduction in species size and reduction in overall productivity. It may also cause more damaging floods, when the river is disconnected from its natural floodplains and must overspill instead into new areas such as urban or agricultural land.

Water quantity and timing

Changes in discharge

Dams and levees are primarily devices for managing flow. Dams retain large quantities of water in the reservoir to be released when needed or to be diverted into other river systems. Levees prevent floodplains from acting as natural dampers by storing floodwater and releasing it slowly over time. In modified systems both the quantity and timing of discharge can be drastically altered from the pre-modified condition. River fish are closely adapted to certain flow regimes. Not only do the adults need adequate quantities of oxygen to maintain them but they also require specific flow characteristics to act as triggers for maturation, breeding and migration. Low water conditions can also pose physical barriers to movement in the main channel and between the channel and the floodplain by disrupting the connectivity of the system. These problems became apparent early on with abstractions from temperate trout streams. As a result minimum flow requirements were established as a basis for management of the water regime. With the extension of water management to large rivers by damming and water transfers, inadequate flow can suppress flood peaks so the floodplain is no longer seasonally flooded. In such cases it may be necessary to plan for releases of water to create artificial floods.

In many species, especially those laying eggs on or in the bottom, well-oxygenated water is needed for their development. Acceleration of flow in rivers during the time when semi-pelagic eggs and drifting juveniles are in the system can mean that these will be swept past the nursery area, and in the case of heavily channelised systems, even out to sea. Ill-considered changes in discharge can, therefore, disrupt the breeding success of vulnerable fishes.

Drawdown

Drawdown is the difference between maximum and minimum levels in lakes and reservoirs and between the total flooded and minimum level in rivers. In reservoirs, lakes and rivers too rapid withdrawal of water can cause stranding of fish, breeding sites and eggs attached to marginal bottom substrates, reducing survival and reproductive success. Equally, accelerated flooding, especially in rivers, will destroy rooted aquatic vegetation and will not allow enough time for the development of the rich supply of food organisms at the edge of the advancing water and on the bottom. It will also prevent nesting species from breeding by drowning nesting sites. For this reason the shape of the flood is as important as its quantity.

Remedial measures

Table 12.2 summarises the effects and possible remedial measures to impacts of interventions in the aquatic ecosystem. Techniques and approaches to remedy adverse impacts of other users are further described in the next sections on enhancement, mitigation and rehabilitation.

Table 12.2 Summary of the impacts of human interventions in river and lake basins on fish and measures for their reduction

Intervention	Impact	Remedial measures
Changes in water quality		
Pollution		
Presence of noxious chemicals in water	Death of fish Reduced survival and growth of juveniles Suppression of breeding Death and disappearance of food organisms	Formulate and rigorously enforce legislation on discharge of noxious substances
Eutrophication		
Increased nutrients	Increase in phytoplankton Increase in phytoplankton feeders Increase in overall productivity up to a critical point Change in nature of phytoplankton from diatom/green algae-dominated communities towards blue–green-dominated communities	Control inputs of nutrients by reducing human waste water and use of fertiliser within basin Biocontrol through introduction or stocking of appropriate species and removal of inappropriate ones
Lowered transparency	Disappearance of bottom vegetation with resulting disappearance of phytophyllic species	Disappearance of bottom vegetation with resulting disappearance of phytophyllic species
Deoxygenation	Shift from species with high oxygen requirements to those supporting lower levels of dissolved oxygen Trend towards increased deoxygenation of bottom waters of lakes and floodplain water-bodies with mortality of bottom faunas	Shift from species with high oxygen requirements to those supporting lower levels of dissolved oxygen. Trend to increase deoxygenation of bottom waters of lakes and floodplain water-bodies with mortality of bottom faunas
Changes in silt loading		
Changes in channel form (due to channelisation or to changes in deposition/erosion process)	Reduction of habitat and community diversity Loss of species	Carry out dredging programmes Stabilise banks and depositional areas with vegetation

(continued)

Table 12.2 (Continued)

Intervention	Impact	Remedial measures
Increased rate of silt deposition in lakes, reservoirs and floodplain structures	Choking of substrates for reproduction, leading to failure to reproduce in lithophils/psammophils Changes in density of vegetation, usually in favour of phytophllis Changes in quantity and type of food available and in the benthos leading to restructuring of the fish community towards illiophages	Control land use (forestry and agriculture) practices in the basin Control upstream mining operations to avoid downstream transmission of silt
Acidification Acid waters	Disappearance of living organisms Changes in fish community Reduction in number of non-visual predators and carnivores	Liming programmes
Decrease in suspended silt load		
Lack of sediment (downstream of dams)	Changes in nutrient cycle and in the nature of the benthos leading to loss of illiophages and increase in benthic limnivores	
Interruptions to connectivity		
Blocking of channels Reservoir creation	Changes in faunal structure with shifts from rheophilic species to limnophilic ones Development of pelagic fish community	Manage reservoir fisheries Introduce new species if required Mitigative stocking programmes
Interruption of migratory pathways by dam walls or by the creation of unsuitable conditions for passage	Elimination of diadromous or obligate migrants by preventing movement to upstream breeding sites by adults and slowing downstream movements of juveniles	Remove dams where possible Install suitable structures for fish passage

(*continued*)

Table 12.2 (*Continued*)

Intervention	Impact	Remedial measures
Changes in morphology		
Loss of habitat diversity through river straightening and bank reinforcement	Loss of species Changes in relative abundance of species	Restore habitat diversity through meander bend restoration pool and riffle restoration vegetation main channel embayments
Prevention of flooding by dams and levees Isolation of the floodplain and draining of wetlands	Loss of floodplain area available for spawning and growth Loss of habitat diversity; change in species composition with loss of obligate floodplain spawners General diminution in productivity of whole system	Reconnection of floodplain through installation of submersible weirs in river channel opening and realignment of levees
Changes in flow		
Temporal changes		
Disruption of spawning patterns through inappropriate stimuli or unnatural short-term flows	Changes in community structure away from seasonal spawners to species with more flexible spawning	Increased releases of flow from upstream control structures at critical times for spawning
In rivers, shift from pulse regulated to stable system dynamics	Diminished productivity at community level	Creation of an artificial pulse by controlled releases of water from upstream control structures
Changes in velocity		
Increases in flow rate (usually due to channelisation)	Young fish in drift swept past appropriate sites for colonisation Local shifts in species composition in tail race with accumulation of predators	Decreased releases of flow from upstream control structures

(*continued*)

Table 12.2 (*Continued*)

Intervention	Impact	Remedial measures
Decreased flow rate	Shifts from rheophilic to lentic communities in reservoir upstream and in controlled reaches downstream	Increased releases of flow from upstream control structures
	Changes in flushing rate resulting in accumulation or low dilution of toxic wastes or anoxic conditions leading to fish mortalities	
Prevention of flooding by dams and levees	Loss of floodplain area available for spawning growth	Reconnect floodplain by installation of submersible weirs in river channel
	Loss of habitat diversity; change in species composition with loss of obligate floodplain spawners	Controlled releases of flow from upstream dams at appropriate times
	General diminution in productivity of whole system	
Drowning of spawning substrates upstream of dams, in channelised reaches of rivers or in lakes	Variable effects usually involving decline of lithophils or psammophils although new wave-washed shore or rock riprap may simulate rhithronic habitats	Reduce flow releases from upstream dams at critical season
Overly rapid filling or drawdown in reservoirs	Flooding of nesting sites	Controlled discharge patterns from reservoir to reduce rapidity of water level change
	Stranding of nests and eggs	
	Drowning of developing vegetation and prevention of development of food organisms	
Increases in phytoplankton in reservoir or downstream due to slower flow and higher water transparency	Increase in abundance of planktivorous fish	Biocontrol through introduction or stocking of appropriate species and removal of inappropriate ones
Changes in temperature		
Stratification in reservoirs	Difficulties of passage for migrant species	
	Elimination of fish in deoxygenated hypolimnion	
	Mortalities downstream of dams due to emission of anoxic waters and hydrogen sulphide	
	Mortalities within the lake or reservoir through deoxygenation of the bottom water	

(*continued*)

232 Inland fisheries

Table 12.2 (Continued)

Intervention	Impact	Remedial measures
Changes in mean temperature caused by low flow regimes or discharge of thermal effluents	Increasing temperature variation can cause shifts in success of spawning due to adverse temperatures for either cold or warm water spawners	
Uptake of water		
Induction of water into power stations or through pumps or irrigation canals	Entrainment of fish into currents diverting them; impingement of fish on turbines and pumps resulting in loss of fish, particularly juveniles	Install efficient screening systems
Water transfers between river systems	Transfer of species and disease organism from one system to another	Install efficient screening systems at inlet and outlet

Table 12.3 Typical conditions requiring resolution in discussions between parties for the conservation of living aquatic resources and fisheries

General category	Condition
Flow	Minimum acceptable flow Timing of flow Speed of change in discharge or water level
Habitat connectivity	Mitigation or removal of obstructions to fish movement Maintenance of access to inflowing tributaries in lakes
Habitat diversity	Maintenance of critical habitats Provision for adequate diversity in main water-body Connectivity to lateral marshes, floodplains, etc. Maintenance of riparian vegetation structure
Water quality	Avoidance of chronic or acute, diffuse or point source pollution by toxic substances Regulation of nutrients within critical limits
Physical disturbance	Limitation on boat wash Limitation on weed cutting
Access	Ensuring access to waterside by interested parties
Basin characteristics	Land-use practice to avoid erosion and uncontrolled runoff Avoidance of inappropriate types of vegetation cover

Basin planning

The large numbers of economic and social activities of growing human populations exert pressure on inland waters. The requirements for the maintenance of healthy stocks of fish and other living aquatic resources and the fisheries that depend on them are frequently secondary to other considerations. It is therefore the task of the fisheries manager and those responsible for the conservation of the environment to negotiate conditions suitable for the maintenance of fish stocks and fisheries. Factors entering into such discussions are set out in Table 12.3.

Where one use impacts on the aquatic systems negotiations may be with the primary user, but more normally the aquatic system suffers a range of abuses from a number of users and discussions should be more broadly based.

Valuation of the living aquatic resource

The essential step in any basin planning process is the recognition that the resources in question are sufficiently valuable to justify protection. This requires some valuation of the resource. Natural resources are notoriously difficult to value. Four major factors enter into such calculations:

- The commercial value of the commodity generated, usually fish or fishing licenses
- The social and economic value of the resource for employment, nutrition, public health and recreation
- The value of the ecosystem service provided by an intact ecosystem, for example self-purification capacity
- The value of the aesthetic or conservation value of the resource and the environment

Commercial and social values are relatively easy to establish because regular costs and benefits can be assigned to products and social benefits. Ecosystem services are more difficult to establish and extraordinarily high estimates have been produced as to the value of intact ecosystems to human society. The valuation of environmental intangibles such as aesthetics or the worth of a single endangered species are even more problematic and are usually assigned by society on the basis of what other benefits it is willing to forgo to achieve the particular goal.

Definition and costing of methods

Methods for the evaluation of fisheries and for assessing the impacts of competing uses of the river basin are discussed later. These methods can be used to establish the actions needed to maintain the aquatic ecosystem and the fishery at an agreed level. These actions have costs as follows:

- The costs of maintaining the aquatic ecosystem have to be calculated on the basis of alternative approaches to protection, mitigation or rehabilitation
- The opportunity costs of other activities that have to be foregone to maintain the aquatic ecosystem in its desired state also have to be established
- The social cost of disease, nutrition, employment and recreational opportunity must also be included.

The fact that the cost–benefit calculation may favour the non-fisheries use in the short term has to be balanced by the longer term interests of the ecosystem. Any other obligations, such as public health, national conservation policies, conformity with international conventions or the interest of other riparian states, have to be taken into account in selecting the final course of action.

Mechanisms

Recently the obligation to produce formalised plans under which the interests of all parties are taken into account has led countries to consider the processes needed for basin management. For instance, the European Community issued a formal directive requiring its members to prepare such plans and the United Nations has been promoting plans for international river and lake basins at a series of regional meetings.

Basin management relies on the setting up of adequate mechanisms to ensure full discussion of the viewpoints of the various stakeholders and to arrive at consensual decisions. Bilateral discussions between fisheries interests and one particular user may be appropriate where a single use dominates the basin. More normally discussions involve many interested parties. In such cases some form of mechanism is needed to bring the various parties together. In the case of larger systems this may be under the umbrella of a lake or river basin commission. In other systems a formal basin management committee has to be established.

Discussions at such forums require institutions that are adequate to represent the position of the various parties. In the case of fisheries such institutions are frequently lacking and the government authority charged with fisheries usually performs this role. Where this is the case efforts should be made to enlarge fisheries representation to include the fishermen and other stakeholders in the fishery.

Education and information

Cross-disciplinary planning of this type can only be effective if all parties are informed about the issues at stake. It is generally thought sufficient to confine such knowledge to the participants in basin management meetings. However, in more complex societies the role of public opinion, often expressed through NGOs, is crucial in determining the allocation of natural resources. In such cases education and information of the general public through the press, radio and television enables decisions to be reached that are consistent with the general will.

Chapter 13
Enhancement

Enhancement is any activity aimed at supplementing or sustaining the recruitment, or improving the survival and growth of one or more aquatic organisms, or at raising the total production or the production of selected elements of the fishery beyond a level that is sustainable by natural processes.

Techniques for enhancement

A variety of techniques ranging from culture-supported capture fisheries to intensive aquaculture can be used to compensate for declines in fisheries due to overfishing, environmental change or inadequacies in the natural ecosystem (Welcomme & Bartley, 1998). The various techniques are often adopted in a stepwise manner leading to a progressive increase in fishery production per unit area of water through increasing human controls on essential parameters of the fish assemblages.

In rough order of magnitude of resources required the techniques are:

- Introduction of new species to exploit underutilised parts of the food chain and habitats not colonised by the resident fauna, or to compensate for loss of species due to environmental disturbance
- Stocking natural waters to improve recruitment, bias fish assemblage structure to favoured species or maintain productive species that would not breed naturally in the system
- Fertilisation to raise the general level of productivity and hence growth of the fish
- Engineering of the environment to improve levels of reproduction, shelter, food resources and vital habitat
- Elimination of unwanted species that either compete with or predate upon target species
- Building up of an artificial fauna of selected species to increase the degree of control and the yield from the system
- Modification of water bodies to cut off bays and arms to serve for extensive and intensive fishponds to increase control over the rearing process
- Introduction of cage culture and parallel intensification of effort of the capture fishery
- Aquaculture through management of the whole system as an intensive fishpond

- Genetic modification or selective breeding of cultured species to increase growth, production, disease resistance and thermal tolerance of the stocked or cultured material.

Introductions

Motives for introductions

The main motives for introducing new species into a lake or river are as follows.

- To create new fisheries that are more resistant to fishing pressure or have greater market value than native fishes. In commercial fisheries such introductions have often involved large predators which were intended to benefit from smaller species of no value to the fishery. In other cases species of known value such as the tilapias have been favoured. New species are introduced into recreational fisheries to improve the variety available to anglers or insert a species of particular trophy or sporting value into an area.
- To fill a 'vacant niche' where existing fish species do not fully utilise the trophic and spatial resources available. In some natural waters geographical accident has resulted in there being few native species, as in some islands, or areas where faunas have been wiped out through glaciation. More commonly the need for introductions arises as a consequence of human activities. For example, in Latin America many new reservoirs lack native species capable of fully colonising lentic waters. In many river basins regulation of flow by dams has eliminated or drastically reduced the native rheophilic faunas, leaving the waters open to colonisation by introduced species.
- To control pests. Several species have been introduced in an effort to control biologically pests and vectors of human disease. Typical of these is the widespread introduction of *Poecilia* and *Gambusia* spp. to control mosquito larvae and several species including *Astatoreochromis alluaudi* for the control of the snail vectors of schistosomiasis.
- To control water quality where suitable phytoplankton-eating species are lacking. Species such as *Aristichthys nobilis*, *Hypophthalmichthys molitrix* and *Oreochromis niloticus* may be introduced to remove excessive algae in eutrophicated systems.
- To develop aquaculture. This remains one of the main motives for the movement of species around the world. Many species have been introduced for culture. These include rainbow trout, common carp and tilapias, which together contribute a large share of inland aquaculture production. Escapes from aquaculture installations have resulted in many accidental introductions into the wild. There has been a tendency in most countries to introduce those few species whose culture is well known rather than to try and develop local species. This tendency can be combated through programmes to identify and develop local species, which in any case often correspond more to local tastes and have higher market values. There is no

doubt that further species and specialised genetic strains will be introduced around the world.
- To fulfil aesthetic and other reasons. Ornamental species are now widely distributed throughout the tropical world through escapes from rearing installations and aquaria. Some species have also been introduced for particular religious or cultural reasons.

Risks from introductions

Introduction of new species is accompanied by the risk of corrupting native fish communities through predation, competition, disease, hybridisation and adverse environmental impacts. Introductions pose special problems in that they insert a totally new element into the fauna whose response is difficult to predict. The following effects can arise.

Environmental disturbance
Introduced species can disturb habitats and in so doing alter ecosystem characteristics to such a degree that native species are threatened. A notable example of such behaviour is the common carp, which roots for food in the muddy bottoms of lakes and rivers, remobilising sediment and biochemical oxygen demand (BOD). This can lead to turbid conditions that reduce light penetration and plankton production. Another form of behaviour is burrowing, and many introduced crayfishes can seriously damage pond banks and river levees.

Predation
The introduction of predatory species into waters from which they were previously absent is one of the major causes of reported negative impacts. Apart from the notorious case of the Nile perch introduction into Lake Victoria, which caused the disappearance of large portions of the cichlid species flock, species disappearances have also been associated with introductions of trout in many areas and of *Cichla ocellaris* in Latin America.

Competition
Competition between the introduced and native species is frequently cited as another cause of potential difficulty. Competition may be for food but in nest building species' breeding sites may play an equally significant role. The disappearance of some tilapias following the introduction of *O. niloticus* is a case in point. While rarely leading to species disappearance the second major cause of negative impacts has been the explosive expansion of populations of small 'r' selected species. This is often accompanied by stunting leading to dense populations of small individuals of little use to a fishery, which compete with and reduce the numbers of more useful species. *Oreochromis mossambicus* shows this behaviour to a certain extent, as do several cyprinids and sunfishes.

Introduction of disease

There have been many examples of diseases and other parasitic organisms that have apparently accompanied introduced fish species to their new home. Often the introduced species is resistant to the disease organism through long co-habitation, but species in the receiving environment are more sensitive and readily transmit the new pathogen outside the original area of the introduction. Diseases can have a disastrous economic impact on aquaculture, as has been repeatedly illustrated in the shrimp, salmon and trout industries.

Genetic contamination/hybridisation

The most extreme genetic effect is hybridisation. Many fish species regularly hybridise with closely related species and frequently with those of greater geographical and taxonomic divergence. Hybrids may be:

- self-fertile and will breed true, in which case an essentially new species is created such as the various 'red tilapia'. Self-fertile strains can eventually revert to their parental forms through backcrossing;
- not self-fertile but capable of producing viable offspring with one or both of the parental species;
- sterile.

Hybridisation among species in the natural environment can pose risks, as valuable adaptive characteristics, such as timing of migration and the ability to locate natal streams, may be lost in the host species. Alternatively, the hybrid can prove more successful and vigorous than the parents, in which case the latter may disappear through competition.

Genetic mixing of different stocks through interbasin transfers of the same species poses similar risks of loss of adaptive characters. Naturalised stocks of exotic species, especially the tilapias and common carp, are also likely to be affected by contamination with genetic material from the various specialised strains being used in aquaculture and the aquarium fish trade.

Co-introduction of nuisance species

Where fish are introduced as juveniles there is a serious risk that fry of other species are also included in the transfer material. If proper precautions are not taken these can readily acclimatise to the receiving habitat. The arrival of *Pseudorasbora parva* and several other species in Europe, for instance, is thought to have originated from contaminated batches of Chinese carp fry imported for aquaculture and weed control (Welcomme, 1988). Some of the recently recorded species from the north of the Mekong basin may also have arrived in Lao PDR by the same mechanism.

Measures to reduce risk

Introductions are generally irreversible and proposals for new introductions call for

Guideline 3 Decision tree for introduction of fish species

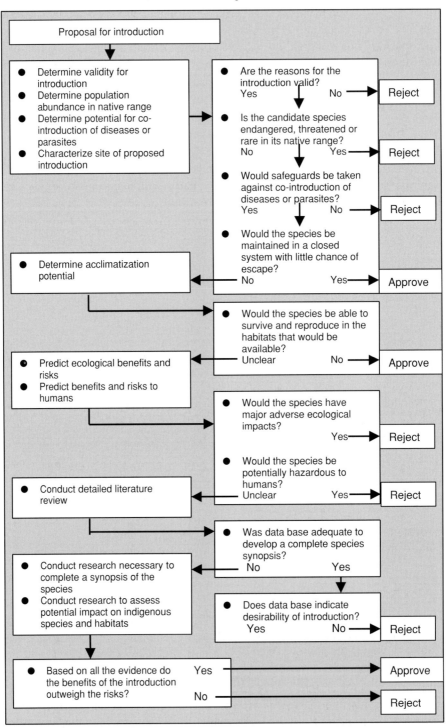

special caution. There is, therefore, strong international pressure to regulate the movement of species to reduce the risks of damage to the environment, to the native fish stocks and to the genetic composition of resident and introduced fish. A general scheme for decision making with regard to introductions is shown in Guideline 3, based on protocols developed by ICES and FAO (International Council for the Exploration of the Sea, 1995; FAO, 1995a).

The precautionary approach to fishery management requires that adverse effects to the aquatic environment be reversible within a time frame of two to three decades (FAO, 1995a). Species introductions into the wild generally fail to meet the requirements of this principle in view of the fact that containment and reversibility have been shown to be nearly impossible. Nevertheless, introductions continue to be viewed as a legitimate management tool given that other elements of the precautionary approach, such as risk analysis, implementing a monitoring system with defined acceptable impact levels, and establishing corrective measures in advance, can reduce their adverse effects.

Stocking[1]

Objectives of stocking

Stocking is the most widespread measure for management of inland fisheries in use today. Most countries report stocking to some degree. As more conventional approaches to management by control of the fishery have proved incapable of limiting effort, compensation for shortfalls in recruitment caused by overfishing and environmental damage have been sought through the addition of young fish to the system. Several major types of stocking can be identified:

- compensation to mitigate a disturbance to the environment caused by human activities;
- maintenance to compensate for recruitment overfishing;
- enhancement to maintain the fisheries productivity of a water-body at the highest possible level;
- conservation to retain stocks of a species threatened with extinction.

Stocking versus natural reproduction

Many stocked fisheries are based on species that do not reproduce to a significant extent in the water-body in question, and rely on stocking for the largest part of recruitment. The population dynamics and assessment of such fisheries are discussed below. Where fish are stocked into self-reproducing populations of the same species, the effectiveness of stocking is particularly difficult to evaluate. Compensatory responses in growth and mortality are likely to reduce the absolute contribution to recruitment from natural stocks, unless this contribution is far below potential.

Furthermore, the contribution to reproduction from stocked fish is often uncertain, it may be inhibited where the stocked fish are drawn from a strain not adapted to the recipient water-body. Empirical studies suggest that stocking may well be superfluous in self-reproducing populations where recruitment is not impaired. Examples are the Finnish coregonid fisheries (Salojarvi & Ekholm, 1990) and the tilapias in Cuban reservoirs (Fonticiella *et al.*, 1995). In other cases, such as the Polish lake fisheries, stocking is believed to have made a long-term positive contribution despite there being naturally reproducing stocks of the target species (EIFAC, 1983). In many cases, for example in Thailand (Bhukaswan, 1988), it has been impossible to evaluate the biological and economic success of stocking because of difficulties in separating stocked fish from natural production. Many salmon fisheries in North America and Europe are composed of both hatchery and naturally produced fish; how they interact and their relative contribution to the fishery provide a topic of much discussion (see Campton, 1995, for discussion and references).

Patterns of stocking

There is a surprising lack of information on the ways in which stocked systems function. The few data sets that exist show an interplay of two main variables: area of the stocked system and stocking rates. They indicate that:

(1) *Yield is related to stocking rate.* Relationships between stocking rate and catch are given in Table 13.1.
(2) Empirical relationships between stocking density and catch may be apparently linear over the range of stocking densities for which data are available (Fig. 13.1). Such relationships must not be extrapolated uncritically, however, as compensatory processes will result in a levelling off or decline of yields at high densities. Linear or quadratic regression models on log-transformed data may be appropriate where variables range over several orders of magnitude and have been used to describe yield–stocking density relationships in water-bodies in tropical Latin America and Thailand (Lorenzen *et al.*, 1998; Quiros, 1999).

Table 13.1 Relationships between stocking rate and catch in reservoirs form China, Sri Lanka and Mexico

Lake	Relationship	Remarks
Nanshahe reservoir, China	$Y = 122.1 + 0.1140D$ ($r^2 = 0.58$)	Data from a time series from a single lake
Sri Lanka	$Y = -1.829 + 0.3714D - 0.0011D^2$ ($r^2 = 0.70$)	
Mexico	$Y = 13.25 + 3.48D$ ($r^2 = 0.66$)	Data from a series of lakes

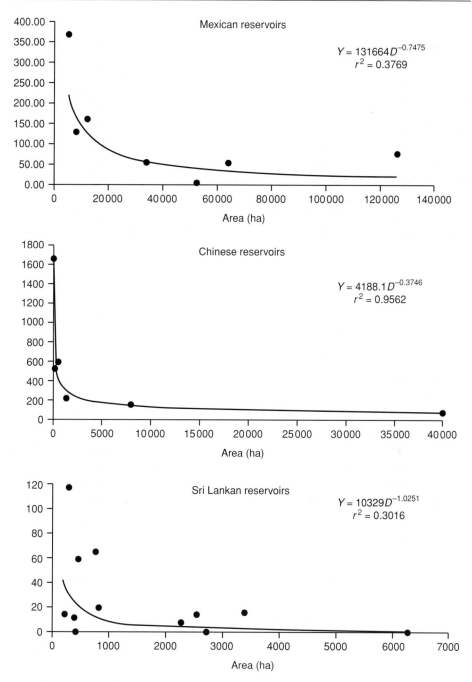

Fig. 13.1 Plots of yield per unit area against area for various sets of stocked reservoirs.

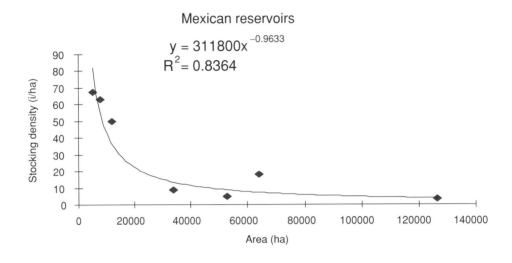

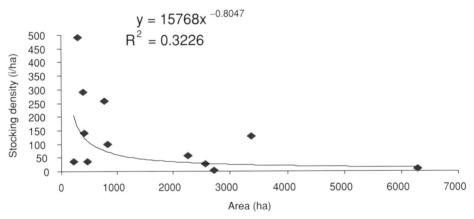

Fig. 13.2 Stocking rates into reservoirs of different areas in Mexico and Sri Lanka.

(3) *Yield per unit area is inversely related to the area of the stocked system.* Stocking has generally proved more effective in small reservoirs in many regions, although larger water-bodies are also stocked (Fig. 13.2). The greater degree of control over small systems and the greater risk of predation and competition in larger ones would seem to favour the efficiency of stocking in smaller water-bodies. Furthermore, larger water-bodies are more difficult to control socially and have a heightened probability of losses through poaching and unreported catch.

This supposition cannot be readily validated as *stocking rate are usually inversely related to lake area*. In nearly all countries practising stocking there is a tendency to stock fish at lower densities into larger water bodies because it is supposed that

Table 13.2 Stocking and production characteristics of reservoirs of different sizes in China

Area of reservoir (ha)	Stocking density	Fish yield
Small (<70)	3000–7500	750–3000
Medium (70–670)	1500–3000	450–750
Large (670–6670)	750–1500	225–450
Super (>6670)	450–750	150–225

(1) competition and predation will be greater in larger water bodies and therefore survival rates will be reduced; and (2) other aspects of enhancement such as fertilisation, control of unwanted species or construction of artificial faunas are more manageable in smaller water-bodies.

Typical stocking densities and resulting yields for Chinese reservoirs of different areas are shown in Table 13.2.

It must be concluded that, at present, the practice of using small water-bodies for intensive and large ones for extensive stocking depends more on the high numbers and cost of fingerlings needed for the larger water-bodies.

Population dynamics of stocking

Rational management of stocked fisheries requires understanding of their population dynamics. This is particularly challenging where populations are supported by a mixture of stocking and natural reproduction. However, many fisheries are mainly or entirely dependent on stocking. Recent studies have elucidated the dynamics of such fisheries and led to the development of practical assessment tools. The key population processes governing the outcomes of stocking in the absence of natural reproduction are density dependence in body growth and size dependence in mortality. In combination, these processes result in density-dependent mortality. Direct density-dependent mortality (other than modulated by growth) may be important where early (pre-fingerling) life stages are used for stocking, or where juveniles are markedly territorial (e.g. in stream-dwelling salmonids).

The following management guidelines can be given for fisheries governed by density-dependent growth and size-dependent mortality (Lorenzen, 1995).

- The optimal stocking regime is dependent on the harvesting regime and vice versa. This is illustrated schematically in Fig. 13.3, where production is shown as a function of stocking density and fishing effort. High fishing effort calls for high stocking densities and vice versa. High stocking densities combined with low fishing effort lead to overstocking, with low production due to slow growth and low survival from stocking to harvest. Conversely, low stocking rates combined with high fishing effort lead to overfishing. Note that both overstocking and overfishing can be alleviated by changes in either stocking density or fishing effort.

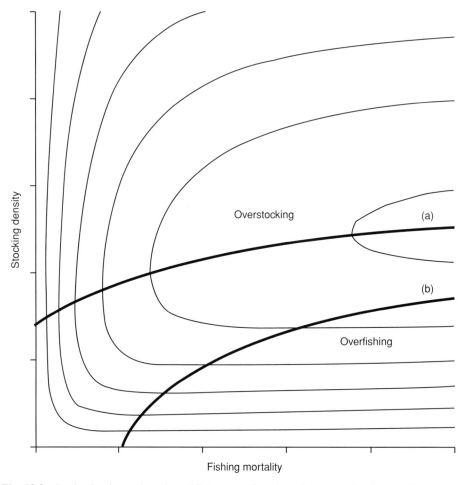

Fig. 13.3 Production in a culture-based fishery as a function of stocking density and fishing mortality. Line (a) denotes the optimal stocking density for a given fishing mortality, line (b) the optimal fishing mortality for a given stocking density. (Modified from Lorenzen, 1995.)

- Potential production from stocked fisheries is inversely related to the size at which fish are harvested. Hence, in combination with the overall ecological productivity of the water-body, the minimum size at which fish are marketable or attractive to recreational fishermen effectively limits the production that can be achieved from stocking. Where large fish are desired, as in certain sport fisheries, stocking densities should be low and overall production will also be low. Where small fish are marketable, high production levels are achieved when stocking densities are high and fish are harvested at the smallest marketable size. Where fish are marketable below their normal size at maturity, culture fisheries can achieve substantially higher levels of production than wild stocks of the same species because large and somatically unproductive spawners can be replaced by a large number of small and somatically productive fish.

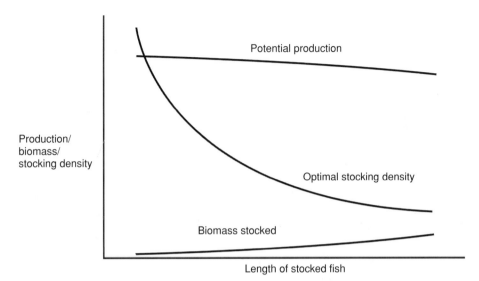

Fig. 13.4 Relationship between the length of stocked fish, potential production and the respective optimal stocking density in numbers and in biomass. (Modified from Lorenzen, 1995.)

- A wide range of different stocking sizes can be used to achieve similar levels of production, but the numbers that need to be stocked decrease in a non-linear way as size increases (Fig. 13.4). This is a consequence of the allometric mortality–size relationship, and the fact that larger seed fish require less time to reach a harvestable size. The biomass of seed that needs to be stocked to achieve a given level of yield increases with increasing seed size, and so does the cost of producing the individual seed fish.

The above list provides general rules that apply to a wide range of stocked fisheries. Where natural reproduction is an important source of recruitment or there is strongly density-dependent mortality after stocking, there are further considerations.

Assessing management regimes in practice

The initial choice of a stocking regime is often based on a best guess, or determined by external constraints (e.g. size and numbers of seed fish available). However, once results from the initial stocking are known, this information should be used to improve stocking (and/or harvesting) regimes of the fishery.

A rough initial assessment of stocking numbers required to achieve a target abundance of harvestable fish can be obtained by inverting the standard mortality formula:

$$N_0 = N_c \exp(z(t_c - t_0))$$

where N_0 is the number to be stocked, N_c is the number desired at age of capture c and z the total mortality between stocking and harvesting, t_c is age at capture, and t_0 is age at stocking (Welcomme, 1976a). The use of this formula in practice requires estimates, or at least good guesses, of the growth of fish (which determines the time between stocking and harvesting), the mortality rate z, and the abundance of harvestable size fish that can realistically be supported by the water-body. In the absence of specific data for the water-body in question, both growth data and potential standing stock can be obtained from studies in other, similar water-bodies. The mortality rate z can be obtained from the allometric mortality equation. In wild fish stocks, $b = 0.29$ and $\mu = 3.0$ year^{-1} (Lorenzen, 1996b). While the allometric exponent b appears to be quite stable between populations, μ is more variable. Mortality in stocked fish may be significantly higher than in wild fish of the same size, and it is therefore recommended to obtain predictions for a range of μ values, from 3 year^{-1} to 30 year^{-1}.

The number of fish desired at harvestable size can be related to the potential productivity of the water-body. Several systems have been used for this, ranging from generalised equations such as the Morpho-Edaphic Index to specialised indexes based on benthos or zooplankton densities. These can be incorporated into the general formula as follows:

$$N = (qp/\bar{w})\exp(-z(t_c - t_0))$$

where N is the number to be stocked, p is the natural annual potential yield of the water-body [Morpho-Edaphic Index (MEI) or alternative estimator], q is the proportion of the yield derived from the species in question, and $\bar{w}$ is the mean weight at capture. Chinese reservoir fishery managers use more empirical expressions for arriving at the number of fish for stocking (Li & Xu, 1995); for instance:

$$d = f/\bar{w}R$$

where d is the annual stocking density (fish ha^{-1}), f is the annual fish productivity (kg ha^{-1}) as estimated from food organism abundance, $\bar{w}$ is the average weight of fish at harvest (kg), and R is the return rate. Food biomass indicators are used to establish the productivity and carrying capacity of the water to be stocked in both China and Russia (Li, 1988; Berka, 1990).

Two main factors influence the size chosen for stocking material: cost and survival. Stocking of fish at too small a size risks high mortality, but the cost of stocking material increases exponentially with length, especially in slow-growing species. The mortality–size relationship provides the basis for a quantitative assessment of the numbers to be stocked of different sizes of seed fish. However, there may also be other considerations, such as features of the biology of the species, that determine the size at which the fish have to be stocked. For example, some migratory and anadromous fishes such as salmonids are usually stocked at a small stage (fry) to acclimate to the natal river and to prepare for migration as their size increases. Cyprinids and other non-migratory forms are generally stocked at a larger stage (fingerlings). In addition, fish stocked at an early stage such as eggs or fry are more likely to be subject to directly density-dependent mortality after stocking than fish stocked at the fingerling or juvenile stage.

Recreational fisheries increasingly tend to rely on even larger fish of cacheable size and to rely less on grow-out in the natural environment.

Once some data on the results of stocking are available, a more rigorous evaluation of stocking regimes should be carried out. This will usually require the use of assessment models, which may be mechanistic (i.e. incorporate explicitly key mechanisms such as growth and mortality) or purely empirical (i.e. describing observed relationships between stocking density and yield by regression equations). Mechanistic models require relatively detailed information on growth, mortality, etc., and are particularly useful where stocking is carried out in larger water-bodies, so that the individual population justifies a study at this level of detail. Because such models are based on an understanding of mechanisms (using well-established growth models, etc.), they can be used to evaluate a wide range of management options in detail. An example of such an analysis that can be adapted to a wide range of fisheries is given in Lorenzen *et al.* (1997). Mechanistic population models can be used in situations of disequilibrium, for example in a developing fishery.

Empirical models are regression models fitted to observations of yield responses to different management regimes and natural conditions. Because they do not embody mechanistic understanding, such models should not be extrapolated beyond the observed range of independent variables. Such models are most useful where data on outcomes under a range of different management regimes are available or can be acquired quickly. This is the case, for example, where a large number of water-bodies is managed in different ways. Information on the natural characteristics, management inputs and yields of these fisheries can be analysed in a comparative perspective. An example is the empirical model developed by Lorenzen *et al.* (1998) for stocked small water-body fisheries in Thailand.

The outcome of stocking depends in part on density-dependent responses, which can only be quantified if data are available for contrasting densities. In many fisheries, stocking densities and/or harvesting regimes have varied in the past as a consequence of fluctuations in seed availability, etc. Where no such information exists, or the contrast in management regimes is insufficient to estimate key parameters, it may be advantageous to vary management on purpose in order to gain crucial information. Such an approach, known as active adaptive management, may bring substantial long-term benefits even if it is associated with short-term loss.

Development of stocked fisheries and decision making

The economic viability of stocking varies widely. Stocking also involves a number of risks that should be assessed before stocking is approved or promoted. Stocking is being carried out under widely different circumstances. On the one side are local communities in developing countries, which may embark on stocking with little research support or government regulation. On the other are government and private bodies in developed countries, which may conduct stocking within a tight and scientifically based regulatory framework. Such different situations call for different

250 *Inland fisheries*

Guideline 4 Decision tree for various types of stocking (from Cowx, 1994b)

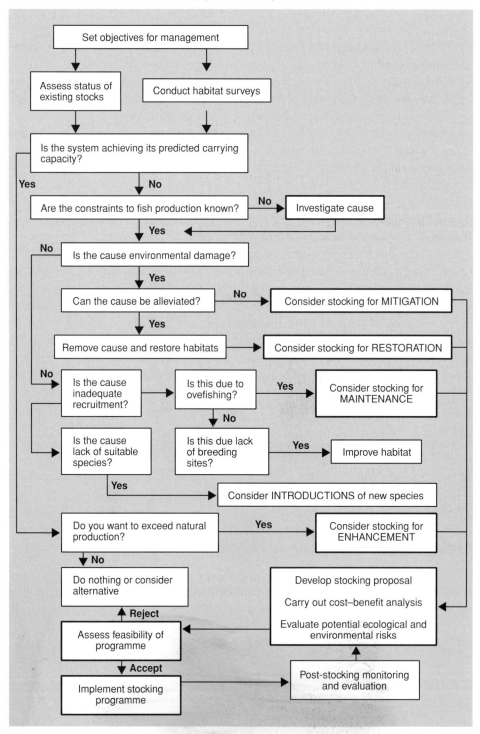

approaches to the development of stocked fisheries. A framework for supporting culture fisheries development in the former situation is outlined in Lorenzen & Garaway (1998).

Conversely, where baseline knowledge is well established and regulations can realistically be enforced, decision making may become more formalised in order to ensure that all relevant considerations are taken into account, and that decisions are transparent and accountable. Flow charts, which serve as checklists to assist in decision making, are given in Guidelines 4 and 5.

Guideline 5 Suggested procedure for planning and executing a stocking exercise (from Cowx, 1994b)

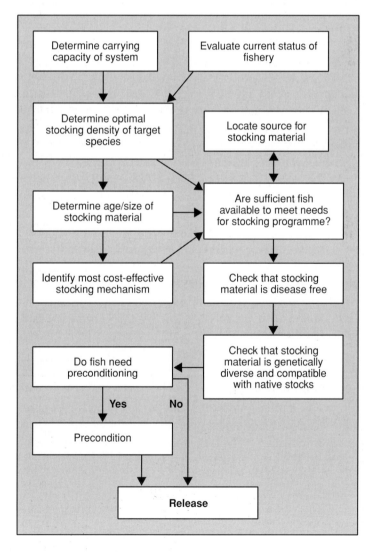

Risks from stocking

Stocking is usually regarded as posing fewer problems than introductions. Inland reservoirs, lakes and rivers are usually stocked with one or more of three categories of fish:

- Native species
- Introduced species that are already well established in the environment
- Introduced species that do not breed in the receiving environment.

In most cases the potential for the gross environmental impacts that may be caused by introductions is not high, but the large quantities involved in stocking and the repeated nature of the operation considerably raises the risks of the following problems.

Genetic effects

Risks of genetic effects on host populations are greater in stocking than with introductions. These arise from two sources: stocking with species that are native to the host water-body, and stocking with introduced species. Where stocking is carried out with species that are already in the receiving water-body there is a risk of genetic swamping whereby the original genetic characters of the host stock are lost to the stocked material. Where particular substocks of fish adapted to the local conditions exist, this may cause problems with some aspects of behaviour such as timing and location of breeding. Escapes of the modified genetic strains from the reservoir or river reach may subsequently dilute the genetic effectiveness of the species over a wide area. In Thailand, where stocking material is drawn from a wide area, and inter-basin transfers of species occur, the risks of such effects are particularly high. For this reason careful selection of broodstock for the production of stocking material is advised. Material to be stocked should be derived only from parents drawn from the receiving basin.

In contrast to the widening of the genetic base beyond that which is adaptive for any stock, there is also a risk from too little diversity. Stocking with material derived from too few breeders can result in a narrow genetic base (low heterozygosity) which will lead to rapid degradation of the material used for stocking of aquaculture ponds and reservoirs, giving poor growth and reproductive potential.

Disequilibrium of fish population

Stocking with one or more target species can produce imbalances in the population, disrupting food chains and threatening the survival of non-target species. This effect is often used deliberately to bias the fishery towards high-value species or to influence the trophic status of the stocked water-body.

Disease

Risks from disease dissemination by stocking are very high and aquaculture has a long history of financial crashes caused by introduction or transfer of diseases along with movements of fish seed. The only way this can be avoided is through greater care and control through regular processes of certification.

Fertilisation

In lakes that are stocked regularly the stocked population may rapidly exceed the natural carrying capacity of the recipient water. Further stocking alone will result in little sustainable increase unless it is accompanied by other measures to increase the productivity of the water. Such measures include the fertilisation of the waters, stocking with balanced groups of species and the modification of the physical form of the environment. Fertilisation is the simplest of these measures and once systematic stocking programmes to raise yield are initiated it appears inevitable that some form of fertilisation will follow. Fertilisation may be through the discharge of nutrient-rich agricultural or urban wastewater into the water body, through the addition of inorganic fertilisers or through the addition of solid organic material of various types. The success of such treatments is demonstrated by increases in yield from 58.5 to 540 kg ha^{-1} in reservoirs of the Wan county, China after the application of inorganic fertilisers (Lu, 1994). Alternatively, 3000–6000 kg ha^{-1} of green manure may be used. Inorganic fertilisers are widely used in Russian reservoirs, where the application rates are determined according to the trophic status of the water body. Fertilisation rates range from 30 kg ha^{-1} of superphosphate and 50 kg ha^{-1} of ammonium nitrate for oligotrophic lakes to 15 kg ha^{-1} and 20 kg ha^{-1}, respectively, in eutrophic ones. Treated lakes show increases of up to five times the number of benthic and planktonic food organisms (Berka, 1990). Schedules for the fertilisation of North American lakes have been proposed which vary according to species, type of lake and climatic region (Moehl & Davies, 1993). Inorganic fertilisers are generally applied at levels of 10 kg ha^{-1} of ammoniated polyphosphate or orthophosphate, 18 kg ha^{-1} of diammonium phosphate or a combination of 18 kg ha^{-1} triple superphosphate and 24 kg ha^{-1} of ammonium nitrate.

Elimination of unwanted species

Much of the production gained by fertilisation is diverted to species other than those stocked. This potential loss of the favoured species can be avoided through the elimination of other elements of the naturally occurring fish assemblages. These may be either competitors, usually small, abundant, fast-growing species which are of no commercial value, or predators which affect the survival rate of the target species. Removal of unwanted fish is common in both recreational and commercial fisheries. It is a keystone of Chinese and Russian management practice but is also attempted in most countries where intensification is reasonably advanced (Berka, 1990; EIFAC, 1991; Cowx, 1994b; Li & Xu, 1995; Sugunan, 1995).

Five main methods are used to control unwanted species (Meronek *et al.*, 1996):

- Chemical applications typically using rotenone or Antimycin
- Physical removal using nets, traps, electric fishing
- Dewatering the reservoir even to the point of complete desiccation
- Stocking with predatory or competing species
- Combinations of the above.

Chemical control is more successful than physical removal, which in turn is more successful than stocking with conflicting species. Combinations of physical and chemical methods may add to the efficiency. A 43% success rate can be anticipated, with complete elimination generally being more successful than partial control of selected species.

It is clear from experiences around the world that 100% control is not easy and ideally the water body should be emptied completely and sealed against subsequent invasions from upstream. However, this is only possible in the smallest of reservoirs and more commonly mechanical and chemical controls are attempted.

Constructing faunas of selected species

Chinese, Russian, Indian and Cuban fisheries managers place great importance on creating a balanced community of species to exploit various trophic levels and spatial niches (Table 13.3).

Table 13.3 Combinations of species used for stocking in extensive and semi-intensive systems

Area	Planktonophage, pelagic	Omnivore–detritivore, benthic	Piscivore, general
China	*Aristichthys nobilis* *Hypophthalmichthys molitrix*	*Ctenopharyngodon idella* *Cyprinus carpio* *Cirrhinus molitorella*	
India	*Catla catla* *Cirrhinus mrigala*	*Labeo rohita* *Labeo calbasu*	
Russia (cold water)	*Coregonus peled*	*Cyprinus carpio* Benthic coregonids	*Perca fluviatilis* *Stizostedion lucioperca*
Russia (warm water)	*Aristichthys nobilis* *Coregonus peled*	*Ctenopharyngodon idella* *Cyprinus carpio*	
Africa	*Oreochromis niloticus*	*Clarias gariepinus*	*Clarias gariepinus*
Latin America (temperate)		*Cyprinus carpio*	*Micropterus salmoides*
Latin America (warm water)	*Oreochromis niloticus* *Oreochromis aureus*	*Cyprinus carpio* *Ctenopharyngodon idella*	

This permits optimisation in the use of the available food resources by the selected species. It also allows gains to be made in trophic efficiency by interactions such as the release of nutrients by bottom-feeding carps, which fertilise the pond for increased phytoplankton production. Balanced mixtures usually include bottom feeders, higher vegetation feeders and pelagic zoo-plankton and phytoplankton feeders. Benthic filter feeders such as freshwater molluscs may also be added, but this is not common at present. Occasionally a low-grade predator may be included. Favoured species are usually Chinese and Indian carps, eels, tilapias and coregonids, with the occasional inclusion of predators such as pike perch or mandarin fish, *Siniperca chautsi*, which have a higher market value.

Introduction of food organisms to supplement the food supply has been mostly confined to the former USSR, where several invertebrate species have been transferred from the east to the west of the country. Berka (1990) claimed impressive increases in the production of fish from reservoirs into which larger, faster growing and more productive mysid and amphipod species were introduced.

Engineering the environment

Physical interventions can be made into stocked lakes and rivers to improve shelter, feeding and breeding grounds. These can include:

- Weed cutting, especially in lakes where high levels of eutrophication encourage the growth of emergent vegetation
- Clearing of accumulated bottom sediments in intensive systems to avoid the build-up of anoxic material
- Grading of bottoms to increase production of benthic organisms and facilitate harvesting
- Installation of spawning gravels and shelter devices
- Deployment of aerators in intensively stocked systems
- Installation of bunds and sluice systems to retain water and to control the inflow and outflow from floodplain pools
- Modification of the water-body through bank straightening, dredging, etc.

A reservoir or lake, or even a river system, may be modified by isolating bays, side-arms or floodplain features by bunds, fences or block nets. The area thus isolated can then be treated as an intensive aquaculture pond, to produce either food fish or stocking material for the main system. Coastal ponds may be similarly isolated by constructing dikes or berms along tidal inlets or around lagoons. Often fish, crustaceans and other organisms are allowed into the ponds during flood, high water, tidal inflow or at other specific times of year by removing or opening part of the dike or berm (Csavas, 1993).

Management as an intensive fishpond

The whole water-body may be treated as a fishpond with full control of all processes. At present this applies mainly to smaller inland water-bodies, for example the Chinese 16 ha ponds which are managed by stocking and fertilisation to attain yields of 5000–6000 kg ha^{-1}. Similar techniques are applied to drain-in ponds such as those of West Africa and Bangladesh. As technologies improve the degree of control over larger reservoirs and lakes may also be increased.

Cage culture

Cage culture is often developed in reservoirs and lakes as an independent process parallel to the enhancement of capture fisheries. In many ways cage culture represents a short cut to aquaculture and as such cage culture associated with capture fisheries is expanding rapidly. It also represents a way for landless people to stake a claim to some area that they can control. The practice originated in China both for the culture of table fish and for the rearing of juvenile fish later used for stocking. Subsequently, other countries have begun using this method as a cheap way to expand aquaculture facilities without making inroads into scarce land resources. For example, Malaysia increased its area of freshwater cages from 2.14 to 4.87 ha and its brackish water cages from 24.29 to 66.82 ha between 1990 and 1993, whereas the area of ponds, although much greater, has increased only slightly (Ferdouse, 1995). Cages deployed in Saguling reservoir in Indonesia for rearing common carp have proved highly cost effective despite stocking densities that, at 1.4 kg m^{-3}, are lower than could be supported by such systems (Rusydi & Lampe, 1990). Cage culture in the lakes of the Pokhara valley, Nepal, has proved similarly attractive as the most direct way to increase the normally low natural production (Swar & Pradhan, 1992). People in Latin America, Chile, Colombia, Brazil and Mexico have begun to use cages to increase the productivity of their reservoirs and lakes.

Installation of cages in a water body will enhance harvests from the wild stock as nutrients and excess food become available to the resident fish. Stocking with selected species will be needed to benefit fully from this, so cage culture may act as a trigger for other forms of enhancement. Eutrophication from cages may eventually exceed the self-purification capacity of the water-body and will then lead to diminished productivity and to the disappearance of species. Strict adherence to guidelines regulating culture densities and water quality should anticipate and avoid such problems.

Genetic modification

Natural water bodies tend to accumulate species and subspecies adapted to the local climatic, abiotic and biotic conditions. Increasing human intervention, including the introduction of exotic species, tends to change the selection pressures of a habitat and may weaken or eliminate local populations. Furthermore, native strains may not be the best adapted to the altered conditions of the water-body under intensive management regimes. New strains of fish are now being developed with desirable characteristics for growth, temperature tolerance and disease resistance for aquaculture.

Genetic modifications are also starting to be used in intensively managed systems. Grass carp *Ctenopharyngodon idella* used for aquatic weed control are rendered sterile by triploidisation (Wynn, 1992), so the chance of them establishing sustaining populations is reduced. Mono-sex populations of tilapia are being produced to reduce unwanted reproduction and the associated reduction in growth rate. Similarly, hybrid crappie (*Pomoxis annularis* Rafinesque × *P. nigromaculatus*) are being used on an experimental basis to avoid their uncontrolled reproduction when stocked as bait-fish in small impoundments (Hooe *et al.*, 1994). Hybrid fishes are also stocked for direct harvest. The saugey, a walleye *Stizostedion vitreum* × sauger *Stizostedion canadense* cross, is very adaptable, displays significant improvement over both of the parent species, is reproductively viable and is being stocked in rivers to support recreational fisheries (White & Schell, 1995). Although specific selection programmes for fish used for inland stocking are not common at present, the genetic structure of many populations in less intensively managed systems is being conserved through hatchery management, reducing stock transfers, and reducing transfers of eggs and fingerlings among hatcheries.

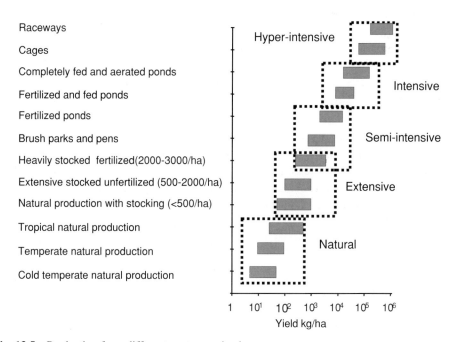

Fig. 13.5 Production from different capture and culture systems.

Table 13.4 Summary of stocking and ancillary enhancement practices at various intensities of input

Type of practice	Stocking density (ha^{-1})	Area	Ancillary practices
Supported natural	< 500	Small to large	Nil
Extensive	500–1000	Small to large	Nil Occasional fertilisation Artificial reefs
Semi-intensive	1000–2000	Small to medium	Some fertilisation Species eradication Artificial fauna Strain selection Hatchery management Habitat modification
Intensive	2000–3000	Small	Intense fertilisation Species eradication Artificial fauna System modification Genetic modification in the short term
Aquaculture	> 3000	Small	Intense fertilisation Intense feeding Aeration System modification Modification of water-body Genetic modification in the short and long term

Summary of enhancement strategies

The increases in productivity associated with the various steps of the intensification process are described in Fig. 13.5 and Table 13.4.

Cost effectiveness

Analysis of the cost effectiveness of enhancements can be carried out at two different levels, that of the individual fishery and that of the society as a whole.

Economics of individual fisheries

Contemporary economic policies maintain that ideally each fishery should be financially self-supporting. In such cases the financial benefit derived from the fishery (B) should equal or exceed the costs of producing the catch (C). Thus:

$$B \geq C$$

In its simplest form, when applied to a stocked fishery B would consist of the price of the fish produced and C would have the following components:

C_1 cost of stocking material, which often amounts to between 40 and 70% of total costs;
C_2 costs of harvesting.

As the fishery is intensified other components are added to C:

C_3 cost of fertilisers;
C_4 cost of removal of unwanted species;
C_5 cost of physically intervening to maintain the environmental quality (draining reservoir, dredging, weed removal, liming, etc.);
C_6 costs of physically modifying the environment (creation of bunds, embankments, creation of spawning and shelter habitats, construction of artificial reefs, etc.);
C_7 costs of genetic manipulation and genetic resource management (selective breeding, hybridisation, polyploidisation, gene transfer or sex manipulation) will increase C_1, cost of stocking material.

In each case, as an additional term is added to C the benefit B should rise to cover the costs involved. This implies that the increasing costs involved in production from enhanced systems can only be met by increases in the productivity of the system with respect to the target species, and by increasing stringency in financial management with the minimisation of support costs to both commercial and recreational fisheries and the elimination of inefficient or unnecessary practices.

In commercial food fisheries the cost–benefit ratios are relatively easy to calculate as the product is readily commercialisable and therefore has a fixed price. Recreational fisheries are less amenable to standard cost–benefit analyses because of the difficulty in assigning an accurate price to the value of a 'recreational fish'. The cost per fish caught in a sport fishery is high relative to the cost of the same fish caught commercially because the recreational fishermen's sport equipment and 'willingness to pay' values, i.e. contingent evaluation, are factored into the price. The cost of maintaining recreational fisheries through stocking and environmental enhancement programmes is also high.

For example, the difference in 'value' between the same fish caught in a sports fishery or a commercial fishery can be substantial. A 2–3 kg barramundi *Lates calcarifer* caught in Queensland, Australia, would yield 1–1.5 kg of fillets worth about A$6–15 kg^{-1} in a commercial fishery, with about half this amount going to the fishermen. The recreational fishermen, however, probably spent about A$150 to catch such a fish (Rutledge *et al.*, 1991; Bartley, 1995).

Economics in the wider context

In considering its cost-effectiveness the fishery should not be treated in isolation. Clearly, at the level of the individual entrepreneur failure to make a profit would prove an obstacle to the continuance of the fishery. However, larger factors have influenced planning at several levels. The component B, while it contains only the price of the fish produced, will lead to strict cost-effectiveness approaches, but in many cases B contains other values, for instance:

B_1 price of fish produced;
B_2 benefit to society of not having to support unemployed fishermen;
B_3 benefit to society of supporting indigenous communities;
B_4 benefit to manufacturing and tourism sectors supporting recreational and commercial fisheries;
B_5 support to the fish farms producing stocking material;
B_6 when enhancement involves endangered species there may be a benefit in that restrictive regulation on fisheries and on water use in order to protect the endangered species may be avoided;
B_7 benefits to society by allowing alternative uses of water, e.g. hydroelectric generation, transportation.

The transition to more intensified systems, therefore, has had impacts on the way in which fisheries are funded. Enhancement of fisheries implies increased costs to the fishery managers in fish seed, fertilisers and labour for species elimination, system modification, etc. One problem here is that the original fishermen are unlikely to be able to mobilise sufficient funds and funding of stocking tends to shift to external financing agencies. The greater predictability, more controlled harvesting season and more concentrated harvest attract investment from businessmen who then take the major part of the profit. This means that the intensification process is frequently accompanied by a loss of independence by the fishermen and a growing dependence on credit or other external forms of funding. As a partial solution to these problems governments may consider subsidies to enhancement either through direct grants or through the creation of state hatcheries, at least in the earlier phases of the adoption of the technology. Governments may also encourage the setting up of rural financing institutions to assist fishermen and rural communities in developing the sector without the intervention of third-party financiers.

In fact, societies may, and often do, subsidise enhancement for the benefit of the larger sector. Such subsidies may take the form of assistance in supply of stocking material, in public works for construction of bunds or restoration of substrates, weed cutting, maintenance of water quality, etc. With the existence of such subsidies the individual entrepreneur can still make a profit and continue the fishery.

Another approach to the socio-economic management of enhanced fisheries is to form co-operatives where the entire sector is vertically integrated. Thus, different members of the co-operative perform habitat improvement, hatchery operation, harvest, processing, distribution, marketing and sales. Losses in one section can be compensated for by large profits in another. Japan is currently employing this strategy in a formal way. Similar approaches to hatchery enhancement of sport fisheries are common. For example, the fishery for white seabass in California is approaching such a system informally by requesting sport-fishing clubs to help with the culture and release of fish hatched in a state-supported hatchery. This in turn receives much of its funding from special licences sold to commercial and sport fishermen (Kent & Drawbridge, 1999).

Note

1. By K. Lorenzen, Renewable Resources Assessment Group, Imperial College of Science, Technology and Medicine, 8 Princes Gardens, London SW7 2QA, UK and R.L. Welcomme.

Chapter 14
Mitigation and rehabilitation

With the high degree of modification of aquatic environments throughout the world there is an increasing trend towards intervening either to improve the functioning of degraded systems or to restore them to conditions that resemble their original condition. Five major strategies for dealing with degraded systems are common:

- *Do nothing*: This is advocated where society places high value on sectors whose processes produce grave and lasting impacts on the living aquatic resources. In such cases costs of any remedial activity are perceived to be so high that it is felt to be a waste of time and resources. This strategy has perhaps been the most frequently used over the past two centuries and has led to the highly modified and polluted situation that many lakes and rivers are in today. Doing nothing is also advocated where the aquatic ecosystem is already functioning adequately within the priorities of society.
- *Mitigation*: Many changes to inland aquatic systems are effectively irreversible and greater value is usually attached to uses of the inland aquatic system other than fisheries. As a result by far the most common tendency today is to attempt to live with modified systems through a series of interventions and management strategies designed to lessen the impact of stress. Actions include the creation of artificial habitat, arranging for flood-simulating water releases or systematic stocking to maintain populations of fish that have no alternative source of recruitment. Mitigation usually involves recurring expenditures, which should be considered part of the cost of the major modifying use of the system.
- *Rehabilitation*: The restoration of modified ecosystems or stocks of fish is possible where the pressures producing the modification have eased or where new technology can be introduced to reduce stresses. Physical rehabilitation is currently being undertaken in Europe, North America and Australia of watercourses modified during the past century for navigation and flood control. There is a concerted attempt to control effluent quality throughout the temperate zones by suppressing point-source pollution and to reverse trends towards eutrophication. Part of this effort has been the reintroduction of vanished species. Rehabilitation usually requires a single major expenditure, although the running costs of treatment plants to improve water quality are not negligible. Because the present state of the river or lake to be rehabilitated is the result of a series of modifications to the landscape, the true objective of restoration is frequently poorly defined. Current land-use patterns or societal

preferences may well determine the result. For example, it would be extremely difficult, and probably undesirable, to return to the fully forested condition prevailing in pre-Roman times in any but the smallest of basins in Europe. In North America, in contrast, a local return to the wilderness state is possible in many areas. Efforts to rehabilitate are frequently thwarted by lack of knowledge of the pre-existing condition or of the requirements of the original fish communities.

- *Protection*: Every effort should be made to preserve inland waters that have not yet been strongly affected by development. Proposals for protection are usually made within the frame of a reserve, site of special scientific interest or other type of protected area. Such an approach needs careful evaluation in view of the opportunity cost to local communities. Where it is determined that there is a net benefit in conserving the resource, protection efforts should seek to discourage physical modifications such as the building of dams, levelling of river channels and the revetting of lacustrine shorelines. Access to the water and fisheries should be limited and levels of exploitation of the fishery, extraction of the water and loading with nutrients and pollutants should be strictly controlled. Should fisheries continue to be permitted within such areas, considerable research in needed to generate sufficient understanding of the limits for exploitation. Greater institutional integration of fisheries management into the more general framework of rural development planning and management is also required. It is recognised that where there is a resource that can be exploited, most developing countries will exercise the option to do so, but damage to the integrity of the system can be limited given the political will. By contrast, many developed economies are exercising the option of conserving water-bodies as an aesthetic value alone.
- *Intensification*: The final strategy implies the distortion of the aquatic resource for fishery purposes. A range of tools is being deployed to this end, including introduction and systematic stocking with desired species, elimination of unwanted species, fertilisation of the water-body and physical modification of the system to facilitate exploitation.

The relationships between these five strategies are shown in Guideline 6.

Objectives of rehabilitation

Rehabilitation does not necessarily involve the restoration of the water-body to its original state but rather the creation of conditions that correspond to societies' needs. The restoration of rivers and lakes to their original condition may or may not be acceptable. For example, nations with large areas of relatively unoccupied land may choose to restore wilderness over part of the territory. Other nations where population pressures are severe would find this process impossible.

Careful definition of the objectives for rehabilitation is necessary and in developing projects the following factors have to be taken into consideration (Guideline 7):

Guideline 6 Choice of strategies for mitigation, rehabilitation or enhancement

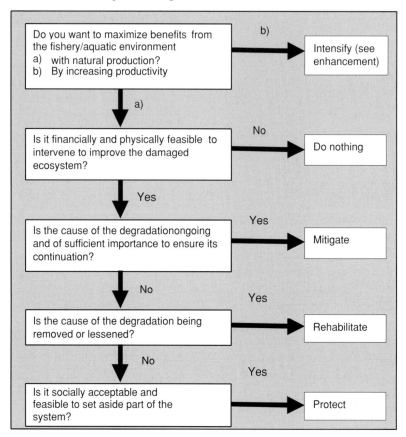

- *The requirements of society*: These may include conservation goals such as the general protection of entire faunas or of a particularly rare species, or use oriented goals such as the maintenance of particular types of fauna for recreation or food fisheries exploitation.
- *The development of the river or lake basin*: No rehabilitation project can be considered in isolation from its basin. Activities upstream of the target water-body that increase sediment loads, change discharge rates and impede migrations can counteract any efforts at a local level. Furthermore, many species are migratory and have feeding and breeding grounds separated by considerable differences. In such cases both feeding and breeding areas will need to be restored and connectivity maintained between them.
- *The minimum area necessary for restoration*: When the area to be restored is limiting to the target species or community rehabilitation will increase their number and well-being. However, as greater areas are restored other factors may intervene to limit the fish population. For example, in a lake, rehabilitation of

Mitigation and rehabilitation

Guideline 7 Scheme for planning rehabilitation programmes

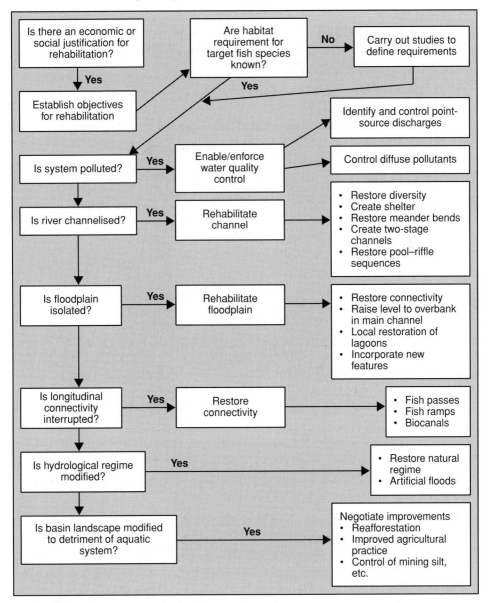

spawning gravels may well increase populations until the availability of food stops further population growth. Similarly, in rivers, floodplains are necessary to many species but the carrying capacity of the environment at low water may be such that only a small proportion of fish spawned can survive to the next year. In such cases only a small amount of the plain needs to be restored for an adequate fish stock to be maintained.

Habitat requirements of fish

The rehabilitation of systems should aim at creating conditions that favour communities of species defined by the objectives. Different types of fish community have differing ecosystem requirements, including:

- *Chemical*: many fish species are sensitive to salinity, pH, nutrient levels or conditions of low dissolved oxygen.
- *Flow*: distinct communities are found in flowing and still waters. Many species also require specific flows as triggers for breeding or to flood the environments they need for breeding and feeding.
- *Structural*: factors such as shelter, appropriate feeding areas and adequate breeding substrate are essential for the survival of individual species.

Ecosystem requirements can be established by simple observation, but more formalised summaries of the conditions required can be generated through sampling procedures such as PHABSIM which establish Habitat Suitability Indexes.

Protection of water quality

Almost all attempts to rehabilitate aquatic environments will fail unless the water is free from toxic pollutants and quality is sufficiently good to allow the fish and other aquatic animals and plants to survive and prosper. For this reason many countries have given high priority to rectifying the poor state of inland waters and have some form of legislation to control pollution. Such legislation often incorporates the 'polluter-pays' principle, which assumes that users of the water and the basin should minimise any deleterious effects, contribute to the mitigation of any impacts of their activities and bear the cost of rehabilitation of the systems when the need for their activity has ceased. The Organization for Economic Co-operation and Development (OECD) definition of the polluter-pays principle is as follows:

(1) The Polluter-Pays Principle constitutes for [OECD] Member countries a fundamental principle of allocating costs of pollution prevention and control measures introduced by the public authorities in Member countries;
(2) The Polluter-Pays Principle means that the Polluter should bear the expenses of carrying out the measures to ensure that the environment is in an acceptable state. In other words, the cost of these measures should be reflected in the cost of goods and services that cause pollution in production and/or consumption.

Uniform application of this principle through the adoption of a common basis for national environmental policies would encourage the rational use and the better allocation of scarce environmental resources and prevent the appearance of distortions in international trade and investment.

In principle, the contribution by those damaging the resource should be total, but this is rarely realisable. Government contributions to mitigation should be seen as a subsidy to the industry. Another generally undesirable form of subsidy is to allow the environment to bear the cost of the abuse, either permanently or temporarily. In this case the cost of the degraded state of the environment will be reflected in the lessened value of the aquatic system for society and a delayed bill for the rehabilitation of the system in the future.

Control and use of vegetation

Vegetation is one of the major components of aquatic ecosystems. It is important to the structure of the environment. It also provides support and refuge for a wide range of organisms, including fish. In normal systems vegetation is maintained in balance with the other components and changes slowly over time through a succession of plant species. Most aquatic ecosystems, river islands, floodplain water-bodies, lakes and reservoirs, are subjected to a natural ageing process which leads to their eventual conversion to dry land through the combined effects of vegetation and siltation. In systems that are influenced by humans or subjected to the introduction of exotic plant species, the balance is no longer controlled and a nuisance occurs through the uncontrolled proliferation of some native element of the community or by the invader.

Vegetation operates at a number of levels.

Vegetation within the drainage basin

Vegetation within the drainage basin is the main regulator of water flows and stabiliser of soils. Forested hill slopes within the drainage retain water falling as rain and release it slowly into river channels. Root masses bind soil and prevent it from being washed into the aquatic system. Removal of vegetation increases the flashiness of the hydrograph and the transfer of silt to the water. Managers of aquatic systems, therefore, should attempt to negotiate with those responsible for forestry and agriculture for the maintenance of adequate vegetation cover within the drainage.

Riparian vegetation

Vegetation growing along the riverbank plays an important role in the aquatic ecosystem by providing shade, cover and structural diversity to the river or lake. This effect is particularly pronounced in small water-bodies where the riparian zone forms a significant part of the whole. Vegetation also stabilises bank structures, contributes to the rain of terrestrial insects that form the basis of food chains in low-order rivers, and retains a large proportion of terrestrial nutrients. Management of riparian vegetation is an important part of mitigation and rehabilitation strategies. Solutions using planting of vegetation to stabilise and shape channels are usually to be preferred to purely mechanical approaches.

Aquatic vegetation

Emergent

Emergent vegetation is rooted in the banks and shallow waters of the system. It can form simple stands along the land–water interface, or floating mats that extend outwards from the shore. It can also fill river channels and shallow lakes to convert them into a swamp. The choking of channels and lakes with native species can follow from eutrophication and reduction of flows. Introduced species, such as alligator weed (*Alternathera philoxeroides*), which have no natural controls tend to become invasive and can cause considerable physical nuisance by preventing navigation near the shoreline. They can also catalyse ecosystem changes by shading and deoxygenation under the vegetation mat. Propagation of such species from water-body to water-body is usually by plant fragments attached to pleasure craft, and strict control is needed to prevent the rapid expansion of infestations.

Submersed and submerged

Submersed vegetation grows on the bottom, which it may cover to form dense meadows. Certain species, such as the water lilies, may also have emergent shoots and floating leaves. Submerged vegetation consists of terrestrial plants which have been flooded and which may or may not survive. Many floodplain and marginal lake grasses have a terrestrial and an aquatic habit, enabling them to withstand long periods of flooding. Excessive expansion of submersed species such *Hydrilla verticullata* can choke the infected water-body, cause rapid fluctuations in temperature, pH and dissolved oxygen, cover open-water feeding and breeding grounds, and reduce plant and animal diversity. Both submerged and submersed vegetation form a substratum for phytophilous fish spawners and a support for food organisms.

Floating

Floating vegetation occupies the surface of the water and is largely independent of the shoreline. It has certain limits to its distribution through wind and wave action. Floating organisms such as the fern *Salvinia* or the higher plants *Pistia stratiotes* and *Eichhornia crassipes* have long histories of causing environmental problems. They are among the fastest growing plants and can proliferate to cover the surface of any water-body. They cause severe problems of shading and deoxygenation beneath the mat, and can physically prevent navigation and fishing.

Control of riparian and aquatic vegetation

Such are the problems caused by invasive plants that a considerable amount of effort has been devoted to their control and elimination. The principal methods are as follows.

Mechanical

Mechanical control consists of cutting, dredging or removal with grabs. It is rarely effective over large areas because of the rapid growth of invasive plants, the expense and effort needed, and the fact that mechanical removal fragments the plants further and most are able to grow vegetatively from the parts that are left. Log booms and barriers may be use to keep the weed out of certain areas. Mechanical methods are used mostly to keep clear river channels and access to boat landings and to water intakes.

Chemical

Spraying with herbicides is commonly used to clear large areas of weed. Chemicals such as 2,4-D, Diquat and Endothal have been used with success. However, large-scale dosages with herbicides can poison fish and the organisms on which they feed. Furthermore, the decaying plants fall to the bottom and create anoxic conditions.

Biological

More recent attempts at vegetation control have concentrated on biological methods. These usually involve the introduction of an animal or plant that acts as a control on the invasive species in its natural range. Some fungi and a range of insects have proved successful in some circumstances (see Table 14.1).

Some fish, particularly grass carp *Ctenopharyngodon idella*, have also been used with success to control soft plants such as *Hydrilla*. In such cases sterile triploid individuals are often used to avoid the risk of the species breeding in its site of activity, although grass carp only breed in certain highly specific riverine environments.

Rehabilitation of lakes and reservoirs

Measures for rehabilitation and mitigation of adverse effects in lakes tend to emphasise water quality and to a lesser extent structure. Lakes world-wide suffer from eutrophication, which brings a series of associated problems principally from low oxygen. In high latitudes lakes tend to suffer from acidification. Physical modifications of lakes are less severe, although urban development or inappropriate water management can damage shorelines. Siltation can also present a problem, especially in areas where there is deforestation of marginal lands for agriculture within the basin.

Table 14.1 Examples of insects used for the biological control of invasive vegetation

Plant	Control agent
Alternathera philoxeroides (Alligator weed)	*Agasicles hygrophila* (Alligator weed beetle)
	Amynothrips andersoni (Alligator weed thrip)
	Vogtia mallori (Alligator weed stem borer moth)
Eichhornia crassipes (Water hyacinth)	*Neochetina eichhorniae* (Weevil)
	Neochetina bruchi (Weevil)
	Saneodes albiguttualtis (Moth)
Hydrilla verticulata (Hydrilla)	*Hydrellia pakistanae* (Leaf-boring fly)
	Hydrellia balcuinasi (Leaf-boring fly)
	Paraoynx diminutalis (Moth)
Pistia stratiotes (Water lettuce)	*Neohydronomous affinis* (Weevil)
	Spodoptera pectinicornis (Moth)

Control of nutrient balance

Eutrophy

Most lacustrine eutrophication originates from the discharge of nutrients into the lake as point-source sewage discharges or by diffuse runoff from agricultural and human activities in the basin. It can also arise through imbalances in the plant or animal communities following fisheries overexploitation. The objective of eutrophication control is not to eliminate all nutrients from the water, since this can be as negative in its effects as some nutrients are always necessary to support life. Rather, control should concentrate first on the reduction of nutrients to a tolerable level, followed by constant fine-tuning to maintain levels at those optimal for the target fish communities.

External controls

Most successful nutrient control programmes depend on removing the inputs. In the case of point sources this is relatively easy where industries, animal-rearing facilities and urban sewage works can be required to adopt such treatment procedures as are necessary to reduce emissions to acceptable levels. In small lakes discharge into the water body can be prohibited entirely. In the case of diffuse sources controls must be sought on fertiliser use in the basin through agreement with farmers. Other sources of

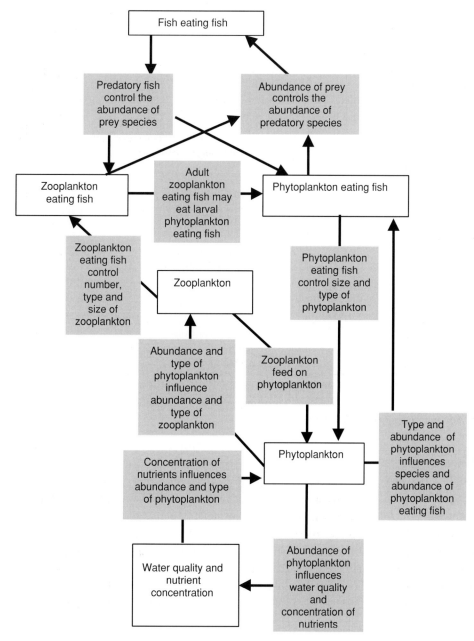

Fig. 14.1 Relationships between different trophic groups controlling the flow of nutrients through the food web.

ground-water contamination such as septic tanks can be strictly controlled or prohibited. In many cases a riparian fringe of trees will remove most of the nutrients from surface and near-surface flow.

Internal controls

Water quality can also be controlled by manipulation of the food webs within any lake or reservoir (Fig. 14.1) (see, for example, Shapiro, 1980; Carpenter *et al.*, 1987). In these instances the ultimate goal of food-web manipulation is to reduce the biomass of phytoplankton. Excessive phytoplankton shades water, preventing the growth of rooted aquatics, increases oxygen demand through its death and decay, and may be toxic if the wrong species are present. Changes in the type of phytoplankton due to eutrophication may also be harmful when blue–greens replace green algae and diatoms, as the former are less suitable as fish food. Such manipulation is based on two principles:

- Pelagic species intervene in planktonic food webs as follows. Species that eat phytoplankton directly alter the abundance of phytoplankton. Therefore increases in numbers of these species will decrease the quantity of phytoplankton. Species that eat zooplankton influence the abundance and size of the zooplankters. Many of these species are selective for the size of zooplankters that they eat. Furthermore, the zooplanktonic organisms themselves prey on different types of phytoplanton. Thus changes in the numbers of zooplankton feeders will indirectly change the abundance and composition of the phytoplankton. Piscivorous species prey on plankton feeders and influence the abundance of both phytoplankton and zooplankton feeders.
- Benthic feeders recycle bottom material by releasing nutrients trapped in the sediment through physical disturbance of the bottom, and through their own excretion. Dense populations of species such as carp can also so muddy the water that bottom macrophytes cannot develop.

The proportional representation of species in a fish community may be disturbed through overfishing or eutrophication in such a manner that predators are unable to control the increase in numbers of lower trophic levels such as cyprinids. Mass removal of such lower trophic-level species has been used in lakes to balance the predator–prey relationship. Selective removal of planktivorous and benthivorous fishes, both by humans and by piscivorous fishes, affects the species composition and biomass of the target fish and, through them, the species composition and biomass of phytoplankton and zooplankton. However, species such as *Aristichthys nobilis* or *Hypophthalmichthys molitrix* can be introduced to eat the unwanted phytoplankton.

In small lakes the deoxygenating effects of eutrophication can be combated by aeration.

Oligotrophy

Oligotrophic lakes and black- and clear-water rivers may contain so few nutrients as to be incapable of supporting any significant biomass of fish. Even where biomass has built up over time, productivity levels are usually so low that recovery from removal of fish may be extremely slow. In cases where improved yields are required the water may be fertilised by the judicious addition of organic or inorganic fertilisers. In planning such operations it should be remembered that the buffer capacity of oligotrophic waters is low and that addition of extraneous chemicals can alter the pH.

Acidification

Acid wastes washed out of the air by rainfall have lowered the pH of many northern lakes in Europe and North America to a point where they do not support aquatic life. The only solutions to this problem are (1) to remove the source of the acid waste and (2) the systematic addition of lime to the lakes to absorb the excess acidity. As most discharges arise from the industrialised lifestyle of the European and North American countries direct reduction is difficult, although measures to improve the quality of emissions from industry and power generation have been adopted. Liming therefore remains the major mitigatory solution. Several Nordic countries and Canada have long-established programmes and have developed special equipment for liming. These programmes have been questioned on the basis of cost.

Physical modification

Shoreline development

The major physical changes to lakes and reservoirs occur at the shoreline. Here urbanisation, road building and wave action from boats can remove many of the original riparian features. These include loss of the sand and gravel beaches that provide spawning substrates and nurseries for many fish species. Loss of riparian trees and emergent vegetation is also common. These provide spawning locations and shelter for fish and other aquatic organisms. In reservoirs a normal riparian zone is lacking as a result of the rapidity of flooding and the lack of time for the aquatic system to develop shoreline features.

Planting of trees and establishing emergent vegetation can rehabilitate denuded shores. Where shorelines have been engineered to destroy the diversity of depths this may require re-grading parts of the shore to create different depth zones in which the vegetation can grow. Where wave action from passing boats is severe, protection by floating booms or artificial offshore islands may be necessary.

Where gravel and sand spawning substrates have been removed or choked with silt, or have disappeared through wave action, import of suitable material may be necessary. This again may need protection with special booms or islands.

Siltation

Lakes and reservoirs are exposed to siltation brought in by inflowing rivers. These often create a depositional delta at their mouths, which becomes colonised with vegetation and converted to marsh or to dry land. In itself this process may be desirable in that it adds diversity and such marginal areas are among the most productive for fish and wildlife in the whole system. Successions of this type are also a natural part of the progress of lakes to dry land. However, when there is excessive siltation from human activities upstream encroachment on the main body of the lake is accelerated and its life may be considerably shortened.

The main check on siltation is to control the pattern of land use within the watershed. Deforestation should be avoided, more conservationist ways of ploughing adopted and any mines in the basin should be equipped with settling ponds.

Mechanical means such as dredging may be adopted in small lakes, and in reservoirs bottom take-off of water encourages the passage of silt. However, in larger water-bodies the volume of silt involved is usually too great for such measures to be effective.

Water-level control

A major problem with reservoirs is the rapidity with which water levels change. These changes are due to abstraction of water for the major purpose of the reservoir, which may be urban supply, irrigation or power generation. Rises in level may also be rapid when floodwaters enter the water-body at a time when initial levels are low and abstraction is minimal. Sudden changes in level of this type are damaging to fish and to their food organisms. Attempts should be made to contain water level changes to rates that are more compatible with the biological needs of aquatic organisms. Unfortunately, because the main function of the reservoir is other than fisheries it is difficult to persuade those responsible to change patterns that are strongly economically detrimental.

Spatial usage within the lake

In many parts of the world, especially in areas of high human population density near to urban centres, lakes constitute a valuable resource for recreation and conservation. As such, demand for the various activities can include angling, boating, swimming, water-skiing and relaxation, as well as conservation for fish, wildfowl and other forms of wildlife. In cases where usage patterns within a lake are numerous and intensive it is generally necessary to create zones for the different uses. Zoning may be permanent as in the case of most wildlife and fishery interests, or seasonal according to the recreational demand placed on the water.

Rehabilitation of rivers

Measures for the rehabilitation of rivers tend to emphasise creating structural diversity and connectivity and to re-establish longitudinal and lateral connectivity. Rivers world-wide suffer from damming and channelisation which have transformed the channels into monotonous embanked reaches isolated by dams and weirs. They have also been separated from their floodplain through drainage and land-reclamation programmes. This section summarises techniques that have been described in more detail in Cowx & Welcomme (1998).

Rehabilitation of channels

The main aim for the rehabilitation of channels is to reintroduce diversity of depth, flow, substrate and riparian structure. Efforts for creating local diversity should con-

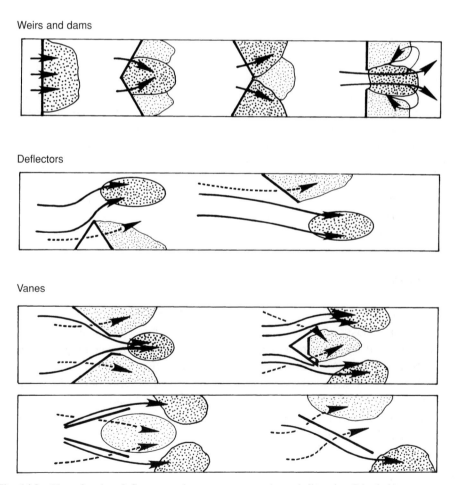

Fig. 14.2 Use of weirs, deflectors and vanes to create channel diversity. Stippled areas = gravel bars; darker areas = pools; solid arrows = main flow; broken arrows = overtopping flows. (Adapted from Cowx & Welcomme, 1998.)

centrate on using locally available materials, which are relatively cheap and harmonise well with the surroundings, rather than on heavy engineering solutions, which are expensive and frequently aesthetically jarring.

Small-scale interventions

Small-scale interventions at a local scale within the existing channel are relatively easy to plan as they have little impact on land and water rights. Major interventions have a much greater impact as they usually involve a widening of the river area and thereby encroach on riparian lands. This means that any large-scale project will involve the active co-operation of riparian landowners or the acquisition of the land by the state.

Activities for restoring channel diversity include the following.

Weirs, deflectors and vanes
Speed and direction of current can be improved through the installation of various types of low weirs and dams, deflectors and vanes. The local changes in current velocity thus produced will create areas of erosion (pools) and deposition (bars and islands) which diversify the types of habitat available (Fig. 14.2). They can also introduce sinuosity into the channel and will cause bottom materials to sort by size according to current velocity. Similar effects can be achieved using obstructions made of wood debris, gabions or stone blocks.

Reintroduction of cover
Cover can be created under riparian vegetation, in half-logs suspended in the water, in pipes let into the bank or placed in mid-channel, on in gabions so placed as to form an artificial undercut bank (Fig. 14.3).

Creation of shallows
Shallows can be created along the channel to encourage growth of emergent vegetation. This serves as a bank stabiliser, provides shelter for fish and serves as a support for food organisms.

Restoration of pool–riffle sequences
Pool–riffle structure can be restored within the channel by placement of submersible dams and weirs made from logs, gabions or boulders. In some cases, for example where severe siltation has buried existing bottoms, gravel will have to be added to the river (Fig. 14.4).

Creation of shallow bays
Shallow bays excavated into the bank can add riparian diversity by creating shallow protected areas. These can serve as spawning sites and nurseries for young fish as well as providing shelter from the current (Fig. 14.5).

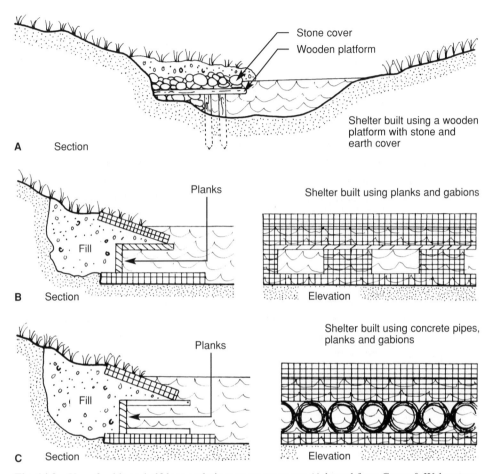

Fig. 14.3 Use of gabions, half-logs and pipes to create cover. (Adapted from Cowx & Welcomme, 1998.)

Larger scale interventions

Other projects may be more large scale, involving many kilometres of river and requiring heavy engineering.

Creation of multi-stage channels

Multi-stage channels should be created to absorb the different flow rates of the river during different seasons (Fig. 14.6). Multi-stage channels usually require the enlargement of the river so that the new submergible stage is large enough to perform its hydraulic function. The space between the main channel and the widened channel can be used to accommodate increased sinuosity and floodplain lake features such as main channel bays and off-channel ponds. In such cases the distance between the new levee walls should be at least seven to ten channel widths.

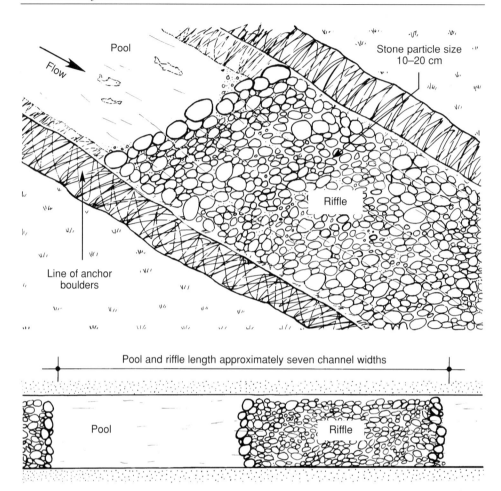

Fig. 14.4 Restoration of pool–riffle sequences in river channels. (Adapted from Cowx & Welcomme, 1998.)

Re-meandering
Re-meandering or the restoration of the former sinuosity of a river channel is one of the best solutions to large-scale reconstruction of channel diversity as it restores all natural habitat types (pool–riffle sequences, point bars, different sediment types, etc.). This can be carried out by artificially digging a new channel, by inducing changes to the channel using deflectors or vegetation, or by reconnecting the old one (Fig. 14.7).

Installation of berms and islands
Shallow water berms and islands (Fig. 14.8) can be used to add diversity to essentially linear structures such as canals. The protected shallow waters are used to establish rooted and emergent aquatic plants which increase bank diversity and provide shelter and food substrates.

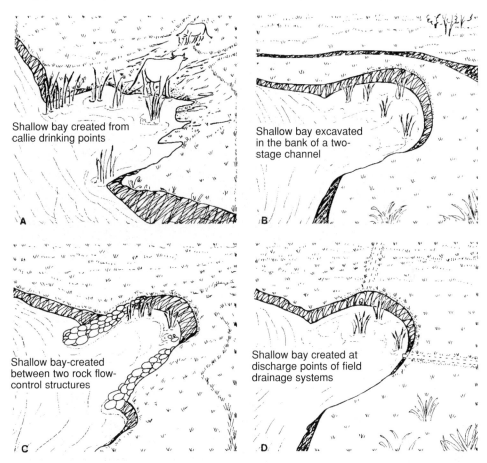

Fig. 14.5 Placement of shallow bays to add riparian diversity. (Adapted from Cowx & Welcomme, 1998.)

Rehabilitation of floodplains

The main aims of floodplain restoration are to reconnect the floodplain to the river and, once connected, to restore original floodplain features. Floodplain restoration projects usually involve the conversion of drained and protected riparian land back to wetland. This implies extensive land purchase. Because of the need to purchase land as well as the scale of the engineering solutions floodplain restoration is very costly. However, it is increasingly apparent that the creation of up-river expansion sites, where waters can be stored during flood peaks for slower release later, can contribute to flood relief downstream. Floodplain restoration projects therefore do not only involve fisheries interests. It is also apparent from the population dynamics of floodplain species that there is an overproduction of young fish in any year in that many of them will be unable to survive the rigours of the dry season. In such cases only portions of the river need to be rehabilitated to play an adequate role in the maintenance

Fig. 14.6 Construction of multi-stage channels. (Adapted from Cowx & Welcomme, 1998.)

of a fish fauna. This approach, which calls for small rehabilitation areas strung out along the channel, can be likened to a string of beads (Fig. 14.9).

Setting back levees

Lowland rivers with ancient floodplains are normally contained by massive levee systems. Humans have heavily occupied such protected floodplains and there is considerable social investment in agriculture, industry and urban development. Depositional processes often involve the channel being raised some distance above the surrounding plains, with attendant risks of flooding and channel jumping. It is, therefore, rarely possible simply to remove the levees and let the river recover its former plain. Rather, a new line of defence from flooding has to be created to replace the old, restrictive levees. The area enclosed by the set-back levees can be used to create two-stage channels with most of the features of the original floodplain (Fig. 14.10). Care should be taken in balancing the needs for flow in such enclosures so as to avoid excessive current, which would scour most of the features and undermine their sheltering function.

Reconnection of relic channels and floodplain water-bodies

Consideration should be given to reconnecting old river and floodplain features such as channels separated from the modified main river or former oxbow and scour lakes. Reconnection can be made through a single channel, usually at the downstream end, or through two connections, one upstream and one downstream (Fig. 14.11). These two approaches can produce very different conditions in the connected water-body. A single downstream connection will produce lentic conditions in the restored water-

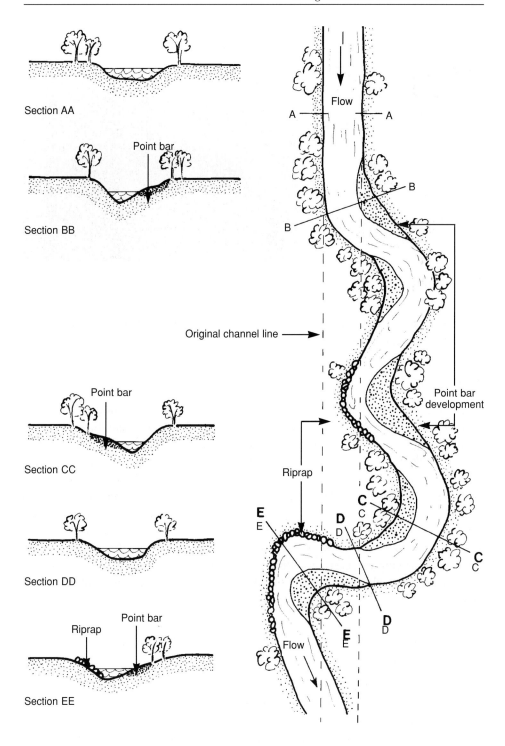

Fig. 14.7 Re-meandering of channels. (Adapted from Cowx & Welcomme, 1998.)

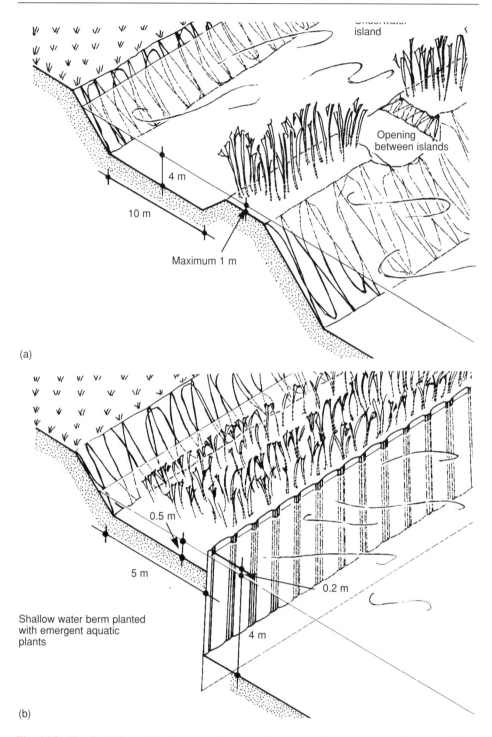

Fig. 14.8 Configuration of shallow-water berms and islands. (a) berm constructed using artificial islands; (b) berm constructed with sheet piling. (Adapted from Cowx & Welcomme, 1998.)

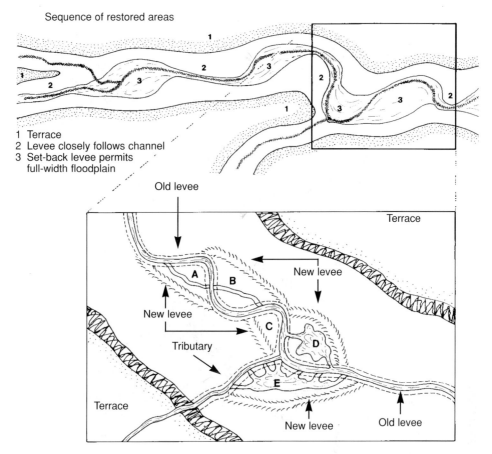

Fig. 14.9 The 'string of beads' approach to large river restoration. (Adapted from Cowx & Welcomme, 1998.)

body with black-fish communities, whereas a double connection will restore flow through it and favour mixed grey-fish and white-fish species. The choice of which approach to adopt will depend on the type of fish fauna desired.

Creation of new floodplain features

Where the original features of the plain have disappeared it may be possible to use artificial features such a gravel extraction sites or borrow pits as substitutes (Fig. 14.12). This means that they should be connected to the river through one or more channels. They can also be engineered to introduce diversity, and provision for such work may be included in the terms of any licence to remove material from the floodplain.

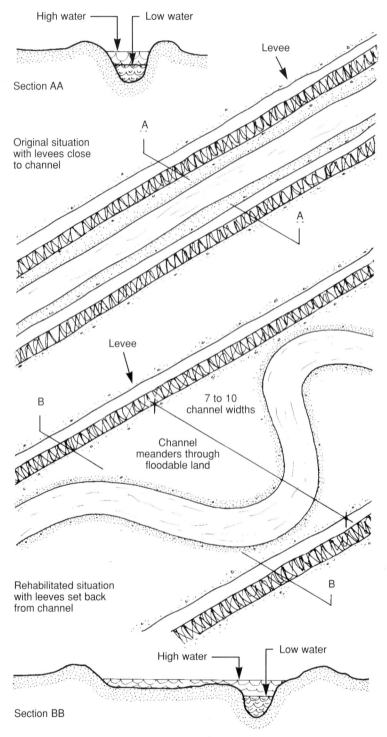

Fig. 14.10 Setting back of levees. (Adapted from Cowx & Welcomme, 1998.)

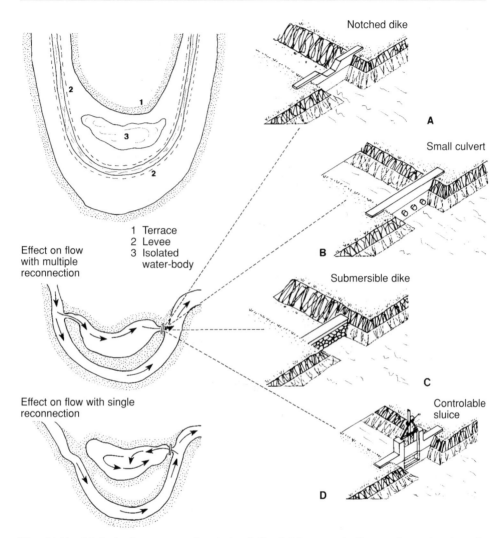

Fig. 14.11 Methods for reconnecting isolated floodplain water-bodies to the main channel. (Adapted from Cowx & Welcomme, 1998.)

Submersible dams

In many cases the bottom of the river channel has become lowered through either dredging or natural erosion. Where this is the case natural overbank flows do not occur and the river is unable to flood its former plain. To overcome this it may prove necessary to install submersible dams within the main channel so as to deflect waters out onto the floodplain (Fig. 14.13). Submersible dams have also been used on some floodplains to slow the evacuation of water.

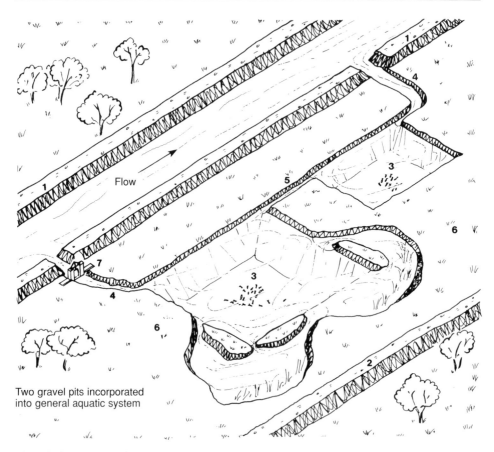

Fig. 14.12 Creation of new floodplain features from gravel and borrow pits: 1, levee; 2, realigned levee; 3, gravel pit; 4, connection to river; 5, connection between pits; 6, potentially floodable land; 7, sluice. (Adapted from Cowx & Welcomme, 1998.)

Protection of fish movement

One of the main problems in rivers is the construction of dams to block the passage of migratory fish. The dam need not be particularly high and even low weirs can impede movements in most species. In some cases it is not possible to provide passage around the blockage by dedicated structures. In such cases fish may be trapped below the dam and trucked to a release site above the dam. Normally, however, some provision is made in the form of a permanent fish pass.

Fish passes

Pool-type fish passes of various designs were an early solution to this problem (Fig. 14.4). These were developed primarily for energetic species such as salmonids in the

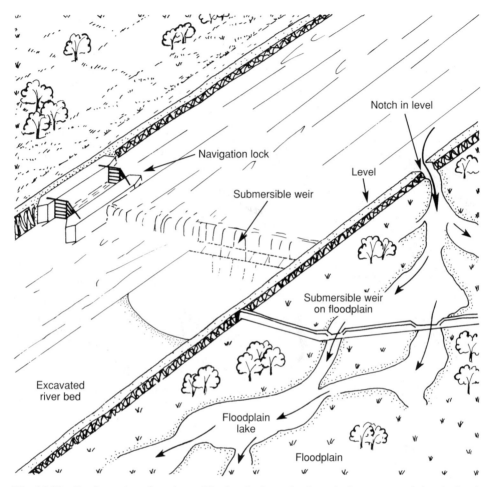

Fig. 14.13 Configuration of a submersible dam in the main channel of an excavated river bed and a submersible dam on the floodplain to conserve floodwater.

northern temperate rivers. Pool fish passes have also been used more or less successfully with other species showing the same type of leaping and sheltering behaviour, but have proved almost totally ineffective with other types of species. As a result there is a continuous search for different kinds of structure that are more compatible with the less energetic migrants.

Fish ramps

Fish ramps (Fig. 14.5) have been developed as an alternative to allow passage of more sedentary species over low obstacles. Fish ramps also have the advantage that they can be made with locally available rocks and can be landscaped into the river to give a more natural appearance. Furthermore, larger rocks set into the ramp can create con-

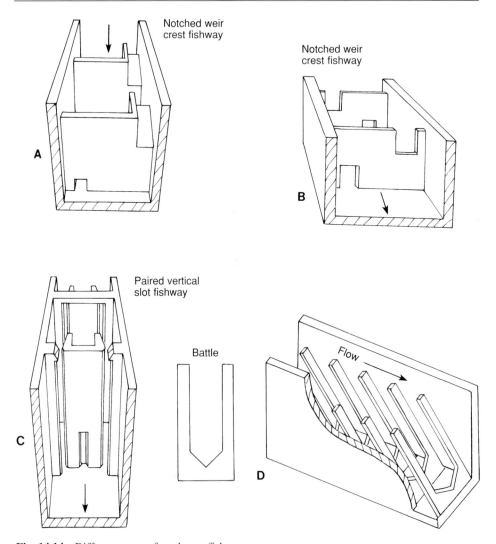

Fig. 14.14 Different types of pool-type fish pass.

ditions analogous to the vanes of a pool-type fishway. These, together with the more grainy texture of the ramp, facilitate the migration of many species that would not negotiate a more traditional structure.

Bypass canals

Dams can be bypassed by biocanals (Fig. 14.6), which are effectively, long, sinuous, artificial rivers that originate above the dam and discharge near the tail waters. These are perhaps the most effective fish pass mechanisms as they also increase the habitat available. They are, however, costly to construct and need a large amount of space.

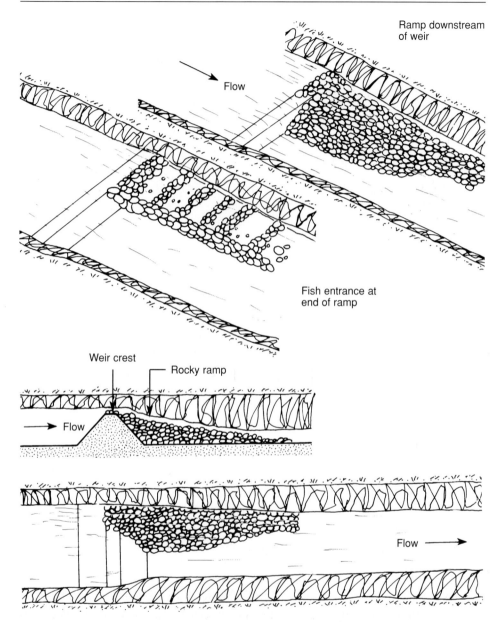

Fig. 14.15 Different types of fish ramp. (Adapted from Cowx & Welcomme, 1998.)

Fish passes of all types are relatively inexpensive when they are incorporated into the original design of a project. They become much more expensive where attempts are made to install them retroactively. Fisheries managers should, therefore, endeavour to ensure that appropriate types of fish pass are planned from the earliest stages of any dam construction project.

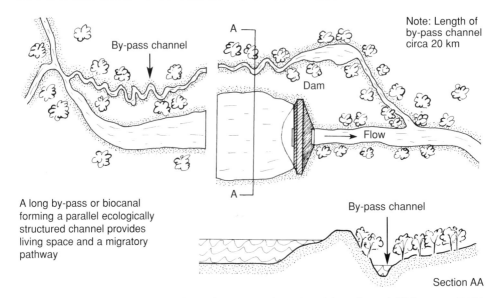

Fig. 14.16 Diagram of a biocanal bypassing a reservoir. (Adapted from Cowx & Welcomme, 1998.)

The efficiency of any type of fishway depends on its capacity to attract fish to its lower end so that they can begin ascent. This depends on the placement of the mouth of the fishway relative to other attractive discharges. Clay (1995) discusses fishway design and placement.

Sluice gates

Some lowland floodplain and wetland systems are poldered and water flow is through canals and rivers where it is controlled by systems of sluices. These sluices inhibit movements of fish. They are often opened at times that do not correspond to periods of migration and may be closed when fish are moving. Sluices of this type are usually operated in the interests of irrigation and flood agriculture, such as rice in Bangladesh. Fisheries managers should negotiate with those controlling the sluices to obtain opening times that correspond more closely to the needs of the fish. Needs may include flood phase, time of day and phase of the moon, and should be studied for the species of greatest interest to fisheries in any region. For example, overshot sluices may favour the passage of pelagic fish eggs into floodplain water-bodies, whereas undershot sluices would favour the passage of bottom-living juvenile and adult fishes.

Removal of dams

There is a growing tendency in countries with a long history of dam installation to remove structures once they have become uneconomic or have reached the end of their useful life. For example, in the USA dam life is reckoned to be about 50 years, after which natural siltation has so reduced the holding capacity of the reservoir as to reduce its effectiveness. Some 25% of dams are currently of greater age. Dam removal is carried out to restore connectivity for upstream migrant species and to restore the natural hydrograph so that former floodplains can be reflooded. Projects for the removal of dams are costly and must be preceded by cost-effectiveness studies to evaluate the gains relative to the loss of services and cost of removal. Several techniques exist for the removal of the dam wall. Portions may be sawn out of concrete structures, earth-filled portions of the wall may be removed by excavation, or the dam may be breached by explosives or mechanically. The removal of a dam is accompanied by activities to restore the original migratory species where these have been permanently damaged by the dam and to restore the vegetation in the former reservoir. The accumulated silt in the reservoir presents a temporary problem. Floods evacuate fine material downstream relatively rapidly, although silt may be deposited in downstream areas causing temporary environmental disturbance and pollution. Coarser material may be evacuated slowly over a number of years or may remain in the former reservoir area. Removal of dams may cause other difficulties, which include:

- remobilising toxic material such as heavy metals and polychlorinated biphenyls (PCBs);
- opening access to invasive, introduced organisms from downstream;
- loss of control of the hydrograph;
- loss of services such as flood control provided by the dam, although these may have become marginal;
- increased risks of flooding of inhabited and agricultural land downstream.

Water regime management

The impoundment of water behind dams and abstractions to divert water to another system disturb the normal hydrological regimes of rivers, fluctuating lakes and reservoirs by causing peak flows to occur at times of the year that are physiologically inappropriate for the fish. They also can cause unseasonable drying out of the river channel, insufficient flows through spawning gravels and failure to flood lateral wetlands. To overcome this, those controlling discharge should be consulted to ensure that there are minimum instream flows in smaller streams and overbank flows large enough to flood the floodplain. This means that water and power-generating companies should be encouraged to reserve part of their water for the conservation of healthy aquatic life downstream of their installation.

It is generally not sufficient simply to restore flows in a haphazard manner. Most river fish species require a particular flow regime to complete their life cycle in the most efficient way possible and there are, therefore, floods of different qualities relative to this reference point. The principal characteristics of a flood regime are illustrated in Fig. 14.17.

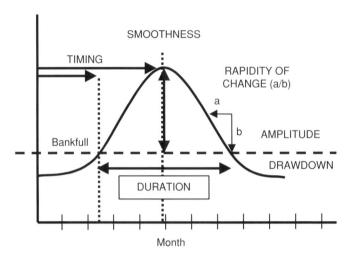

Fig. 14.17 Characteristics of a flood curve.

The characteristics of a flood regime may be described by a series of hydrological indices (HIs) representing the overall intensity of the flood. Various measures have been used to indicate the relative intensity of a flood in any one year. These are often derived from a measure of discharge or water height such as rainfall, discharge rates or measurements at a gauging station. Individual measurements can be combined to form the index which, for example, may represent the area under the flood curve above bankfull in the flood and the area below the bankfull level during the dry season. An index of relative intensity says little on its own because two floods having similar indices may well differ in the quality of the flood. The principal considerations determining quality, which should also be considered in re-creating flow regimes for biological purposes, are:

- *Timing*: Most river fish time their reproduction to coincide with the maximum supply of food and shelter for their young during the flood. Spawning usually occurs at bankfull in grey and black fishes. Migratory white fishes have to anticipate the flood to migrate and arrive at upstream spawning sites on the rising flood. In either case fish depend on a series of signals to prepare physiologically for the event so as to be in spawning condition in time.
- *Amplitude*: The amplitude of the flood determines the area of floodplain covered by water. This is important for the release of nutrients by the advancing aquatic/terrestrial ecotone. In general, the greater the amount inundated the greater the release of nutrients and the deposition of silt for the terrestrial phase. However, in some systems such as the Nile Sudd advancing river floods may encounter sterile, rain-flooded areas at the floodplain margin. In such cases there is no advantage to increasing flooding over a greater area.

Mitigation and rehabilitation 293

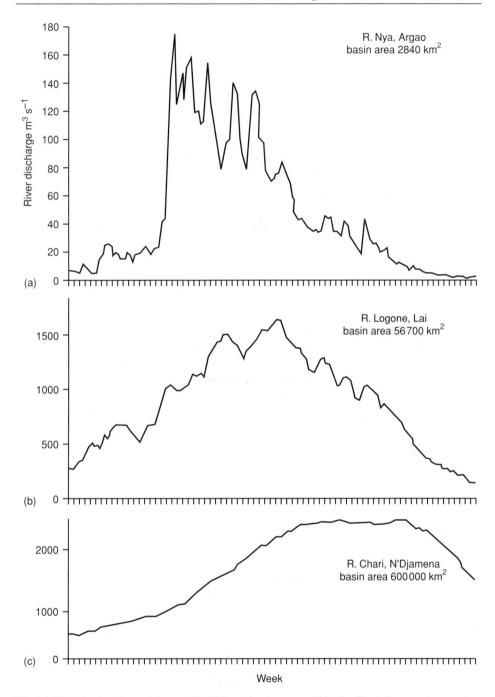

Fig. 14.18 Flood regime of rivers with different basin areas within the Chari–Logone river system (Africa) showing increasing smoothing with increasing area of the drainage basin.

- *Duration*: The duration of the flood determines the exposure of young fish to the food-rich floodplain environment. In general, long floods should result in better growth and survival than short ones.
- *Rapidity of change*: The speed with which floodwaters rise and fall on the plain [represented by the slope of the ascending and descending limbs of the flood curve (a/b)] is critical for a number of reasons. Too rapid rises in floodwater disturb nests and submerge attached eggs to too great a depth. Furthermore, floating and emergent vegetation is not able to accommodate readily to rapid changes in level. Overly rapid falls in level result in increases in mortality through greater numbers of fish being stranded on the plain or in temporary water-bodies. The greater overall energy of floods that change level rapidly also disturbs the deposition–erosion cycle.
- *Smoothness*: In natural flood regimes hydrographs become smoother with increasing river order (Fig. 14.18). In general, the smoother the hydrograph the more suitable the environment for fish. Frequent reversals in the rise and fall of floods increase the risk of failure of spawning by exposing spawning substrates and nests. They also increase the risk of stranding mortality, interrupt the drift of fry onto the floodplains and increase the risk of juveniles being swept back into the main river.
- *Drawdown*: The hydrograph during low water can be as critical to the dynamics of the fish community as can the high-water regime. The population size in some river systems with especially high drawdowns is conditioned by the amount of water remaining in the system at low-water. This is demonstrated by a high correlation coefficient for a relationship between year-to-year fluctuations in a low-water hydrological index and catch (see, for instance, the Kafue Flats fishery prior to impoundment of the river; University of Michigan, 1971). Because of the dynamics of fish assemblages in fluctuating systems the ratio between high- and low-water hydrographs can be of great significance. This ratio can be expressed as a percentage of the water area remaining in the system at low water relative to the total flooded area.

Management of aquatic vegetation

In contrast to its invasive and nuisance role, aquatic and riparian vegetation can contribute positively to the rehabilitation of the environment. Vegetation can be used to stabilise riverbanks, and create shelter and habitat diversity. Trees and emergent aquatics can be used to train river courses by encouraging siltation and thus deflecting water in much the same way as artificial wing dykes and deflectors. Wherever possible vegetation should be used to replace heavy engineering solutions, as it is more environmentally friendly and adds aesthetic value to the system. The differences between a vegetated and rehabilitated river and one in which no reconstruction has taken place are shown in Fig. 14.19.

Mitigation and rehabilitation 295

(a)

(b)

Fig. 14.19 Contrast between (a) a channelised system (the River Ouse in England) and (b) a natural river (the River Lot in France).

Vegetation can be planted as whole plants. Usually, however, this approach is costly and time consuming and cuttings are used instead. These are laid under bio-matting which serves to hold them in the required position and protect them during their earlier growth. Log booms that cut down on wave action may provide additional protection on the exposed outside of bends in the river.

Chapter 15
Biodiversity and conservation issues[1]

Importance of biodiversity

Since the adoption of the Convention on Biodiversity, the conservation and sustainable use of living aquatic resources has become a primary goal for fisheries management. This section examines the role of biodiversity in the functioning of aquatic ecosystems. It also describes some of the measures used to quantify biodiversity so that it can be monitored, documented and sustainably managed in the face of growing human impacts.

The theory

The role of biodiversity in the functioning of ecosystems is currently being debated among scientists. Experimental and theoretical studies often give conflicting answers. However, some trends are emerging following careful analysis of experimental systems. Diverse systems are more productive than simpler ones, but only up to a certain level of diversity; past this level species may become redundant (Baskin, 1995). Similarly, the stability of ecosystems increases up to a point, and then becomes independent of diversity. However, if apparently redundant species have different tolerances to stress or other environmental changes, they will not be equivalent in a dynamic ecosystem as they may have a role to play if conditions change (Chapin *et al.*, 1997). Some species appear to be more important members of a community than others, for example, top predators such as the Nile perch, species that modify habitat such as crayfish, or species that link ecosystems such as migrating salmonids that bring nutrients from the ocean to inland streams.

Diversity of communities of living aquatic organisms has been stated as being important for the productivity, stability (resistance and resilience) and aesthetics of inland water ecosystems.

Productivity

At the gene level, diversity is important in order to avoid inbreeding depression (i.e. the loss of fitness-related traits owing the expression of deleterious recessive genes) and to provide positive heterosis (hybrid vigour). For example, Kincaid (1983) assessed effects of reduced variability due to inbreeding in rainbow trout, *Oncorhynchus mykiss*, and determined that per cent hatch and survival were affected at an inbreeding level of 12.5% (the level from one generation of half-sib matings), whereas levels greater than 25% reduced fecundity, growth and survival. In farmed populations of tilapia, *Oreochromis niloticus*, a highly heterogeneous mixed-strain stock performed better than less diverse strains (Eknath *et al.*, 1993).

Theory states that genetic diversity is important for viability, disease resistance and continued adaptation (Allendorf & Leary, 1986). It is extremely difficult to quantify or document such dependence in natural populations where critical research has not been done. Leberg (1990) created semi-natural ponds that were stocked with either a full-pair (brother–sister) or an unrelated pair of mosquito fish, *Gambusia holbrooki*, where a 25% reduction in genetic heterozygosity led to a 56% reduction in final population size (however, all populations did increase from the original founding pair of fish). In another laboratory study, survival, growth, early fecundity and developmental stability of *Poeciliopsis occidentalis* were all higher in the most heterozygous stocks and lowest in the least variable stock (Quattro & Vrijenhoek, 1989).

Productivity of inland water systems and fisheries is dependent on biological diversity in that fish of commercial interest must have prey items to consume and a functioning ecosystem. But are diverse systems more productive than simplified ones? In farming, monoculture is extremely productive but requires high levels of external input. In fishery systems, input is possible in the form of nutrients or hatchery-raised fish, but in the absence of these interventions, how do complex systems compare with simplified ones?

Species diversity of chydorid Cladocera in Danish lakes decreased with increasing primary productivity (Krebs, 1978). The diversity of runs of Pacific salmon allows this group of species to exploit different riverine habitats at different life-history stages. The productivity of the mixed fishery in the ocean may be increased through this diversity (see also below on stability of Pacific salmon); however, it appears that the ocean habitat may be limiting production (Bigler *et al.*, 1996). Water development was shown to reduce fish diversity but not productivity over an entire drainage system (Winemiller, 1994).

Lévêque (1995) cites three African lakes that have similar levels of productivity, 100–200 kg ha^{-1} (as measured by fishery production and admittedly a crude measure of productivity), but differing levels of species diversity. This is contrasted to Lake Nakuru in East Africa that has only one fish species of tilapia, *Oreochromis alcalicus grahami*, but that contributes 625–2436 kg ha^{-1} to pelican populations. Fisheries management focuses on that part of secondary productivity useful to humanity, i.e. the harvest, but there are other components to productivity that are important and are not considered in catch statistics.

The Nile perch simplified the food web in Lake Victoria by eliminating perhaps hundreds of species of haplochromine cichlids, yet the productivity of the lake has increased (Reynolds & Gréboval, 1988) and more recently the haplochromine populations are showing signs of partial recovery. This increase may not be due to reduced diversity but because the lake is receiving increased nutrients and sediment from the surrounding degraded landscape. Lake Chapala in west central Mexico, another lake highly impacted by pollution, water diversion and the introduction of exotic fishes, has also experienced increased fish production with the addition of exotic species. However, production is now highly erratic, 'suggesting less secure livelihood for fishers'. Desirable native fishes such as the pescado blanco, *Chirostoma* spp., popoche, *Algansea popoche*, and catfishes, *Ictalurus* spp., have declined (Lyons *et al.*, 1998).

The productivity of systems is influenced by a variety of factors and inputs. As Lévêque (1995) states, 'There is no simple relationship between biodiversity and ecological processes such as productivity'. Inputs or manipulation can increase productivity. Clearly, when stocking fish in a new habitat, such as a newly filled reservoir, forage fish should also be stocked, as well as species that can exploit various food sources, such as the planktivorous *Limnothrissa miodon* introduced to Lake Kariba in southern Africa. In species recovery or other stocking programmes use of genetically diverse stock is preferable, as suggested in the endangered species recovery programme described by Quattro & Vrijenhoek (1989).

Stability: resistance and resilience

Theory on the relationship of diversity and stability of natural systems is also contradictory. Elton (1958) and MacArthur (1955) stated that complex (higher biodiversity) ecosystems were more stable, i.e. better able to withstand impacts. The theory is that with more connections in the food web, disturbance to any particular path will be compensated for by another path. However, evidence for this was hard to produce from natural systems and other theories stated that diversity led to instability (May, 1971). Recent work by McCann *et al.* (1998) has smoothed out some of the controversy by looking at communities not in a stable equilibrium and by using non-linear approaches to effects of disturbance. The diversity of certain groups of fishes has enabled them to cope with some environmental change. Pacific salmon are known for their diversity of migration patterns and their homing ability. Historically, spring runs of chinook salmon in California were most abundant, but required the most difficult inland migrations. Today, because of water-development projects and other forms of habitat degradation, the spring runs have almost been entirely eliminated from the state. The state still enjoys wild chinook salmon because of the prevalence of fall runs; the diversity of spawning migration pattern kept at least some of the salmon spawning.

An organism's ability to tolerate different environmental conditions is partially determined by its genes. Differences in pollution (e.g. heavy metal) tolerance, heat resistance, salinity tolerance, etc., are under a degree of genetic control. For example, wild populations of guppies, *Poecilia reticulata*, from diverse stocks were more adaptable to full-strength seawater than were domesticated stocks that had reduced variability through inbreeding (Chiyokubo *et al.*, 1998). When different domesticated saline-sensitive stocks of guppies were crossed, salinity tolerance increased in the progeny to a level similar to wild stocks, indicating positive heterosis.

Genetic studies on stream-dwelling darters found that genetic variation was an indicator of water quality (Heithaus & Laushman, 1997). Specific alleles were associated with water pollution levels in the stoneroller minnow, *Campostoma anomalum*; fish possessing alleles sensitive to low water quality were completely absent from the most polluted streams. Although other confounding factors determine the relationship between genetic variation and species distribution or niche width, the stoneroller minnow had the highest levels of variation and was able to inhabit the highest diversity of stream habitats in this study. Species diversity did not change as polluted streams in Sweden were restored to higher water quality. Rather, the types of species inhabiting improved rivers changed: trout, *Salmo trutta*, stone loach, *Barbatula barbatula*, and eel, *Anguilla anguilla*, increased in numbers and presence in relation to improvements in water quality (Eklöv *et al.*, 1998).

Moyle & Light (1996) examined how communities with differing levels of diversity resisted invasion by exotic species. Their results, similar to work on terrestrial plant communities, were that diversity did not protect a community from invasion and that physical parameters influenced invasion success. Many communities are diverse because the physical conditions are suitable for the growth of many species, even exotic ones.

A healthy functioning ecosystem is not necessarily static or stable in its species composition. In the marine environment there are 'regime shifts' where the relative abundance or presence of species changes over time as a result of natural events (Botsford *et al.*, 1997). These shifts are possible due to presence of diversity in other areas and to the connections in the marine environment. In inland waters similar events occur through colonisation of water-bodies. Ecosystems with alien species undergo fluctuations between native and alien species over time (Moyle & Light, 1996). Thus, ecosystems are constantly drawing on the pool of aquatic genetic resources to withstand both natural and anthropogenic impacts.

At the level of ecosystem function Hollig & Meffe (1996) criticised current resource management that often seeks to control or modify natural systems through reduced diversity (e.g. monocultures), reduced variation (e.g. flood control) and increased control over natural phenomena (e.g. wildfire suppression). Hollig & Meffe (1996) promoted a golden rule of natural resource management '… to retain critical types and ranges of natural variation in resource systems in order to maintain their resiliency'.

Aesthetics and the existence value of biological diversity

The diversity of fishes has played a role in human culture for thousands of years. The walls of the Pharaohs' tombs *c.* 3200 BC, the totem poles of Pacific North-west Indians and the Japanese Gyataku (images of fish pressed onto rice paper) show humanity to have used images of fishes to symbolise important aspects of our life, such as creation, success, power, good, evil, fertility and spirituality (Kreuzer, 1974; Moyle & Moyle, 1991). The detailed paintings of ancient Egypt are thought by some to act as reminders to the Pharaohs that, in death, they still had the responsibility to provide Egyptian people with fish (Klingerder, cited in Moyle & Moyle, 1991). The diversity of common carp and goldfish has graced porcelain, sculpture and brush paintings for centuries in Asia. A sacred pond contained many species of fish and the water was thought to be the source of all life in Ancient Sumaria (Kreuzer, 1974). The image of the fish was one of the earliest symbols of Christianity and continues as a symbol of faith and devotion today. Thus, biological diversity contributes to humankind in ways not associated with 'use' or with commercialisation, except in the case of tourism and recreational fisheries and, therefore, tends to be underappreciated or undervalued in fisheries management.

This value can have profound effects on other development projects. The Tennessee darter, a small perch protected under the US Endangered Species Act (ESA), temporarily stopped the construction of the Tellico Dam on the Little Tennessee River, USA, because the dam would flood critical habitat. Because of the financial implications of conserving such endangered species, the US Congress created the Endangered Species Committee, composed of political appointees and agency leaders that could overrule decisions made under the Act in light of economic considerations. Ironically, this Committee also ruled in favour of the darter, but the dam was built anyway following passage of an energy and water development bill that included exemption from the ESA for Tellico Dam (Moyle, 1993).

Apparently the Tennessee darter did not have enough of the right kind of 'value'. The aesthetic and existence value of freshwater diversity continues to be important to society, but the value is difficult to quantify. Loomis & White (1996) placed economic value on threatened species in the USA by calculating the 'willingness to pay', i.e. how much a household would pay to save a threatened species. The results were partially encouraging in that Pacific salmon and steelhead scored second on the ranked list behind the northern spotted owl, but at the bottom of the list were the squawfish and striped shiner. Moyle & Moyle (1995) review several strategies for placing market value, ecosystem value, existence value and intergenerational value on endangered species in order to promote their conservation.

Some measures of biodiversity

Biological diversity is defined as the 'variability among living organisms from all sources including, *inter alia*, terrestrial, marine and other aquatic ecosystems and the ecological complexes of which they are apart: this includes diversity within species, between species and of ecosystems' (CBD, 1994). As acknowledged by the CBD, biological diversity is a complex concept and several indices have been developed to quantify it. Each index has its advantages and limitations and it is not the purpose of this chapter to review them critically.

Owing to the nature of many freshwater systems, i.e. many are discrete waterbodies often isolated from other water systems, much of the biological diversity present is endemic or found in very limited areas. This isolation is also responsible for the genetic differentiation and speciation of many groups of inland fishes such as salmonids, catastomids, cichlids and cyprinids. However, diversity can also be generated within a contiguous water body, as in the African Great Lakes where species radiation within certain families, e.g. the cichlids, has occurred without obvious restrictions in gene flow and in an evolutionarily short time-span (Fryer, 1996).

Genetic diversity

Indices have been created to measure the genetic diversity of biological species. Common measures are listed below. Heterozygosity (H) is a measure of genetic diversity derived from the number of loci with more than one gene (polymorphic loci) and the gene frequencies:

$$H = 1 - (p^2 + q^2)$$

where p and q are the gene frequencies at a given locus with two genes (more genes can be added into the formula). Average heterozygocity is then found by averaging the heterozygocity of each locus studied. Proportion of polymorphic loci (P)

$$P = \text{number of variable loci/total number of loci}$$

measures the loci with more than one gene; however, it treats all variable loci equally, regardless of whether there is only one rare gene or several genes causing the variation at a locus.

Because not all individuals in a population contribute genetic material to the next generation, the effective population size,

$$N_e = \frac{4 N_m N_f}{N_m + N_f}$$

where N_m and N_f are the numbers of males and females, respectively, is a useful measure of the genetic resource of a group that can be passed to the next generation. Effective population size is usually much smaller than the actual number of individuals in a group. Consider the extreme example of a population of all males: the effective population size would be 0 no matter how many individual males were present. Effective population size can be useful in setting conservation goals and monitoring wild and hatchery populations (Bartley *et al.*, 1995).

The way in which genetic diversity is distributed between subgroups of a species is also important. The measure F_{ST} defines how much difference exists between groups compared with the variation within groups

$$F_{ST} = 1 - \frac{H_S}{H_T}$$

where H_S is the average heterozygosities of all subpopulations studied and H_T is the total heterozygosity derived from the overall gene frequencies (Chakraborty &

Leimar, 1987). Freshwater species in general have more variation due to between-group variation than marine species. This is presumably due to the potential for migration and therefore gene flow of both adults and larvae in marine systems (Gyllensten, 1985). Population size is a main determinant of genetic diversity in that small populations are often subject to loss of genes through inbreeding depression or random loss of genes due to small numbers of reproducing adults.

There is a growing awareness of the importance of population diversity, or stocks, in ecosystem function and fishery management. The concept of a stock will not be treated here as it has been widely debated in fishery science and should be operationally defined for a given fishery situation, e.g. stock of ground fish (multi-species), stock of chinook salmon (multiple runs of a single species) or stock of winter run of chinook salmon in the Sacramento River (specific stock with spatial and temporal characters within a species). A key concept in protection of endangered biodiversity by the US ESA is the evolutionarily significant unit (ESU) that has been operationally defined for Pacific salmon (Waples, 1991). For a population or stock of fish to be considered an ESU, and hence protectable under the ESA, it must represent an important genetic legacy of the species and must be substantially reproductively isolated from other conspecific population units (Waples, 1991). Thus, the diversity indices listed above could also take into account discrete stocks and ESUs.

Species diversity

Species richness can be measured by simply recording the number of species in a given area (N), but this treats rare and common species identically. More meaningful estimates can be provided by weighting the number of species by their relative abundance indices such as the Shannon–Wiener function, H', (which predicts the abundance of the next species), and the Simpson's index of diversity, D (which assesses the probability that two collections will be different):

$$H = \sum_{i=1}^{S} P_i \log_2 P_i$$

where p_i is the proportion of individuals of species i in the community of S species. Equity measures can be defined as H/H_{max}, where H_{max} is where are all species are equally abundant and is given by $H_{max} = \log_2 S$.

The index of endemism, E, is an important attribute of inland aquatic biodiversity. Tropical islands are most famous for their coral reef diversity, but their freshwater fauna is usually more unique, endemic and often extremely threatened. Sisk *et al.* (1994) developed a simple index as:

$$IE_c = 100 \left[\frac{1}{2} \left(\frac{EM_c}{M_t} + \frac{EB_c}{B_t} \right) \right]$$

where EM_c is the number of endemic species in area C, M_t is the total number of species in the database, EB_c is the number of endemic other taxa in the area and B_t is the total number of other relevant species in the database.

Indices of biotic integrity (IBI) (Angermeier & Schlosser, 1987) incorporate a broader array of community attributes such as species composition, trophic habits, presence of exotic species and fish health to describe not only species diversity, but also ecosystem health. Karr & Dudeley (cited in Moyle & Randall, 1998) defined biotic integrity as 'the ability to support and maintain a balanced, integrated, adaptive community of organisms having a species composition, diversity, and functional organisation comparable to that of the natural habitat of the region'. Several attributes, usually around 12, can be combined into the index. Not unexpectedly, such a complicated index was shown to be a better indicator of degraded water systems than the Shannon–Wiener function.

The watershed IBI of Moyle & Randall (1998) included native Ranid frogs, anadromous fish, exotic species, roads and dams, i.e. factors indicating the state of larger areas. However, Moyle & Randall (1998) noted that numbers of native resident fish and abundance of native fish captured much of the same information as the more complicated index for California stream communities. Such indices can be developed for other areas based on site-specific criteria. The important consideration would seem to be to include measurements that are appropriate for watershed assessment similar to those of Moyle & Randall; The frogs give an indication of the state of the terrestrial habitat; the anadromous fish indicated the level of habitat fragmentation; introduced and native fish indicated biotic interactions; and road and dam development were indicative of specific human interventions that have or are likely to have an impact.

In concluding this section it must be noted that not all genetic diversity or species diversity is equal and numeric indices cannot capture the whole essence of a dynamic ecosystem. For example, adding one large predator, such as the Nile perch, to a system will probably have more of an effect than adding several small herbivores. We should also remember that the genetic diversity of humans is more than 98% identical to chimpanzees (Gibbons, 1998); the remaining 2% must be extremely important.

Management for biodiversity

Conservation of biological diversity is now the obligation of all nations ratifying the Convention on Biological Diversity (CBD, 1994). The mode in which the provisions of the Convention can be applied to fisheries is set out in the FAO Code of Conduct for Responsible Fisheries (FAO, 1995a) and in the guidelines to the Code for Inland Fisheries (FAO, 1997a, b). While the Convention aims at striking a balance between conservation and use its general tenor tends to be conservationist, whereas the Code of Conduct is aimed more at containing fisheries within sustainable limits. It is unrealistic to demand that valuable sources of food and recreation such as are found in rivers throughout the world remain unexploited solely to protect their biodiversity. Unfortunately, the process of fishing, together with management practices for the improvement of the yield or recreational value of chosen elements of fish assemblages, do erode a biodiversity that is also under threat from pollution and general environmental deterioration. Because of this, freshwater fishes are the most threatened group of animals utilised by humans (Bruton, 1995), with 20% either extinct, threatened or vulnerable (Moyle & Leidy 1992). Management of inland waters now must consist of managing the diversity of the fish populations of all types of water, as well as the original task of managing the yield. Furthermore, the management function is no longer only the domain of those responsible for the fishery but also concerns all other users who can impact the water and the living organisms inhabiting it. River management has now become a national (and international in the case of transnational river basins) enterprise in which all interests should negotiate to maintain the quantity and quality of water in the rivers, the connectivity and diversity of fluvial habitats and the sustainability of the fish communities.

Indicators of change

Management of biodiversity depends on an adequate evaluation of its status. In lakes and rivers with few species it is relatively easy to identify one that is endangered and to make appropriate arrangements for its protection. However, in water-bodies with large numbers of species it is almost impossible to keep track of any but the few very large species that are normally the first targeted by the fisheries. In such cases the general health of the biodiversity has to be measured using indicators of trends within the population (Table 4.6), as well as the indicators suggested in the watershed IBI of Moyle & Randall (1998).

Management of the gene pool

Two major types of management are recognised for the conservation of biodiversity.

- *Ex situ* management, whereby genetic material, eggs, sperm, embryos and living representatives of the species are conserved in specialised places apart from the natural habitat. In the case of fish, living individuals are usually maintained in aquaria, often by enthusiastic hobbyists or aquarium societies. *Ex situ* collections of germplasm may be conserved in frozen genebanks. Conservation of genetic material in frozen genebanks is less common in fish than in plants and domestic animals, although there are several areas now where this is being done, for example in Canada, Norway, Iceland, Brazil, Russia, Finland, the USA, the Philippines and India (Harvey *et al.*, 1999). The Convention on Biological Diversity (CBD, 1994) and others (see papers and discussions in Harvey *et al.*, 1999) see a positive role for *ex situ* conservation in support of *in situ* conservation, research, insurance against loss of genetic resources, and education.

- *In situ* management, whereby fish species and assemblages are protected in their natural habitat, except in the case of domesticated organisms, in which case it is the environment where they acquired their special characteristics, such as a farm or pond in a given environment. Endangered species may also be introduced into non-native, but natural, habitats that are similar to their native ones and thus kept alive when the native habitat is under threat. In fisheries, *in situ* conservation is the major approach that has been adopted so far and the following sections discuss the main mechanisms that can be used.

Management of the fishery and the fishers

One of the major causes of deterioration in the diversity of inland fish populations is the fishery. Heavy fishing pressure has degraded stocks of many species to a point where they are heavily compromised. The main policy for fishery managers must be to reduce access to the overburdened fisheries and thus to allow them to recover. In temperate areas fishing pressure is mainly exerted through recreational fisheries which, because of the demand for specific species, can also deform populations through management to favour the preferred target. In such cases education can produce shifts in perception on the part of the fishing community. In both cases, current thinking is that success in conserving wild fish species can only be achieved by fully involving the fishing communities through co-management systems. In its turn, co-management involves a transfer of ownership to the people who actually fish the resources, giving them an interest in maintaining the diversity of the fishery. It would be simplistic to view this type of approach in isolation, as it must be accompanied by incentives for the fishing communities and supported through education, research and other infrastructure (see Chapter 11, p. 195).

Rehabilitation of the environment

Fishing is not alone in its capacity to damage biodiversity. In most areas of the world by far the greatest harm has been done by non-fishery activities. The economic power of activities such as power generation, navigation, agriculture and industry is very difficult to counterbalance, although increasingly the preservation of a fish species has been sufficient motive to question and even halt possibly detrimental activities in the basin. If this is not possible the fishing community has to make the best of a bad job and adopt such measures as are possible to mitigate the negative impacts. When pressure on the natural resource is released through better conduct of industry and agriculture, steps should be taken to rehabilitate aquatic environments (see Chapter 14, p. 262).

Protocols, guidelines and codes of practice

Guidelines and codes of practice have been developed to minimise adverse impacts on natural ecosystems, i.e. to promote, among other things, *in situ* conservation. The International Council for the Exploration of the Sea (1995) and regional fisheries bodies have adopted codes of practice on the use of alien (exotic) species and genotypes (genetically modified organisms). The main elements of these codes are:

- preparation of a proposal planning and justifying the introduction;
- evaluation of the proposal by an independent body and environmental impact assessment;
- refinement or rejection of the proposal;
- establishing fish health management (i.e. quarantine) and monitoring programmes if the proposal is approved;
- adapting fishery management to new situation(s).

Inland water systems, as most other systems in today's world, are increasingly under the influence of humanity (Vitousek *et al.*, 1997) and will need to be managed to some degree. Can inland waters be managed for biodiversity? Such a management scheme would place priority on the conservation of natural levels of diversity rather than on production and employment and, as stated by Hollig & Meffe (1996), would allow natural cycles of floods, fires, etc., to take their course. However, it should be stressed that management for biodiversity and management for sustainable development are not mutually exclusive.

Parks and reserves

Reserves, parks and other types of protected areas may be used for a wide variety of objectives, including nature conservation, recreation or sustainable use (IUCN, 1994). The principle of establishing aquatic reserves for fish has recently been further reinforced by the Convention on Biological Diversity and by Ramsar's decision that the conservation of fish species alone is sufficient to justify the setting up of a Ramsar site. Traditional management often prohibits fishing in designated portions of rivers, floodplains or lakes for all or part of the year. It is not uncommon in such circumstances for religious or traditional values to be invoked as an authority for such closure. In reality, such areas serve as reserves and play a role in protecting fish stocks. Under modern management practice several types of conservation area exist. Parks are generally large features of the landscape managed holistically for a number of purposes. Reserves are smaller features of the landscape managed for a few particular conservation objectives. Other types of protected area such as sites of special scientific interest (SSSIs) are even smaller local features. The following sections describe the management aspects of two categories of reserve that may be useful in inland fisheries; two main types, conservation reserves and harvest reserves, are discussed.

Conservation reserves

Conservation reserves serve the following objectives:

- To preserve a whole faunal complex which depends on a certain type of landscape (wetland, floodplain, mountain lake, etc.). This calls for the designation of a large area, often an entire lake, river or tributary system.
- To preserve a single species. This requires somewhat less resources in terms of area than the full faunal complex although it is frequently difficult to isolate the species from others with which it customarily interacts.
- To conserve essential habitat needed for completion of some aspect of the life cycle. Many species are threatened by modifications to their ecology that endanger one or more of their critical habitats. These are usually spawning or nursery areas, although summer or winter refuges may also be much diminished.

Conservation of species through parks and reserves requires that the protected areas include habitat corresponding to all the main needs of the fish (feeding, breeding, nursery and shelter habitats) as well as the connectivity between them. It also requires the conservation of companion species, vegetation and trophic resources necessary for the complex of food webs. This may be relatively easy in lakes and small streams where habitats are usually fairly congruent. It is more difficult in large rivers where the different types of habitat may be separated by considerable distances and where upstream landscape effects may prove critical in the maintenance of water quality and quantity downstream. Locally resident black and greyfishes may be easily accommodated in a single-channel–floodplain reserve, but longer distance whitefish migrants are extremely difficult to provide for within conventional reserve structures.

Conservation reserves are usually permanent features of the allocation of the landscape although, in some cases such as areas that are only briefly occupied by seasonal migrants, seasonal protection may be adequate. However, in such cases the use that is made of the reserved area outside the season should not damage the environment so as to diminish its usefulness for the fish. Conservation reserves usually depend on total exclusion of fishermen from the designated area. In order to achieve this some form of policing is usually necessary. Ideally, some incentive can be provided for local people to respect the reserve. This may be achieved by a clear demonstration that the setting aside of part of a potential fishing ground has benefits for the fishing communities. The direct involvement of the fishermen in the setting up and policing of the area may also contribute to success. Where such benefits have become apparent there is often a spontaneous movement to create further reserves. By contrast, reserves that are set up and policed by central governments without the involvement of the fishermen stand less chance of success.

Harvest reserves[2]

Harvest reserves have many of the characteristics of IUCN Category IV 'Habitat/

Species Management Areas'. They may be used to protect fish in locations where they are particularly vulnerable to overexploitation. Protecting fish in a few key locations may lead to more breeding fish each year and hence maintain the annual production of new recruits to the fishery. In rivers fish are easiest to catch in floodplain pools and river channels during the dry season and also during their spawning aggregations and migrations. By the time the dry season arrives, floodplain fish stocks have usually been greatly depleted. Those few fish remaining must produce next year's stock. Resisting the temptation to take these easy catches at this time may mean that many more survive to spawn at the start of the floods and lead to many more fish for capture in the next flood season.

Harvest reserves differ from other types of protected areas in two important ways. Firstly, harvest reserves are intended to increase the harvest of fish (i.e. to benefit the fishing community) and not just to protect fish for their own sake. When fishers understand this objective, they have a much clearer incentive to support their local reserve and to resist the temptation to poach. Since communities tend to prioritise their own interests, harvest reserves are likely to be more attractive to fishermen than biodiversity sanctuaries or closed conservation areas would be. Harvest reserves will only achieve their objective if consideration is given to where and when the extra fish produced by the reserve will be caught, and by whom. Harvest reserves will thus only provide benefits to fishers if either: the reserve is located in a water-body from which fish can emigrate to fished areas; or some fishing is allowed inside the reserve, either in limited seasons, or with non-threatening gears.

The second important difference between fishery harvest reserves and most other protected areas is that harvest reserves may not always need to be closed for the whole year. They should be managed instead with a flexible combination of regulations, adapted to maximise local benefits. In some situations (e.g. where the reserve water-body is the main fishing place for a village), more benefits may be achieved by closing the reserve for only part of the year, and allowing fishing in the reserve at other times. Closing the reserve all year may leave virtually no fishing opportunities for the villagers, who may then be irresistibly tempted to fish illegally in the reserve. In other situations, year-round closure may be more effective (e.g. in particularly vulnerable habitats, or where fully closed 'taboo' areas are traditional practices).

Of the various possible fishery management tools (effort limitations, quotas, size limits, etc.), harvest reserves are particularly suitable for co-management approaches for the following reasons:

- They are conceptually *simple*, with easily understandable effects on fish stocks
- They are *traditional* management tools in many places, with proven local acceptability
- Their high *visibility* makes illegal fishing relatively easy to detect (it is easier to see a poacher fishing in a reserve area than to see who is using illegally small mesh sizes, or using too many units of gear)
- Most importantly, they may be specifically designed to give *benefits* to local communities.

310 *Inland fisheries*

The distribution of people who benefit from a harvest reserve depends on the dispersal patterns of the extra fish produced. Reserves inhabited by relatively non-migratory species such as the riverine blackfish or lacustrine species inhabiting isolated bays or rock reefs will mainly increase fish catches within a small local area. Reserves protecting more migratory species may give benefits to whole catchments, owing to their much wider dispersal patterns. In rivers blackfish reserves should generally be located in dry-season water-bodies in floodplain areas. Whitefish reserves should be located in spawning grounds, usually in upstream parts of the catchment. In lakes some migratory species will need reserved access to tributary rivers for breeding. Migratory species may also need additional management restrictions (e.g. controls on barrier traps) to ensure that some fish can return to the reserve each year to spawn. Blackfish reserves are more likely to be supported by local communities than whitefish reserves, since the extra fish produced by their management efforts will stay mainly within their own waters.

Co-management strategies involving such harvest reserves and other management tools should be developed using the social development guidelines given in Chapter 11 (p. 201). Local stakeholders should thus be directly involved in the decision to introduce a harvest reserve, and in its selection from the various possible local water-bodies. Villagers are more likely to know the hydrology of their local area and to support the reserve if they agree that the best water-body has been selected. Real participation involves much more than just saying 'yes' to the outside expert's choice, and will require good social facilitation skills.

While recognising the importance of such participatory principles and the need for local adaptation and flexibility, the following notes provide general technical guidelines for the selection and co-management of harvest reserves.

- Several small reserves should be selected, scattered around a lake or river catchment, rather than one large one. In river systems, several small harvest reserves may give more benefits to fishers, owing to the limited dispersal patterns of blackfish. To spread out the management costs between the beneficiaries, one or more reserves may be established in each village where suitable social structures and reserve habitats exist.
- Reserves should be selected in several different habitat types to protect different fish species and their various life stages. In rivers some reserves may, for example, be selected in river sections, with others in floodplain lakes. In lakes individual rocky structures, small islands, isolated bays and tributary rivers are particularly suitable. Some reserves may include all such habitats.
- 'Blackfish' reserves should be selected in deep, permanently flooded water-bodies in floodplain areas. Such water-bodies are more likely to protect blackfish over the full dry season, while other shallower pools dry up, resulting in fish kills.
- The use of particularly dangerous dry-season gears (e.g. de-watering, poison, electric fishing, fish drives) should be restricted in blackfish reserves to ensure the survival of fish over the dry season (most floodplain fish spawn at the start of the flood).

- 'Whitefish' reserves should be selected in the upstream stretches of rivers used as spawning grounds for these migrant species. These should be matched with floodplain reserves where the juvenile and adult feeding grounds are located. Care should be taken to ensure that there is adequate communication between the two areas by negotiating with other developers of the river to ensure that adequate bypass structures are in place round any dams and that pollution levels are low enough to permit the passage of fish. Additional controls may also be needed on barrier traps to ensure that the whitefish can reach the protected spawning grounds. Some species of fish may cross national frontiers in completing their life cycle. This means that spawning and feeding reserves may be in different countries (or different states of the same country). International collaboration and agreement are needed in such cases.
- Harvest reserve water-bodies should be selected to leave enough alternative fishing grounds to maintain the fishing opportunities of nearby villages. A village's only water-body or fishing ground should not be set aside as a fully closed reserve, as it is likely to be badly poached.
- Where possible, a new reserve should be close to the village(s) involved in its management. Such closeness maximises the visibility of fishing activities and minimises opportunities for illegal fishing.
- Water-bodies should be selected that have good connections to surrounding fished areas (e.g. through water channels or across flooded land), to ensure that the extra fish produced in the reserve may be caught. Where whole disconnected water-bodies, such as lakes, dams or isolated ox-bow lakes are permanently closed as reserves, they may give no benefits to any fishable waters. Only partial closures should be made in such water-bodies, either as one or more small partial reserves, or as a closed season. In large lakes, a reserve may be introduced in the local waters of each fishing village around the lake.
- Water channels should be maintained to enable the passage of fish in and out of the reserve, and to ensure the transfer of fish stocks to fished areas. In floodplain areas, this may involve the removal of silt or vegetation, or restrictions on the use of barrier traps in channels around reserves, especially during the rising and falling water seasons.
- The location of the reserve should be made as clear as possible, by defining boundaries at easily recognisable local features, such as bridges, well-known buildings (mosques, schools, etc.) and river confluences. Marker posts may also be used if these are made clearly visible. Grid references, which are invisible on the ground, should be avoided. It is also helpful if modern reserves can be located in areas that are protected by traditional or local religious sanctions.
- Either permanent or seasonal closures may be used to ensure protection of critical life-cycle phases. Closed seasons may be set to prevent the capture of juvenile fish in the high-water season, brood-fish during the early flood spawning season, or migrating fish during the rising/falling water seasons. The actual period for a closed season may need to change between years depending on the dates of flooding.

Size, location and area

Conservation reserves are usually established on the basis of a perceived need to preserve a particular landscape, community or species. Such choices are usually dictated by higher level policies determined by scientific advice, public demand or conformity to the requirements of an international treaty.

In harvest reserves adaptive management practices should be used to determine the optimal combination of regulations (number and sizes of reserves, seasonality of closures and use of other management tools) for each locality.

The area needed for conservation and harvest reserves can be defined through biological or social criteria. The heavy demand for land and water in most areas of the world means that social and political standards dominate. Scientific advice to governments on the types of habitat that need to be conserved or to local communities on the extent and nature of the local set-aside may influence policy. At present, biological criteria are usually fairly arbitrary, although in both lakes and rivers current models may serve as a starting point for these more empirical approaches. For example, in rivers, the idea of reserves for rehabilitation sufficient to repopulate the river from a limited area is expressed as the 'string of beads' approach to river restoration. However, the optimal area of each bead has yet to be established.

Notes

1. By D.M. Bartley, FIRI, FAO Fisheries, Rome, Italy.
2. By D. Hoggarth, SCALES Inc., 6 Highgate Gardens, St Michael, Barbados.

Chapter 16
Legislation[1]

The purpose of this section is to provide a brief survey of the treatment of inland fisheries in international and national law. The importance of inland fisheries has been recognised in a number of international instruments. The subject has received important attention from the Subsidiary Body on Scientific, Technical and Technological Advice (SBSTTA) of the Convention on Biological Diversity (CBD), which focused on living aquatic resources among other priority areas. Several other international instruments are relevant to inland fisheries, including the Food and Agriculture Organisation (FAO) Code of Conduct for Responsible Fisheries. Such international attention has increased awareness of crucial issues concerning inland fisheries, and provided 'moral authority' for countries and institutions to take action to conserve biological diversity within inland waters, wetlands and other environments. However, the real power for conservation and sustainable use must come from national governments (Welcomme, 1999b). It is at this level that policies and legislation must be formulated that address the complex issues of preservation versus use, trade and equity, and financial support for needed progress.

International instruments

Various conventions and other international instruments apply either directly or indirectly to inland fisheries. This section will briefly describe the relevant provisions of five such instruments: Agenda 21, the Convention on Biological diversity, the FAO Code of Conduct, the UNESCO Convention Concerning the Protection of the World Cultural and Natural Heritage, and the Ramsar Convention.

Agenda 21

The United Nations Conference on Environment and Development (UNCED) took place in June 1992 in Rio de Janeiro, Brazil. During the meeting five documents were signed, among which was the main substantive work of the Earth Summit, Agenda 21.

Agenda 21 is a non-binding document. Instead, it was intended to serve as a set of normative principles that will determine appropriate international behaviour in the twenty-first century. The programme areas that constitute Agenda 21 are described in

313

terms of the basis for action, objectives, activities and means of implementation. The multisectoral nature of water resources development in the context of socio-economic development is recognised. So too is the multi-interest character of the utilisation of water resources for water supply, sanitation, agriculture, industry, urban development, hydropower generation, inland fisheries, transportation, recreation, lowland and flatland management and other activities.

Chapter 18 of Agenda 21 recognises that freshwater resources are an essential component of the Earth's hydrosphere and an indispensable part of all terrestrial systems (18.1). It calls for the protection of aquatic ecosystems with respect to the following programme areas (18.5):

- Integrated water resources development and management
- Water resources assessment
- Protection of water resources, water quality and aquatic ecosystems
- Drinking-water supply and sanitation
- Water and sustainable urban development
- Water for sustainable food production and rural development
- Impacts of climate change on water resources.

Part 40 of Chapter 18 sets forth the activities to be implemented by all the states, according to their capacity and available resources, and through bilateral or multilateral co-operation, including United Nations and other relevant organisations as appropriate. The activities fall into the following categories:

(a) Water resources protection and conservation
(b) Water pollution prevention and control
(c) Development and application of clean technology
(d) Groundwater protection
(e) Protection of aquatic ecosystems
(f) Protection of freshwater living resources, which include:
   (i) control and monitoring of water quality to allow for the sustainable development of inland fisheries
   (ii) protection of ecosystems from pollution and degradation for the development of freshwater aquaculture projects
(g) Monitoring and surveillance of water resources and waters receiving wastes
(h) Development of national and international legal instruments that may be required to protect the quality of water resources, as appropriate.

While only item (d)(i) makes explicit mention of inland fisheries, it is clear that the entirety of the above framework has direct relevance to the management of these resources.

Convention on Biological Diversity

The CBD, another outcome of the UNCED process with important ramifications for inland fisheries, came into force in December 1993. The CBD has three principal objectives set forth in its first article. These are (1) the conservation of biological diversity; (2) the sustainable use of its components; and (3) the fair and equitable sharing of the benefits arising out of the utilisation of genetic resources (Art. 1).

The CBD set out various obligations towards achieving these objectives. For example, Contracting Parties are required to develop national biodiversity strategies (Art. 6.a); to integrate biodiversity conservation and sustainable use of its components into sectoral and cross-sectoral plans, etc. (Art. 6.b); and to take steps to identify and monitor biodiversity (Art. 7). Under Article 8, countries are obligated to establish *in situ* conservation strategies. This includes the development of protected area systems (Art 8.a); however, the obligation to regulate and manage 'biological resources important to the conservation of biological diversity' applies both within and without protected areas (Art. 8.c), including such requirements as the maintenance of viable populations of species in natural surroundings (Art. 8.d).

The CBD is a framework agreement. The details of its implementation are, for the most part, left to Contracting Parties to handle through domestic legislation or in further regional and international agreements. Indeed, from a lawyer's point of view, the CBD is impressive both for its comprehensive scope and for the extensive amount of legal work that it leaves to be done elsewhere. It may be, as the CBD Secretariat states, that the CBD 'is unique in the sense that it is the only multilateral, legally binding instrument that covers all the world's ecosystems, thereby taking a comprehensive rather than sectoral approach' (UNEP/CBD, 1997). The job of giving detailed, substantive legal content to this multisectoral vision, however, will be the task of laws and agreements drafted on specific subjects, often shaped by the perspectives of particular sectors.

The CBD is intended to deal comprehensively with the subject of biodiversity; as such, it does not tend to deal specifically with different types of resources. The CBD's obligations with respect to biodiversity apply generally, whether or not the resource in question is marine, terrestrial or otherwise. Thus, living aquatic resources are not singled out for special treatment in the text of the CBD but become the subjects of specific programmes, such as the Jakarta Mandate for marine and coastal biota.

With respect to inland aquatic biodiversity, the Conference of the Parties approved a work plan on inland water ecosystems with Decision IV/4: 'Status and Trends of the Biological Diversity of Inland Water Ecosystems and Options for Conservation and Sustainable Use'. This document is divided into four sections. Section II (Status and Trends of Biological Diversity of Inland Water Ecosystems) recognises three key characteristics of the biological diversity of inland waters:

- The biological diversity of inland waters relies on ecosystems and habitats containing high diversity and large numbers of endemic and threatened species, which are unique or associated with key ecological processes.
- Inland water ecosystems perform valuable ecological functions, and inland water species, genomes and genes are of social, scientific and economic importance.

- The essential ecological functions performed by inland water ecosystems include, *inter alia*, maintenance of the hydrological balance, retention of sediments and nutrients, and provisions of habitats for various animals, including migratory birds and mammals. Other ecosystem functions are the breakdown of anthropogenic pollutants and the sequestering of excess nutrients.

This document constitutes a programme of work on the biological diversity of inland water ecosystems. Elements of this programme include: identifying information gaps and needs to be addressed in order to obtain a global assessment of the biological diversity of inland waters; developing regional guidelines for assessments; applying the ecosystems approach; integrating the consideration of the biological diversity of inland waters into sectoral planning; restoring and rehabilitating ecosystems; education and public awareness; traditional knowledge; and the development of indicators.

FAO Code of Conduct for Responsible Fisheries

The FAO Conference, at its 28th Session on 31 October 1995 adopted a Code of Conduct for Responsible Fisheries along with Resolutions 4/95 and 5/95 calling on FAO, member states and all those involved in fisheries to implement the Code. The Code is a voluntary instrument and a key one in that it deals, among other things, with fisheries management and development, embracing conservation and environmental issues while taking into account social and economic considerations.

The Code also constitutes an important contribution to the implementation of relevant international instruments, because it was formulated in such a way that its interpretation and application conform to the relevant rules of international law. It is a comprehensive document that addresses all those involved in fisheries and applies to all types of fisheries, both within the exclusive economic zone and on the high seas, in inland waters, as well as aquaculture. In addition to its general principles, six of the Code's articles address substantive technical areas including fisheries management, fishing operations, aquaculture development, integration of fisheries into coastal area management, post-harvest practices and trade and fisheries research.

Technical guidelines have been prepared to implement the Code for inland fisheries (FAO, 1997a). The guidelines have no formal legal status. Instead, they are intended to provide general advice in support of the implementation of Articles of the Code pertinent to the development and management of inland fisheries, which are: Article 6: General Principles; Article 7: Fisheries Management; Article 9: Aquaculture Development; and Article 10: Integration of Fisheries into Coastal Area Management.

The guidelines include some crucial definitions. 'Inland fisheries', for example, is defined as 'any activity conducted to extract fish and other aquatic organisms from inland waters'. The guidelines note a special characteristic of inland fisheries, namely, that most impacts to the resource originate from outside the sector. Implementation of the Code with respect to inland fisheries, therefore, inevitably involves negotiation with external sectors. Development of inland fisheries shares many of the same problems found in aquaculture, in that efforts are made to increase productivity of the fishery through enhancements in order to meet societal demands (Bartley & Pullin, 1999).

UNESCO Convention

The UNESCO Convention Concerning the Protection of the World Cultural and Natural Heritage (1972) protects both 'cultural heritage' and physical and biological formations and areas which are of 'outstanding universal value'. Parties to this Convention commit themselves to do 'all (they) can' to accomplish the goals of: identification, protection, conservation, presentation and transmission to future generations of cultural and nature sites (whether listed or not). To this end, a party must act 'to the utmost of its own resources and, where appropriate with … international assistance and co-operation'.

Each state, in fulfilling these commitments, shall endeavour:

(a) to adopt a general policy which aims to give the cultural and natural heritage a function in the life of the community and to integrate the protection of that heritage into comprehensive planning programmes;
(b) to set up within its territories, where such services do not exist, one or more services for the protection, conservation and presentation of the cultural and natural heritage with an appropriate staff and possessing the means to discharge their functions;
(c) to develop scientific and technical studies and research and to work out such operating methods as will make the state capable of counteracting the dangers that threaten its cultural or natural heritage;
(d) to take the appropriate legal, scientific, technical, administrative and financial measures necessary for the identification, protection, conservation, presentation and rehabilitation of this heritage; and
(e) to foster the establishment or development of national or regional centres for training in the protection, conservation and presentation of the cultural and natural heritage and to encourage scientific research in this field.

Ramsar Convention

The Ramsar Convention on Wetlands of International Importance Especially as Waterfowl Habitat (1971) is designed to regulate the deterioration of wetlands by human activity (drainage, pollution or saltwater intrusion). For this Convention, the term 'wetlands' includes swamps, bogs, salt-water marshes and other wet areas, including marine areas with a depth at low tide of 6 m or less. This Convention requires the member states to:

- designate areas for listing as wetlands of international importance (based primarily on their use by waterfowl). A state is required to list at least one wetland, by submission of a precise decision or map at the time it becomes a party to the Convention;
- promote conservation of listed wetlands;
- promote conservation of unlisted wetlands;
- promote the conservation of waterfowl.

In addition, the Convention requires parties to protect non-listed wetlands 'by establishing nature reserves ... and providing adequately for their wardening'. Loss of wetlands should be avoided and minimised. Where all or part of a listed wetland is lost, other areas should be listed and protected in its stead (Young, 1993). It is also important to note that whereas the Ramsar convention has historically used criteria for listing sites based on their importance to birds, fish are now recognised as a basis for listing new sites (Ramsar Convention, 1996).

National legislation

National legislation bears directly on the relationship between the fishery and society. Its role is threefold: to ensure that the benefits of the fishery are distributed to the society as a whole, to protect the fishery and ensure its sustainability, and to protect the fishermen by providing a legal framework in which they can operate. Traditionally, these functions have been performed by central governments with a tendency to strong centralised regulations applicable to the fishery as a whole. Modern trends to decentralisation require changes to this model. It is not the purpose of this section to provide a comprehensive survey of legislation applicable to inland fisheries but rather to highlight some of the principal issues and trends by referring to specific examples.

Defining inland fisheries and their placement within the legislative framework

One characteristic that emerges from a broad sampling of national laws is the lack of uniformity with respect to the definition of inland fisheries, and the location of the subject within the legislative framework. In some countries, inland fisheries are dealt with in specific legislation. In France, for example, the subject is covered by the law of 1984 regarding 'freshwater fisheries and the management of fisheries resources in inland waters',[2] which recognises that the protection of the aquatic environment and the fisheries patrimony is of general interest.[3] In many other countries, however, there is no separate legislative treatment of inland fisheries, and no clear definition of the term. At times the subject is treated as a subcategory of maritime fisheries law, while in other cases it may be nested within a country's general environmental legislation.

As a recent review of relevant legislation in Latin America indicates,[4] most of the legislation in that region does not include a specific definition of inland fisheries. Most of the countries regulate inland fisheries in the maritime fisheries law. This is the case of Chile, for example, where the Fisheries and Aquaculture Law of 1989,[5] as amended, applies to all fishing activities undertaken in terrestrial waters, inland waters, territorial sea and the Exclusive Economic Zone (EEZ) or in adjacent areas to the EEZ where there is national jurisdiction in conformity with laws and international treaties. Similarly, in Peru, Decree no. 25.977 enacts the General Fisheries Law, which is applicable to both inland and marine fisheries.

It is also noteworthy that when referring to inland fisheries, lawmakers use different terms such as terrestrial waters,[6] river and lagoon waters,[7] and freshwaters.[8] Thus, there is a tendency for legislation in several countries to treat the subject in a fragmented, location-specific manner, rather than treating the resource in an integrated way. There are some exceptions such as Mexico, where the Fisheries Regulation of 1992, in its Article 2 applies the concept of inland fisheries to 'the streams, rivers, lakes, lagoons and reservoirs in the federal jurisdiction'.

Creating institutional linkages between inland fisheries and water management: the case of the United Kingdom

Ideally, management of inland fisheries requires close co-ordination between fisheries institutions on the one hand, and institutions responsible for water management on the other. Not infrequently, however, this type of co-ordination is hampered by the fact that the institutions involved are embedded within their own legislative traditions; that is, there are separate and distinct bodies of fisheries law and water law, between which the linkages are often weak or unclear.

A good example of an attempt to deal with this problem is provided by the UK, where inland fisheries is principally regulated by the Salmon and Freshwater Fisheries Act,[9] a 1975 consolidation of the Salmon and Freshwater Fisheries Act 1923 and certain other enactments relating to salmon and freshwater fisheries. The Act vests the power to regulate the fishing of salmon and trout in water authorities [later succeeded by the National Rivers Authority (NRA), now renamed and restructured into the Environment Agency (EA), see below]. Similarly, water authorities may regulate other freshwater fisheries including fish of any description within their area.

Although the legal basis for the regulation by the NRA of fisheries is provided in the 1975 Salmon and Freshwater Fisheries Act, the institutional set-up for inland fisheries can be found in the Water Resources Act 1991. This Act vests power to maintain, improve and develop salmon fisheries, trout fisheries, freshwater fisheries and eel fisheries in the NRA (S.114), with these activities to be undertaken with direct reference to the 1975 Salmon and Freshwater Fisheries Act. The Water Resources Act provides that the Minister, on application by the Authority, may by Order (a published Statutory Instrument), for any area described in the Order, modify any provisions of the 1975 Act in relation to the regulations of fisheries, or any provision of a Local Act relating to any fisheries in the area (apparently, there are some powers vested in local authorities with respect to fisheries). Such a modification is subject to quite heavy procedures, and it shall not apply to or affect any fish-rearing licences.

The NRA was replaced by the EA. The EA's mandate was extended to include waste management and pollution control functions, but its organisational structure and fisheries mandate were left unchanged, and its operations are carried out in much the same manner. The EA is a national para-statal agency or quango (quasi non-governmental agency), with a large degree of independence from the government. In all, the Agency is responsible for water management (flood protection, water abstraction, groundwater, water pollution, land drainage), pollution control, waste management, inland and coastal fisheries and navigation. There is an arms-length relation to the Secretary of State for the Environment, who can give directions to the Agency. With respect to inland fisheries, both the Secretary for the Environment and the Minister of Agriculture, Food and Fisheries may give directions. It appears that the two ministers have to agree on such directions. Some of the regulatory powers are vested in the Minister or Secretary.

The Board of the Agency consists of between 8 and 15 members, two of whom are appointed by the Minister for Fisheries. Eligible appointees are persons of experience and shown capacity in areas relevant to the functions of the Agency. Significantly, there is no legal requirement to appoint representatives from various sectors or local government.

Whilst being a national agency, its operations are carried out on the basis of river basins, as was the case with the predecessor water authorities. Management decisions and regulation are as a matter of operational practice based on a system of integrated resource planning. Typically, such plans designate priority uses of water resources – including fisheries – and specific measures to accommodate such uses or to harmonise multiple use of water resources. There is no legal obligation to prepare such plans and there is also no obligation to consult on such plans as a result. In practice, there is an extensive mechanism of public consultation on the plans, including public hearings and several rounds of comments from industry, local government, non-governmental organizations (NGOs) and others.

The 1991 Water Act does provide for the establishment of local and regional fisheries advisory committees (S.8). The committees are to be consulted by the Agency in respect of its inland fishery functions. Members are those people who appear to be interested in trout, salmon, eel or other fisheries'.

The problem of inland fisheries in a federal system: the case of Sudan

The management of inland fisheries encounters special problems in the context of federal systems, where jurisdiction over river basins or other water-bodies may be divided among competing state governments. An example of such problems, and an ongoing attempt to deal with them through legislative reform, can be found in Sudan.

From 1991 onwards, Sudan[10] has gone through a number of important constitutional developments. A series of Presidential Decrees has established a federal system of government, most notably by the creation of 24 Federal States in 1994.[11] These developments are to be consolidated in a new Constitution that has been approved by the General Assembly and is now going through a series of state-level referenda.

The creation of a state level of government entails that a number of legislative powers and competencies is devolved to that level of government, yet at the same time certain functions are retained at the federal level. With regard to inland fisheries, the Federal Ministry of Animal Resources is responsible for the promotion and development of fisheries resources and aquatics, as well as policies and arrangements for their conservation and exploitation. General policy development is thus vested in federal institutions.

At state level, however, inland fisheries matters are within the purview of state government institutions, usually in the hands of state ministries for animal resources. In the area of inland fisheries, as well as in other areas, national fisheries legislation remains in force unless and until it has been replaced by state legislation. So far, three states (Khartoum State, White Nile State and Upper White Nile State) have enacted state fisheries laws. The state-level laws replace the existing national Freshwater Fisheries Ordinance from 1954, but are generally similar to the 1954 Ordinance.

The 1954 Ordinance contains a number of regulatory instruments, including the requirement for each fisherman to have an annual licence, and for each fishing boat to be registered and licensed annually. The ordinance further prohibits the introduction of non-indigenous species and the use of poison or electricity as means to capture fish, unless under a special licence. The Minister may make regulations with respect to types of nets and gear to be used in fishing and, again by regulation, may declare closed areas and seasons. Recent recommendations made by FAO to modernise the ordinance include the introduction of more transparency in licensing procedures, provisions on data collection, a framework to guide policy making and planning, and improved enforcement mechanisms.

There remains, however, a degree of uncertainty as to which level of government has competence to enact fisheries legislation. This uncertainty has not been clarified yet, which is mostly a result of the fact that the federalisation process is of such recent nature.[12] From a fisheries management point of view, the main problem with a state-level approach is that the fish resources are almost exclusively within the Nile river system, which is a transboundary river system crossing many state boundaries. The institutional issue described above is thus highly relevant if an effective fisheries management regime is to be established that addresses the river system as a whole. Such an approach would be hampered by the existence of a large number of autonomous states that may have varied approaches to fisheries management.

In contrast, outside the Nile river system, inland fishing takes place for the most part in lakes, or tanks, whether perennial or seasonal. In these instances, there are few transboundary issues to be addressed as the fisheries resources are not located in rivers that cross many administrative boundaries. Some types of inland fishery are thus truly 'local' activities that could be managed effectively at a level other than the central government level.

In light of these issues, therefore, recent draft legislation seeks to apply different governance approaches to different types of inland water. The draft Inland Fisheries Management and Conservation Bill distinguishes the inland waters of Sudan between those confined within the boundaries of individual states and those that cross one or more state boundaries. The latter largely comprise the Nile river system. According to the draft, responsibilities for fisheries management for the 'Inter-state fishery waters' is vested in the Federal Government, in particular the Director of Fisheries. Other fishery waters remain within the purview of state fisheries authorities.

Creating a legal framework for co-management of inland fisheries: the case of Zambia

In many natural-resource sectors, increasing attention is being paid to the importance of involving local people in the management of local resources on which they depend. Yet traditional natural resources laws typically place virtually all management rights and control in government, and fail to provide a suitable framework within which local communities and governments can negotiate mutually beneficial and legally secure partnerships. An increasing number of countries is addressing these shortcomings by amending laws to provide clearer mechanisms for co-management of local resources. Such a reform effort in the case of inland fisheries is currently underway in Zambia.

In Zambia,[13] the Fisheries Act vests responsibility for the control and management of fisheries in the Ministry of Agriculture, Food and Fisheries, in particular the Director of Fisheries, head of the Department of Fisheries (DOF). The administration of the Act is carried out through four subdivisions within the Ministry, i.e. fisheries research, aquaculture research, fisheries management and aquaculture extension. In carrying out its management functions the fisheries management division is supported by fisheries officers in various provinces, who are in turn supported by fisheries officers at the District level.

The Fisheries Act No. 21 of 1974 and its four implementing regulations and orders from 1986 provide the legal framework for fisheries management. The Act prohibits certain methods of fishing and the introduction of any species of fish or the importation of any live fish without the written permission of the Minister. The Act empowers the Minister to declare any area of water as a prescribed area for purposes of recreational, subsistence or research fishing, to determine the method of fishing and to issue a Special Fishing Licence with conditions for specific activities in a specified area. The Act also regulates commercial fishing through the declaration, by the Minister, of any area of waters as a commercial fishing area for which regulations may be made to prohibit, restrict or regulate fishing, the method of fishing, prescribe closed fishing seasons, prescribe the type of licences for commercial fishing and prescribe records to be kept and information to be provided by persons fishing in a commercial fishing area. In addition, the Minister may require the registration of fisherman and appoint a committee to co-ordinate and improve commercial fishing activity for any commercial fishing area.

The basic management tools under the Act are the minimum necessary to manage fisheries, but the experience in Zambian fisheries demonstrates that the legal mechanisms are inadequate, largely because of the administration's inability to implement them.[14] This is the result of a lack of financial resources and logistical support, lack of staff and inadequate training, among other factors. In addition, it is worth noting that following a restructuring at the ministerial level in which the DOF was separated from the Ministry of Parks and Wildlife, the Department of Fisheries has lost access to a significant enforcement apparatus – including strong legal powers and human resources – that existed within the Ministry. Under the new structure, enforcement is distinctly weaker.

In 1994, the Government of Zambia established and put into operation a community-based management programme, or co-management, of the fisheries in Lake Kariba. In view of many of the implementation and enforcement problems surrounding fisheries management, it is now intended to apply this form of management more widely, and proposals are being made to provide the necessary basis for the community approach. The basic components of co-management, based on Lake Kariba fishery's experience, are: (1) the relocation, regrouping and organisation of the numerous fishing villages and camps that were established by fishers of the inshore fisheries on Lake Kariba into fewer and bigger villages and camps; (2) the setting up of Village Management Committees (VMCs) for these villages and camps; (3) the division of the lake and lake shore area into zones; and (4) the establishment of zonal committees chaired by traditional Chiefs that supervise, assist and co-ordinate the work of VMCs of the recognised fishing villages and camps.

Co-management is aimed at monitoring the implementation of fisheries management programmes and enforcement of regulations, and the facilitation and promotion of development projects in the fishing villages through its machinery.

The proposed framework for co-management envisages the VMCs to operate under the supervision of the Zonal Committees who are, in turn, supervised by a Fishery Management Board for the fishery area. The Fishery Management Board will report to a management authority whose establishment is under a new Fisheries Act. The other notable features of the co-management system are the powers vested in the VMCs to collect fees, the maintenance of accounts by the Fishery Board,[15] the need for authorisation from the Director of DOF to build houses, the issuance of fishing rights and the VMCs' ability to initiate socio-economic projects.

In support of the proposed changes described above, a Fisheries Bill is currently under consideration that solidifies a framework for decentralised fisheries management in line with the Lake Kariba experiences. The Bill provides for the declaration of Fishery Management Areas (S.12), for which area a fishery management plan is to be prepared. The plan shall be prepared in consultation with the Fisheries Management Board, and must include indications as to how local communities will be involved in the implementation of the plan. The Fisheries Management Boards consist of 12 members, nine of whom are 'members representing traditional rulers, fishing communities and non-governmental organisations in the fishery area' (S.21). In addition, such members must all be residents in the fishery area.

The Bill also provides for the possibility further to fine-tune community-based management through regulations. The Minister may make regulations to 'establish and regulate community based or other decentralized fisheries management and any aspect of their operation including the granting or delegation of powers consistent with this Act to any person, classes or groups of persons to enforce or implement fisheries conservation and management measures and regulations' [S.59 (2)(l)]. Regulations may also provide for the control, development and management of any fishery management area and the development, establishment and implementation of integrated community projects within fishery management areas.

Regional legislation

In international rivers and lakes, authority for all or part of the legislation guiding the fishery of that particular water may be assigned by the riparian countries to an international fishery body. The oldest fishery commission in the world is that for Lake Constance, which was formed in the nineteenth century with the specific task of managing the lakes fisheries. Similar commissions exist for the Great Lakes of North America, the Caspian Sea and Lake Victoria. These Commissions usually have a role in apportioning catch among member countries, in deciding on mesh and other restrictions, and on the licensing of fishermen. They may also play a role in negotiating and setting environmental criteria. A similar role may be played in rivers which also have basin commissions such as the Danube, the Uruguay and the Niger. Usually such bodies are more preoccupied with navigation, water rights and total discharge. However, they have a valuable potential role to play in protecting fisheries through guaranteeing passage of river migrants, keeping pollution within acceptable limits, controlling access to the fishery, etc.

Notes

1. By C. Leria, LEG, FAO, Via delle Terme di Caracalla, Rome, Italy.
2. Law no. 84–512, 29 June 1984, *Journal Officiel de la Republique Francaise*, **2039**, 30 June 1984.
3. Article 2.

4. Lería, 'Preliminary Analysis of Legislation referring to inland fisheries in Latin America', COPESCAL–Seventh Session, 16–20 January 1995.
5. Chile, Law no. 18.892 of 1989 'Ley General de Pesca y Acuicultura'.
6. Aquaculture and Fisheries Law of Chile, No. 18.892 of 1989.
7. Uruguay, Decree 711/971, Article 3.
8. Guatemala, Governmental Decree 1235, Article 117.
9. 1 August 1975.
10. Information on Sudan is almost entirely derived from S. Hodgson, Preliminary Report on Inland Fisheries Legislation, prepared under FAO project TCP/SUD/6611.
11. S. Hodgon, Preliminary Report on Inland Fisheries Legislation, p. 5.
12. See S. Hodgson, Preliminary Report on Inland Fisheries Legislation, for a detailed discussion.
13. The information on Zambia is largely derived from B. Kuemlangan, Zambia Draft Fisheries Legislation, Preliminary Report, prepared under TCP/ZAM/6613.
14. B. Kuemlangan, Zambia Draft Fisheries Legislation, Preliminary Report, p. 25.
15. Funds for these accounts are portion of licence fees and fines retained by the Board, voluntary contributions and grants.

Chapter 17
Conclusion

This review of inland fisheries and of approaches to their conservation and management is based on knowledge accumulated over many decades. Advances in inland water science have not been uniform. Lakes and small temperate streams were studied relatively early and gave rise to the science of limnology around the beginning of the twentieth century. These systems were already being managed through stocking and environmental enhancement in the latter half of the nineteenth century. Research into reservoirs followed the wave of dam construction of the 1970s. Investigation of large rivers is even more recent and was systematically presented for the first time at the Large River Symposium in 1985 (Dodge, 1989). Throughout most of the history of inland water science, efforts concentrated on the biological aspects and in most cases worked with contemporary ecological theory and fish population dynamics.

Changing patterns for the resource

More recently, however, a vision has emerged of inland waters as functional parts of the landscape subject to societal pressures and values. Policies for management of any natural resource are conditioned by a number of factors that are liable to change. These include:

- *Public perception*: This depends on the political climate of the time. Until recently public attitudes were governed by strongly centrist systems of organised religion and centralised nation states and on the whole adhered to a coherent world-view whereby man was the master of all. This world-view has given way, surprisingly rapidly, to one in which the natural world is perceived as having 'rights' that humankind is obliged to conserve and support, a process formalised through the Biodiversity Convention. This view has arisen coincident with the rise in awareness of our dependency on the harmony and diversity of the natural world. Part of this process is the intent to share equitably the benefits of fisheries and to devolve control of the fishery from central government to local authorities and even fisher communities.
- *Use patterns*: The classic conflict between use and preservation is resolved in different ways depending on human pressures. In poor areas there is a tendency to exploit rather than conserve, whereas in wealthier parts of the world the emphasis

is reversed. In some circumstances this balance is determined by population, although some of the most densely populated areas are also the most conservation minded. Temporary political instabilities and warfare can also disrupt normally protective systems, inflicting local and temporary damage. Decisions can also be linked to pressure from special-interest groups. A typical example of this is the move from food to recreational fisheries encouraged by the emergence of affluent and influential urban groups.

- *Demography*: Apart from the sheer pressure of numbers, use patterns are liable to change with population distribution. Recent trends towards urbanisation have decreased diffuse rural pollution but increased point-source urban-generated discharges. Shifts of populations away from the hinterland to the coasts and to the riparian areas of lakes or rivers have also placed pressure on aquatic environments. For example, the elimination of the host of river blindness in West Africa has allowed people to occupy river valleys that had hitherto been shunned. The increased population density has in some cases resulted in rapid removal of the riparian gallery forests, increases in siltation and pressure on the fishery.
- *Changes in the nature of the resources*: While the generalised ecological principles governing the functioning of aquatic ecosystems are constant, fishery resources are constantly changing in response to changing land-use patterns in the river or lake basin and to climatic variation. Introductions of new species can change the composition and type of the resource, as in Lake Victoria, where the introduction of a large predator changed the fishery from a local subsistence activity to a commercial operation for export. Eutrophication of a water body can alter the whole population pattern. Damming can produce far-reaching changes in the nature of the upstream and downstream ecosystems, and so on.

These changes are not slow. There has been a rapid acceptance of the Biodiversity Convention and an application of its provisions in many parts of the world. Demographic shifts can happen very quickly following an enabling change such as the removal of a disease vector. Shifts in land-use pattern through deforestation can equally alter the whole nature of an ecosystem in a matter of years. These changes are not permanent. Policies for decentralisation, demographic shifts and land-use patterns can be reversed and fisheries can collapse.

All this means that any decisions for the management of any fishery can no longer be regarded as long term. Management today involves uncertainty and requires corresponding flexibility and responsiveness to change. For this reason there should be less reliance on centralised legislation with rigid enforcement. Instead, more emphasis should be placed on consultative and adaptive systems that are able to accommodate change and where regulation becomes self-imposed. To do this the fishery needs to be kept under review through monitoring of the resource, and of the social and economic climate under which it is pursued. The results of the monitoring have to be incorporated into future plans taking into account past history, present performance and future circumstance.

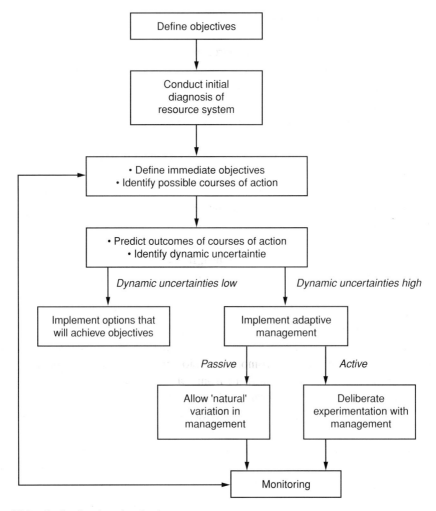

Fig. 17.1 Cycle of actions in adaptive management.

Monitoring

This process becomes a continuous cycle of formulation, implementation, monitoring and review, which should involve all stakeholders (Fig. 17.1). Monitoring at this stage can be carried out at different levels. At the most basic level landing committees, parish or village councils should give their interpretation of the success or failure of the policies as formulated. They should also be involved in the collection of statistics on landings, the gear used and the number of people fishing. At the district level consultations among a number of villages should enable some broader appreciation at a larger geographical scale as well as providing a level of scientific assessment. Finally, central government should support and validate more local efforts through more formalised, periodic scientific and socio-economic surveys.

Needs for participatory management

The current philosophy for natural resources management is to weaken centralised control and to place the task of management more with local communities. Clearly, this shift in emphasis will work only if the devolution of control proceeds in an orderly manner with the establishment of appropriate infrastructure. Furthermore, in order for all to participate fully in the shift towards participatory management certain preconditions are needed.

Firstly, the level of education has to be raised. The level of knowledge required for participatory management and the operation of high-input, enhanced systems implies a greater level of knowledge than that which is needed for a simple extractive activity. In some countries the development of the new intensified systems of management has been accompanied by the emergence of research, often empirical, on the most appropriate ways to manage the systems. For example, there is already a wealth of information in countries such as China, Cuba, India and Poland on stocking, feeding and fertilisation ratios. Research alone is not enough, as the extension of the results and the application of the formula by the fishermen also require a higher level of preparation at their level. This trend reinforces the drift from simple fishermen to a more complex organisation of the fishery. Governments can support the spread and more effective application of conservation, management and enhancement strategies through research. They can also promote extension and training to upgrade the capacity of fishermen to use this approach to management. The level of knowledge and capacity on the part of local government level is also higher in that it is now called upon to provide advice and co-ordinate activities in sectors of which it had little prior knowledge

Secondly, finance has to be made available through either government or private financiers to fund the increased investment in the fishery and the natural environment. The transition to more locally managed systems has impacts on the way in which fisheries are funded. In developed economies where the emphasis has been on recreational fisheries, fishing clubs and associations are required to an increasing degree to take over the costs of stocking from the government agencies which previously carried out this function. Enhancement of fisheries implies increased costs to the fishery managers in fish seed, fertilisers and labour for species elimination, system modification, etc. The original fishermen are unlikely to be able to mobilise sufficient funds and funding of stocking tends to shift to external financing agencies. The greater predictability, more controlled harvesting season and more concentrated harvest attract investment from businessmen who then take the major part of the profit. This means that the intensification process is frequently accompanied by a loss of independence by the fishermen and a growing dependence on credit or other external forms of funding. As a partial solution to these problems governments should consider subsidies to enhancement, either through direct grants or through the creation of state hatcheries, at least in the earlier phases of the adoption of the technology. That such a transition can be successfully negotiated was demonstrated when the centrally planned economies of Eastern Europe and the Russian Federation collapsed and the role of central government in subsidising the fishery by producing stocking material from government hatcheries disappeared. As a result, the private sector and fishermen's co-operatives have had to develop their own hatcheries.

On the whole, existing financial institutions, especially in developing countries, do not support investment in fisheries and aquaculture. As a consequence, alternative sources of funding need to be established. Governments should also encourage the setting up of rural financing institutions to assist fishermen and rural communities in developing the sector without the intervention of third-party financiers. An alternative is to form co-operatives where the entire sector is vertically integrated. In this way habitat improvement, hatchery operation, harvest, processing, distribution, marketing and sales can all be performed by different members of the co-operative. Losses in one section can be compensated for by large profits in another. Japan is currently employing this strategy in a formal way. Hatchery enhancement of white seabass in California is approaching such a system informally by requesting sport-fishing clubs to help with the culture and release of fish hatched in a state-supported hatchery. This, in turn, receives much of its funding from special licences sold to commercial and sport fishermen (Kent & Drawbridge, 1996).

In areas with intensive food fisheries involving large numbers of fishermen sustainability may only be achieved if effort is withdrawn from the fishery. This means that alternative occupations will have to be found for dispossessed fishermen. Here the answers rarely lie with subsistence agriculture; rather, intensified culture of specialised crops for urban supply or for export will be more appropriate. This will need financial inputs through appropriately constituted rural banks or foreign investment.

Habitat rehabilitation is also an area needing considerable amounts of funds. In many areas local governments and property developers are supporting small-scale rehabilitation schemes for aesthetics and marketing. Additional value can be added to such schemes by building in features that favour fish and other forms of wildlife with little additional cost. Larger schemes, however, involve considerable costs in the form of engineering for the realignment of levees and modifying of river channels, as well as the cost of buying back occupancy of the floodplain. This requires a major commitment on the part of society, which is only recently becoming possible as the additional advantages in the form of flood protection and other ecosystem services become apparent. In some cases, where countries are too poor to allow for this kind of undertaking, costs are borne by the international community through the Global Environment Program.

While further research is always desirable, knowledge of inland waters, their fish populations and fisheries is currently at a technical level that is sufficient to allow humanity to manage these resources according to the aspirations expressed in the Convention on Biological Diversity and the Code of Conduct for Responsible Fisheries. It is also adequate for the management of inland waters for the sustained satisfaction of human needs. Currently, however, social, economic, policy and institutional issues impede application of the technical know-how. Efforts are now focusing on ways in which to resolve them so that inland fisheries can be better integrated into the whole field of natural resource management.

References

A

Aass, P., Nielsen, P.S. & Brabrand, A. (1989) Effects of river regulation on the structure of a fast-growing brown trout (*Salmo trutta* L.) population. *Regulated Rivers: Research and Management*, **3**, 255–66.

Abbot, J. & Gujit, I. (1998) *Changing Views on Change: Participatory Approaches to Monitoring and Evaluation*. International Institute for Environment and Development (IIED) Sustainable Agriculture and Rural Livelihoods, London.

Aguilar-Manjarrez, J. & Nath, S.S. (1998) A strategic reassessment of fish farming potential in Africa. *CIFA Technical Paper 32*, FAO, Rome.

Allan, J.D. & Johnson, L.D. (1997) Catchment-scale analysis of aquatic ecosystems. *Freshwater Biology*, **37**, 107–11.

Allan, J.D., Erickson, D.L. & Fay, J. (1997) The influence of catchment land use on stream integrity across multiple spatial scales. *Freshwater Biology*, **37**, 149–61.

Allen, K.R. (1971) Relation between production and biomass. *Journal of the Fisheries Research Board of Canada*, **28**, 1573–81.

Allendorf, F.W. & Leary, R. (1986) Heterozygosity and fitness in natural populations of animals. In: *The Science and Scarcity of Diversity* (ed. M.E. Soule), pp. 57–76. Sinauer, Sunderland, MA.

Angermeier, P.L. & Schlosser, I.J. (1987) Assessing biotic integrity in the fish community in a small Illinoise stream. *North American Journal of Fisheries Management*, **7**, 331–8.

Arnold, W.S. & Norris, H.A. (1998) Integrated resource management using GIS: shellfish aquaculture in Florida. *Journal of Shellfish Research*, **17**, 318.

Arnold, W.S., Norris, H.A. & Berrigan, M.E. (1996) Lease site considerations for hard clam aquaculture in Florida. *Journal of Shellfish Research*, **15**, 478–9.

B

Backiel, T. & Le Cren, E.D. (1978) Some density relationships for fish population parameters. In: *The Ecology of Freshwater Fish Production* (ed. S.D. Gerking), pp. 279–302. Blackwell, Oxford.

Bagenal, T. (ed.) (1978) *Methods for the Assessment of Fish Production in Fresh Waters*, 3rd edn. Blackwell Scientific Publications, Oxford.

Banarescu, P.M. (1965) *Pisce-Osteichthyes*. Fauna R.P.R., Bucuresti.

Baras, E. (1991) A bibliography on underwater telemetry. *Canadian Technical Report of Fisheries and Aquatic Sciences* **1819**: 1–55.

Baras, E. & Lagardère, J.P. (1995) Fish telemetry in aquaculture: review and perspectives. *Aquaculture International* **3**, 1–26.

Baras, E. & Philippart, J.C. (Eds.) (1996) *Underwater Biotelemetry*. University of Liège, Liège.

Baras, E. *et al.* (2000) A critical review of surgery techniques for implanting telemetry devices into the body cavity of fish. In *Proceedings of the 5th European Conference on Wildlife Telemetry*, (ed. Y. Le Maho). CNRS Éditions, Paris.

Barrett, G.W., Van Dyne, G.M. & Odum, E.P. (1976) Stress ecology. *BioScience*, **26**, 192–4.

Barthem, R.B., Lambert de Brito, M.C. & Petrere, M. (1991) Life strategies of some long distance migratory catfish in relation to hydroelectric dams in the Amazon basin. *Biological Conservation*, **55**, 339–45.

Bartley, D.M. (1995) Marine and coastal area hatchery enhancement programmes: food security and conservation of biological diversity. Thematic paper presented at the *International Conference on the Sustainable Contribution of Fisheries to Food Security*, Kyoto, Japan, 4–9 December 1995.

Bartley, D.M., Kent, D.B. & Drawbridge, M.A. (1995) Conservation of genetic diversity in a white seabass (*Atractoscion nobilis*) hatchery enhancement programme in southern California. *Special Symposium Proceedings of the American Fisheries Society*, **15**, 249–58.

Bartley, D.M. & Pullin, R.S.V. (1999) Towards policies for aquatic genetic resources. In *Towards Policies for Conservation and Sustainable Use of Aquatic Genetic Resources* (eds R.S.V. Pullin, D.M. Bartley & J. Kooiman), pp. 1–16. ICLARM, Penang.

Baskin, Y. (1995) Ecosystem function of biodiversity. *BioScience*, **44**, 657–60.

Bayley, P.B. (1983) *Amazon fish populations. Biomass, production and some dynamic characteristics*. Dissertation, Dalhousie University.

Bayley, P.B. (1988) Factors affecting growth rates of young tropical floodplain fishes: seasonality and density-dependence. *Environmental Biology of Fishes*, **21**, 127–42.

Bazigos, G.P. (1974) The design of fisheries statistical surveys – inland waters. *FAO Fisheries Technical Paper No. 133*, FAO, Rome.

Beamesderfer, R.C.P., Rien, T.A. & Nigro, A.A. (1995) Differences in the dynamics of potential production of impounded and unimpounded white sturgeon populations in the Lower Columbia River. *Transactions of the American Fisheries Society*, **124**, 857–72.

Beddington, J. & Cooke, J.G. (1983) The potential yield of fish stocks. *FAO Fisheries Technical Paper No. 242*, FAO, Rome.

Begon, M. & Mortimer, M. (1986) *Population Ecology. A Unified Study of Animals and Plants*, 2nd edn. Blackwell Scientific Publications, Oxford.

Belk, M.C. (1993) Growth and mortality of juvenile sunfishes (*Lepomis* sp.) under heavy predation. *Ecology of Freshwater Fish*, **2**, 91–8.

Berka, R. (1990) Inland capture fisheries of the USSR. *FAO Fisheries Technical Paper No. 311*, FAO, Rome.

Berkes, F. (1995) Community-based management of common property resources. In: *Encyclopaedia of Environmental Biology*. Vol. 1, pp. 371–3, Academic Press, New York.

Beverton, R.J.H. & Holt, S.J. (1957) On the dynamics of exploited fish populations. *MAFF, Fisheries Investigations Series* II, **19**, 1–553.

Beyer, J. & Sparre, P. (1982) Modelling exploited fish stocks. Danish Institute for Fisheries and Marine Research Charlottenlund Slot, Charlottenlund.

Bhukaswan, T. (1988) The use of cyprinids in fisheries management of the larger inland water bodies in Thailand. In: Indo-Pacific Fisheries Commission, Papers contributed to the Workshop on the Use of Cyprinids in the Fisheries Management of Larger Inland Water Bodies of the Indo-Pacific (ed. T. Petr). *FAO Fisheries Report 405*, 142–50.

Bigler, B.S., Welsch, D.W. & Helle, H.J. (1996) A review of size trends among North Pacific salmon (*Onchorhynchus* spp.). *Canadian Journal of Fisheries and Aquatic Sciences*, **53**, 455–65.

Bninska, M. (1985) The effect of recreational uses upon aquatic ecosystems and fish resources. In: *Habitat Modification in Freshwater Fisheries* (ed. J.S. Alabaster), pp. 223–35. Butterworth, London.

Bonetto, A.A. (1976) Calidad de las aguas del Rio Parana – introduccion a su estudio ecologico. Corrientes Argentina, Instituto Nacional de Ciencias y Tecnica Hidrica (INCYTH), 202 pp.

Botsford, L.W., Castilla, J.C. & Petersen, C.H. (1997) The management of fisheries and marine ecosystems. *Science (Washington)*, **277**, 509–15.

Bouvet, Y., Pattee, E. & Meggouth, F. (1985) The contribution of backwaters to the ecology of fish populations in large rivers. Preliminary results on fish migrations within a side arm and from the

side arm to the main channel of the Rhone. *Verhandlungen des Internationalen Vereins fur Theoretisce un Angerwonte Limnoogie*, **22**, 2576–80.
Bromley, D.W. (1992) The Commons, Property and Common Property Regimes. *Making the Commons Work: Theory Practice and Policy* (ed. D.W. Bromley), pp. 3–16. ICS Press, San Francisco.
Bronte, C.R., Elgeby, S.J.H. & Swedberg, D.V. (1993) Dynamics of a yellow perch population in western Lake Superior. *North American Journal of Fisheries Management*, **13**, 511–23.
Brown, T.G., Barton, L. & Langford, G. (1996) The use of a geographic information system to evaluate terrain resource information management (TRIM) maps and to measure land use patterns for Black Creek, Vancouver Island. Canadian Manuscript Report of Fisheries and Aquatic Sciences/Rapport. Manuscrit canadien des sciences halieutiques et aquatiques. Imprint varies, No. 2395.
Bruton, M.N. (1995) Have fishes had their chips? The dilemma of threatened fishes. *Environmental Biology of Fishes*, **43**, 1–27.
Burton, T.M. & Prince, H.H. (1995) A landscape approach to wetlands restoration research along Saginaw Bay, Michigan: baseline data collection and project description. Proceedings of the 38th Conference of International Association of Great Lakes Research, International Association for Great Lakes Research, p. 90.

C

Caddy, J.F. & Bazigos, G.P. (1985) Practical guidelines for statistical monitoring of fisheries in manpower limited situations. *FAO Fisheries Technical Paper No. 257*, FAO, Rome.
Caddy, J.F. & Mahon, R. (1995) Reference points for fishery management. *FAO Fisheries Technical Paper No. 347*, FAO, Rome.
Cambray, J.A. (1990) Fish collections taken from a small agricultural water withdrawal site on the Groot River, Gamtoos River system. *South African Journal of Aquatic Science*, **16**, 78–89.
Campton D.E. (1995) Genetic effects of hatchery fish on wild populations of Pacific salmon and steelhead: what do we really know? *American Fisheries Society Symposium*, **15**, 337–53.
Carpenter, S.R., et al. (1987) Regulation of lake primary productivity by food web structure. *Ecology*, **68**, 1863–76.
Casley, D.J. & Kumar, K. (1987) *Project Monitoring and Evaluation in Agriculture*. Johns Hopkins University Press, Baltimore, MD.
Casley, D.J. & Kumar, K. (1988) *The Collection, Analysis and Use of Monitoring and Evaluation Data*. Johns Hopkins University Press, Baltimore, MD.
CBD (1994) Convention on Biological Diversity. Text and Annexes. Interim Secretariat for the Convention on Biological Diversity, Chatelaine, Switzerland.
Chadwick, E.M.P. (1982) Dynamics of an Atlantic salmon stock (*Salmo salar*) in a small Newfoundland river. *Canadian Journal of Fisheries and Aquatic Sciences*, **39**, 1496–501.
Chakraborty, R. & Leimar, O. (1987) Genetic variation within a subdivided population. In: *Population Genetics and Fishery Management* (eds N. Ryman and F. Utter), pp. 89–120. University of Washington Press, Seattle, WA.
Chambers, R. (1992) Rural Appraisal: Rapid, Relaxed and Participatory. IDS Discussion Paper No. 311. Institute of Development Studies, University of Sussex.
Chambers, R. (1994) Participatory rural appraisal (PRA): challenges, potentials and paradigm, *World Development*, **22**(10), 1437–54.
Chapin, F.S., Walker, B.H., Hobbs, R.J., et al. (1997) Biotic control over the functioning of ecosystems. *Science (Washington)*, **277**, 500–504.
Chapman, D.W. (1967) Production in freshwaters. In: *The Biological Basis of Freshwater Fish Production* (ed. S. Gerking), pp. 3–29. Blackwells, Oxford.

Charles, S. (1996) Etude experimentale de la qualité des alevins de saumon, *Salmo salar* L. destines au repeuplement. Contribution a l'analyse hydraulique de l'habitat (Quality of hatchery salmon, *Salmo salar* L., juveniles for restocking. Contribution to the habitat hydraulics study). Bordeaux 3 University, France.

Chevey, P. & Le Poulain, F. (1940) La pêche dans les eaux douces du Cambodge. *Memoires de l'Insitut Oceanographiuque de l' Indochine*, **5**, 193.

Chitradivelu, K. (1972) Growth, age composition, population density, mortality, production and yield of *Alburus alburnus* (Linneaus, 1758) and *Rutilus rutilus* (Linneaus, 1758) in the inundation region of the Danube – Zofin. *Acts of the University of Carolingea (Biology)*, **1972**, 1–76.

Chiyokubo, T., Shikano, T., Nakajima, M. & Fujio, Y. (1998) Genetic features of salinity tolerance in wild and domestic guppies (*Poecilia reticulata*). *Aquaculture*, **167**, 339–48.

Christensen, M.S. (1993) The artisanal fishery of the Mahakam River floodplain in East Kalimantan, Indonesia. III Actual and estimated yields, their relationship to water levels and management option. *Journal of Applied Ichyology*, **9**, 202–209.

Christensen, V. & Pauly, D. (eds) (1993) *Trophic Models of Ecosystems*. ICLARM, Manila.

Chu-Fa Tsai & Ali, L. (1987) The changes in fish community and major carp populations in beels in the Sylhet-Mymensingh basin, Bangladesh. *Indian Journal of Fisheries*, **34**, 78–88.

Clark, C.W. (1985) *Bioeconomic Modelling and Fisheries Management*. Wiley, New York.

Clay, C.H. (1995) *Design of Fishways and Other Fish Facilities*. CRC Press, Boca Raton, FL.

Corbley, K.P. (1996) National GIS fills 'gaps' in biological diversity. *GIS World*, **9**(8), 50–53.

Cordiviola de Yuan, E. (1992) Fish populations of lentic environoemnts of the Parana river. *Hydrobiologia*, **237**, 159–73.

Cowx, I.G. (1991) *Catch Effort Sampling Strategies*. Fishing News Books, Oxford.

Cowx, I.G. (ed.) (1994a) *Rehabilitation of Freshwater Fisheries*. Fishing News Books, Blackwell Science, Oxford.

Cowx, I.G. (1994b) Stocking strategies. *Fisheries Management and Ecology*, **1**, 15–30.

Cowx, I.G. (ed.) (1996) *Stock Assessment in Inland Fisheries*. Fishing News Books, Blackwell Science, Oxford.

Cowx, I.G. & Welcomme, R.L. (1998) *Rehabilitation of Rivers for Fish*. Fishing News Books, Blackwell Science, Oxford.

Craig, J.F. & Kipling, C. (1983) Reproduction effort versus environment; case histories in Windermere perch, *Perca fluviatilis* L., and pike *Esox lucius* L. *Journal of Fish Biology*, **22**, 713–27.

Crisp, D.T. & Mann, R.H.K. (1991) Effects of impoundment on populations of bullhead *Cottus gobio* L. and minnow, *Phoxinus phoxinus* L. in the basin of Cow Green reservoir. *Journal of Fish Biology*, **38**, 731–40.

Crisp, D.T., Mann, R.H.K., Cybby, P.R. & Robson, S. (1990) Effects of impoundment upon trout (*Salmo trutta*) in the basin of Cow Green Reservoir. *Journal of Applied Ecology*, **27**, 1020–41.

Cross, D. (1991) Reservoirs, GIS, and bass habitat. General Technical Report, USDA Forest Service, RM, **207**, 121–5.

Crozier, W.W. & Kennedy, G.J.A. (1995) The relationship between a summer fry (0+) abundance index, derived from semi-quantitative electrofishing and egg deposition of Atlantic Salmon in the River Bush, Northern Ireland. *Journal of Fish Biology*, **47**, 1055–62.

Csavas, I. (1993) Aquaculture development and environmental issues in the developing countries of Asia. In: *Environment and Aquaculture in Developing Countries* (eds R.S.V. Pullin, H. Rosenthal & J.L. MacLean). *ICLARM Conference Proceedings*, **31**, 74–101.

Cunningham, S., Dunn, M.R. & Whitmarsh, D. (1985) *Fisheries Economics. An Introduction*, 1st edn. Mansell Publishing, London.

Cushing, D.H. (1974) The possible density-dependence of larval mortality and adult mortality in fishes. In: *The Early Life History of Fish* (ed. J.H.S. Blaxter), 1st edn. pp. 103–11. Springer, Berlin.

Cushing, D.H. (1988) The study of stock and recruitment. In: *Fish Population Dynamics*, (ed. J.A. Gulland), 1st edn. pp. 105–28. Wiley, New York.

D

Daget, J. & Ecoutin, J.-M. (1976) Modeles mathematiques de production applicables aux poissons subissant un arret annuel prolonge de croissancee. *Cahiers ORSTOM (hydrobiologie)*, **10**, 59–70.

Daget, J., Planquette, N. & Planquette, P. (1973) Premieres données sur la dynamique des peuplements de poissons du Bandama (Cote d'Ivoire). *Bulletin du Musee Nationale de l'histoire naturelle (3me serie)(Ecologie generale 7), Paris*, **151**, 129–43.

Dahl, G. (1971) *Los peces del Norte de Colombia*. INDERENA, Colombia.

Dansoko, F.D. (1975) *Contribution a l'etude de la biologie de Hydrocyon dans le delta central du Niger*. Thesis, Bamako, Ministere de l'Education Nationale.

Das, S.M. & Pande, J. (1980) Studies on annual winter fish mortality in Lake Naintal. *Indian Journal of Animal Science*, **50**, 561–4.

Delmendo, L.M. (1969) Lowland fishponds in the Philippines. *Current Affairs Bulletin of the IPFC*, **54/55**, 1–11.

Dodge, D.P. (ed.) (1989) Proceedings of the International Large River Symposium (LARS). *Canadian Special Publication of Fisheries and Aquatic Sciences*, **106**, 629 pp.

Dudley, R.G. (1972) *Growth of Tilapia of the Kafue floodplain, Zambia: predicted effects of the Kafue Gorge Dam*. PhD dissertation, University of Idaho, Moscow, USA.

Dynesius, M. & Nilsson, C. (1994) Fragmentation and flow regulation of river systems in the northern third of the world. *Science*, **266**, 753–62.

E

Eckmann, R. von (1983) Zur situation der fischerei im peruanischen Amazonsgebiet (the fisheries situation in the Peruvian Amazon region). *Archiv fur Hydrobiologie*, **97**, 509–39.

Eckmann, R., Gaedke, U. & Wetzlar, H.J. (1988) Effects of climatic and density-dependent factors on year-class strength of *Coregonus lavaretus* in Lake Constance. *Canadian Journal of Fisheries and Aquatic Sciences*, **45**, 1088–93.

EIFAC (1983) Report of the EIFAC Working Party on Stock Enhancement. *EIFAC Technical Paper No. 44*.

EIFAC (1991) Report of the International Seminar on mass removal of (unwanted) fish in large inland water bodies, Lahti, Finland, 10–12 June 1991. *EIFAC Occasional Paper No. 26*.

Eklöv, A.G., Greenberg, L.A., Brönmark, C., Larsson, P. & Berglund, O. (1998) Response of stream fish to improved water quality: a comparison between the 1960s and 1990s. *Freshwater Biology*, **40**, 771–82.

Eknath, A., Tayamen, M.M., Palada de Vera, M.S., *et al.* (1993) Genetic improvement of farmed tilapia: the growth performance of eight strains of *Orechromis niloticus* tested in different farm environments. *Aquaculture*, **111**, 171–88.

Elliott, J.M. (1985) Population regulation for different life-stages of migratory trout, *Salmo trutta*, in a Lake District stream, 1966–83. *Journal of Animal Ecology*, **54**, 985–1002.

Elton, C.S. (1958) *Ecology of Invasions by Animals and Plants*. Chapman & Hall, London.

F

Falk, J., Graefe, A., Drogin, E., Confer, J. & Chandler, L. (1992) Recreational boating on Delaware's inland bays: implications for social and environmental carrying capacity. Technical Report. Delaware University Sea Grant College Program.

FAO (1993) Fisheries management in the South-East arm of lake Malawi, the Upper Shire river and lake Malombe, with particular reference to the fisheries on chambo (*Oreochromis* sp.). Committee for the Inland Fisheries of Africa (CIFA) Technical Paper 21, 113 pp. *FAO Fisheries Technical Paper No. 294*.

FAO (1995a) *Code of Conduct for Responsible Fisheries*. FAO, Rome.

FAO (1995b) Precautionary Approach to Fisheries. Part I: Guidelines on the precautionary approach to capture fisheries and species introductions. *FAO Fisheries Technical Paper No. 350/1*, FAO, Rome.

FAO (1997a) *FAO Technical Guidelines for Responsible Fisheries*. No. 4. *Fisheries Management*, FAO, Rome.

FAO (1997b) *FAO Technical Guidelines for Responsible Fisheries*. No. 5. *Aquaculture Development*, FAO, Rome.

FAO (1999a) Review of the state of world fishery resources: inland fisheries. *FAO Fisheries Circular No. 942*, FAO, Rome.

FAO (1999b) Management guidelines for Asian floodplain river fisheries: Part 1. A spatial heirarchy and adaptive strategy for co-management; Part 2. Summary of DFID research. *FAO Fisheries Technical Paper Nos 384/1*, p. 384/2.

FAO (1999c) Guidelines for the routine collection of capture fishery data. *FAO Fisheries Technical Paper No. 382*.

Ferdouse, F. (1995) Malaysian aquaculture industry – investment opportunities and export potentials. *INFOFISH International*, **6/95**, 18–23.

Fernando, C.H. & Holcik, J. (1982) The nature of fish communities: a factor influencing the fishery potential and yields of tropical lakes and reservoirs. *Hydrobiologia*, **97**, 127–40.

Fisher, W.L. & Toepfer, C.S. (1998) Recent trends in geographic information systems education and fisheries research applications in U.S. universities. *Fisheries*, **23**(5), 10–13.

Fonticiella, D.W., Arboleya, Z. & Diaz, G. (1995) La repoblación como forma de manejo de pesquerias en la acuicultura de Cuba. *COPESCAL Documento Ocasional*, **10**.

Fox, P.J. (1976) Preliminary observations on fish communities of the Okavango Delta. In: *Proceedings of the Symposium on the Okavango Delta and its Future Utilisation*, pp. 125–30. Gaberones, National Museum, Botswana Society.

Fremling, C.R., Rasmussen, J.L., Sparks, R.E., Cobb, S.P., Bryan, C.F. & Clafin, T.O. (1989) Mississippi River fisheries: a case history. In: *Proceedings of the Large Rivers Symposium*, Vol. 106 (ed. C.P. Dodge), pp. 309–51. Canadian Special Publication of Fisheries and Aquatic Sciences.

Fryer, G. (1972) The conservation of the Great lakes of East Africa: a lesson and a warning. *Biological Conservation*, **4**, 256–62.

Fryer, G. (1996) Endemism, speciation and adaptive radiation in great lakes. *Environmental Biology of Fishes*, **45**, 109–31.

Fryer, G. & Iles, T.D. (1972) *The Cichlid Fishes of the Great Lakes of Africa*. Oliver & Boyd, Edinburgh.

Fuentes, C.M. (1998) *Deriva de larvas de sábalo*, Prochilodus lineatus, *y otras especies de pesces de interés comercial en et río Paraná inferior*. Tesis Doctoral. Facultad de Ciencias Exactas y Naturales, Universidad de Buenos Aires.

G

Garcia, S.M. & Newton, C. (1997) Current situation, trends and prospects in world capture fisheries. In: *Global Trends: Fisheries Management* (eds E.L. Pikitch, D.D. Huppert & M.P. Sissenwine), pp. 3–27. American Fisheries Society Symposium 20. Bethesda, MD, USA.

Gaugush, R.F. (1994) Combining field sampling, a geographic information system, and numerical modelling to analyze sediment distribution in a backwater lake of the Upper Mississippi River. *Lake and Reservoir Management*, **9**(2), 75–6.

Gehrke, P.C. (1992) Enhancing recruitment of native fish in inland environments by accessing alienated floodplain habitats. In: *Australian Society for Fish Biology Workshop Recruitment Processes*, Hobart, 21 August 1991 (ed. D.A. Hancock), pp. 205–9. Australian Government Publishing Service, Canberra.

Genovese, P.V. & Emmett, R.L. (1997) Desktop geographic information system for salmonid resources in the Columbia River Basin. Report: NOAA-TM-NMFS-NWFSC-34 NTIS accession number: PB98118383.

Gibbons, A. (1998) Which of our genes make us human? *Science*, **281**, 1432–4.

Gomez, A.L. & Monteiro, F.P. (1955) Estudo da populacao total de peixes da represas da sestacao Experimental de Biologia e Piscicultura de Pirassununga, Sao Paulo. *Revista Biologica Maritima de Valparaiso*, **6**, 82–154.

Goodchild, G.A. (1991) Code of Practice and Guidelines for safety with Electric Fishing. *EIFAC Occasional Paper No. 24*.

Goodnight, W.H. & Bjorn, T.C. (1971) Fish production in two Idaho streams. *Transactions of the American Fisheries Society*, **4**, 769–80.

Gordon, W.R. (1994) A role for comprehensive planning, Geographical Information System (GIS) technologies and program evaluation in aquatic habitat development. *Bulletin of Marine Science*, **55**, 995–1013.

Gowan, C., Young, M.K., Fausch, K.D. & Riley, S.C. (1994) Restricted movements in resident stream salmonids: a paradigm lost. *Canadian Journal of Fisheries and Aquatic Sciences*, **51**, 2626–37.

Grainger, R.J.R. & Garcia, S.M. (1996) Chronicles of marine fishery landings (1950–1994): trend analysis and fisheries potential. *FAO Fisheries Technical Paper No. 359*, FAO, Rome.

Grey, D.L. (1986) The development and management of the Northern Territory barramundi (*Lates calcarifer*) fishery. In: *Proceedings of the First Asian Fisheries Forum*, Manila, Philippines, 26–31 May 1986 (eds J.L. Maclean, I.B. Dizon & I.V. Hosillos), pp. 375–80. ICLARM, Manila.

Grimble, R. (1998) Stakeholder methodologies in natural resource management. *Socio-economic Methodologies, Best Practices Guidelines*. Naturals Resources Institute, Chatham, UK.

Gross, M.R., Coleman, R.M. & McDowall, R.M. (1988) Aquatic productivity and the evolution of diadromous fish migration. *Science*, **239**, 1291–3.

Gulland, J.A. (1983) *Fish Stock Assessment. A Manual of Basic Methods*, 1st edn. John Wiley & Sons, Chichester.

Gyllensten, U. (1985) The genetic structure of fish: differences in the intra-specific distribution of biochemical genetic variation between marine, anadromous, and freshwater species. *Journal of Fish Biology*, **26**, 691–700.

H

Halls, A.S. (1998) *An assessment of the impact of hydraulic engineering on floodplain fisheries and species assemblages in Bangladesh*. PhD thesis. University of London.

Halls, A.S. (1999) Spatial models for the evaluation and management of inland fisheries. Report to the Food and Agricultural Organisation of the United Nations, Rome.

Halls, A.S., Debnath, K., Kirkwood, G.P. & Payne, I.A. (2000) Density-dependent recruitment of *Puntius sophore*, in floodplain waterbeds in Bangladesh. *Journal of Fish Biology*, **56**, 905–14.

Halls, A.S., Hoggarth, D.D. & Debnath, D. (1999) Impacts of hydraulic engineering on the dynamics and production potential of floodplain fish populations in Bangladesh. *Fisheries Management and Ecology*, **6**, 261–85.

Halls, A.S., Jones, C.J. & Mees, C.C. (2000) Information systems for the co-management of artisanal fisheries. Final Report to the Department for International Development.

Hanson, J.M. & Leggett, W.C. (1985) Experimental and field evidence for inter- and intraspecific competition in two freshwater fishes. *Canadian Jounal of Fisheries and Aquatic Science*, **42**, 280–6.

Harris, J.H. (1988) Demography of Australian bass, *Macquaria novemaculeata* (Perciformes, Percichthyidae), in the Sydney Basin. *Australian Journal of Marine and Freshwater Research*, **39**, 355–69.

Harvey, B., Ross, C., Greer, D. & Carolsfeld, J. (1999) *Action Before Extinction*. World Fisheries Trust, Victoria, BC.

Heithaus, M.R. & Laushman, R.H. (1997) Genetic variation and conservation of stream fishes: influence of ecology, life history, and water quality. *Canadian Journal of Fisheries and Aquatic Science*, **54**, 1822–36.

Henderson, B.A., Collins, J.J. & Reckahn, J.A. (1983) Dynamics of an exploited population of lake whitefish (*Coregonus clupeaformes*) in Lake Huron. *Canadian Journal of Fisheries and Aquatic Science*, **40**, 1556–67.

Hilborn, R. & Walters, C.J. (1992) *Quantitative Fisheries Stock Assessment. Choice, Dynamics and Uncertainty*, 1st edn. Chapman & Hall, New York.

Hoggarth, D.D. (1985) *The feeding and growth of Welsh yellow eels (*Anguilla anguilla *L.)*. MSc thesis, University of Wales.

Hoggarth, D.D. & Kirkwood, G.P. (1996) Technical interactions in tropical floodplain fisheries of South and South-east Asia. In: Cowx, I.G. (ed.) *Stock Assessment in Inland Fisheries* (ed. I.G. Cowx), pp. 280–92. Fishing News Books, Oxford.

Hoggarth, D.D., Halls, A.S., Dam, R.K. & Debnath, K. (1999a) Recruitment sources for fish stocks inside a floodplain river impoundment in Bangladesh. *Fisheries Management and Ecology*, **6**, 287–310.

Hoggarth, D.D., Cowan, V.J., Halls, A.S., *et al.* (1999b) Management guidelines for Asian floodplain river fisheries. *FAO Fisheries Technical Paper Nos 348/1, 348/2*, FAO, Rome.

Holcik, J. & Kmet, T. (1986) Simple models of the population dynamics of some fish species from the lower reaches of the Danube. *Folia Zoologica*, **35**(2), 183–91.

Holden, M.J. (1963) The populations of fish in dry season pools of the River Sokoto. *Fisheries Publication of the Colonial Office, London*, **19**, 58 p.

Hollig, C.S. & Meffe, G.K. (1996) Command and control and the pathology of natural resource management. *Conservation Biology*, **10**, 328–37.

Hooe, M.L., Buck, D.H., & Wahl, D.H. (1994) Growth, survival, and recruitment of hybrid crappies stocked in small impoundments. *North American Journal of Fisheries Management*, **14**, 137–42.

Hopkins, C.L. (1971) Production of fish in two small streams in the North Island of New Zealand. *New Zealand Journal of Marine and Freshwater Research*, **5**, 280–90.

Horton, R.E. (1945) Erosional development of streams and their drainage basins: hydrophysical approach to quantitative morphology. *Bulletin of the Geographical Society of North America*, **56**, 275–370.

Huet, M. (1949) Apercu des relations entre la pente et la population des eaux courantes. *Sweizer Zeitung Hydrobiologie*, **11**, 333–51.

Hugueny, B., Camara, S., Samoura, B. & Magassouba, M. (1996) Applying an index of biotic integrity based on fish assemblages in a west African river. *Hydrobiologia*, **331**, 71–8.

Hutchinson, E.G. (1975) *A Treatise on Limnology*. Vol. 1, Part 1. *Geography and Physics of Lakes*. Wiley and Sons, New York.

Hutchinson, N.J., Dodington, V., Stirling, M., *et al.* (eds) (1995) Shoreline alterations on the Muskoka Lakes, Ontario – mapping techniques, methods development, preliminary results and interpretive framework. *Lake and Reservoir Management*, **11**(2), 151.

Hyslop, E.J. (1980) Stomach contents analysis: a review of methods and their application. *Journal of Fish Biology*, **17**, 411–29.

I

IIED (1998) PLA Notes 31: *Participatory Monitoring and Evaluation Series*. International Institute for Environment and Development (IIED) Sustainable Agriculture and Rural Livelihoods, London.

Ingendahl D., Bach, J.M., Larinier, M. & Travade F. (1996) The use of telemetry in studying downstream migration of Atlantic salmon smolts at a hydro-electric power plant in South-West France: preliminary results. In: *Underwater Biotelemetry* (eds E. Baras & J.C. Philippart), pp. 121–8. University of Liège, Liège.

International Council for the Exploration of the Sea (1995) ICES Code of Practice on the Introductions and Transfers of Marine Organisms – 1994. *ICES Co-operative Research Report No. 204*.

Irwin, E.R. & Noble, R.L. (1996) Effects of reservoir drawdown on littoral habitat: assessment with on-site measures and geographic information systems. In: *Multidimensional Approaches to Reservoir Fisheries Management* (eds L.E. Miranda & D.R. DeVries), *American Fisheries Society Symposium*, **16**, 324–31.

Isaac, V.J. & Ruffino, M.L. (1996) Population dynamics of tambacqui, *Colossoma macropopum* (Cuvier), in the lower Amazon, Brazil. *Fisheries Management and Ecology*, **3**, 315–33.

Isaak, D.J. & Hubert, W.A. (1997) Integrating new technologies into fisheries science: the application of geographic information systems. *Fisheries*, **22**(1), 6–9.

IUCN (1994) *Guidelines for Protected Area Management Categories*. CNPPA with the assistance of WCMC. IUCN, Gland, Switzerland; Cambridge, UK.

J

Janauer, G.A. (1997) Macrophytes, hydrology, and aquatic ecotones: a GIS-supported ecological survey. *Aquatic Botany*, **58**, 379–91.

Johanesson, K.A. & Mitson, R.B. (1983) Fisheries acoustics: a practical manual for biomass estimation. *FAO Fisheries Technical Paper No. 240*.

Jones, G.P. (1984) Population ecology of the temperate reef fish *Pseudolabrus celidotus* (Bloch & Schneider) (Pisces:Lambridae). 1. Factors influencing adult density. *Journal for Exploration of the Sea; Biology and Ecology*, **75**, 277–303.

Jones, R. (1973) Density-dependent regulation of the numbers of cod and haddock. *Rapport. R-V. Reunion du Conseil. International pour l' Exploration de la Mer*, **75**, 277–303.

Junk, W.J., Bayley, P.B. & Sparks, R.E. (1989) The flood pulse concept in river–floodplain systems. In: *Proceedings of the International Large Rivers Symposium*, Vol. 106 (ed. D.P. Dodge), pp. 110–27. Canadian Special Publication of Fisheries and Aquatic Sciences.

K

Kapetsky, J.M. (1974) *Growth mortality and production of five fish species of the Kafue River floodplain, Zambia*. PhD thesis, University of Michigan.

Kapetsky, J.M. (1998) Geography and constraints on inland fishery enhancements. In: *Inland Fishery Enhancements* (ed. T. Petr), pp. 37–63. *FAO Fisheries Technical Paper No. 374*, FAO, Rome.

Kapetsky, J.M. (1999) The development and training requirements for geographic information systems (GIS) and remote sensing (RS) applications in relation to inland fisheries (including aquaculture) in the lower Mekong Basin. Report prepared for the Fisheries Unit, Mekong River Commission Secretariat, Phnom Penh, Cambodia, May 1999.

Kapetsky, J.M. (2000) Present applications and future needs of meteorological and climatological data in inland fisheries and aquaculture. *Journal of Agricultural and Forest Meteorology*, **2789**, 1–9.

Kapetsky, J.M. & Chakalall, B. (1999) A strategic assessment of the potential for freshwater fish farming in the Caribbean Island States. *COPESCAL Technical Paper No. 10* (Suppl.), FAO, Rome.

Kapetsky, J.M. & Nath, S.S. (1997) A strategic assessment of the potential for freshwater farming in Latin America. *COPESCAL Technical Paper No. 10*, FAO, Rome.

Kapetsky, J.M., Hill, J.M. & Worthy, L.D. (1988) A geographical information system for catfish farming development. *Aquaculture*, **68**, 311–20.

Kapetsky, J.M., Hill, J.M., Worthy, L.D. & Evans, D.L. (1990) Assessing potential for aquaculture development with a geographic information system. *Journal of the World Aquaculture Society*, **21**, 241–9.

Kapetsky, J.M., Wijkstrom, U.N., MacPherson, N., Vincke, M.M.J., Ataman, E. & Caponera, F. (1991) Where are the best opportunities for fish farming in Ghana? The Ghana aquaculture geographical information system as a decision-making tool. FI: TCP/GHA/0051 Field Document. *Field Technical Report No. 5*, FAO, Rome.

Karenge, L.P. & Kolding, J. (1994) On the relationship between hydrology and fisheries in lake Kariba, Central Africa. *Fisheries Research*, **22**, 205–26.

Karr, J.R., Fausch, K.D., Angermeier, P.L., Yant, P. & Schlosser, I.J. (1986) Assessing biological integrity in running waters: a method and its rationale. *Illinois Natural History Survey, Special Publication 5*.

Keleher, C.J. & Rahel, F.J. (1996) Thermal limits to salmonid distributions in the Rocky Mountain region and potential habitat loss due to global warming: a geographic information system (GIS) approach. *Transactions of the American Fisheries Society*, **125**, 1–13.

Kennedy, M. & Fitzmaurice, P. (1969) Factors affecting the growth of coarse fish. In: *Proceedings of the 4th British Coarse Fish Conference*, Liverpool, pp. 42–9. University of Liverpool, Liverpool.

Kent, D.B. & Drawbridge, M.A. (1996) Developing a marine ranching program: A multidisciplinary approach. In *Marine Ranching: Global Perspectives with Emphasis on the Japanese Experience*, pp. 66–78. FAO Fisheries Circular No. 943.

Kerr, S.R. & Ryder, R.A. (1988) The applicability of fish yield indices in freshwater and marine ecosystems. *Limnology and Oceanography*, **33**, 973–81.

Kieser, R., Langford, G. & Cooke, K. (1995) The use of geographic information systems in the acquisition and analysis of fisheries acoustic data. International Council for the Exploration of the Sea. International Symposium on Fisheries and Plankton Acoustics, Aberdeen, UK, 12–16 June 1995 (summary). ICES, Copenhagen.

King, M. (1995) *Fisheries Biology Assessment and Management*, 1st edn. Fishing News Books, London.

Kincaid, H.L. (1983) Inbreeding in fish populations used for aquaculture. *Aquaculture*, **33**, 215–27.

Kirkwood, G.P., Beddington, J.R. & Rossouw, J.A. (1994) Harvesting Species of Different Lifespans. In: *Large-Scale Ecology and Conservation Biology*, 35th Symposium of the British Ecological Society with the Society for Conservation Biology, University of Southampton, 1993 (ed. P.J. Edwards, R. May & N.R. Webb), pp. 199–227. Blackwell Scientific Publications, Oxford.

Kolding, J. (1993) Population dynamics and life history styles of Nile tilapia, *Oreochromis niloticus*, in Furguson's Gulf, Lake Turkana, Kenya. *Environmental Biology of Fishes*, **37**, 25–46.

Koutnik, M.A. & Padilla, D.K. (1994) Predicting the spatial distribution of *Dreissena polymorpha* (zebra mussel) among inland lakes of Wisconsin: modeling with a GIS. *Canadian Journal of Fisheries and Aquatic Sciences*, **51**, 1189–96.

Krebs, C.J. (1978) *Ecology: The Experimental Analysis of Distribution and Abundance*, 2nd edn. Harper & Row, New York.

Kreuzer, R. (1974) Fish and its place in culture. In: *Fishery Products* (ed. R. Kreuzer), pp. 22–47. FAO and Fishing News Books, Surrey.

L

Lackey, R.T. (1979) Options and limitations in fisheries management. *Environmental Management*, **3**, 109–12.

Laé, R. (1992) Influence de l'hydrobiologie sur l'evolution des pêcheries du delta Centrale du Niger, de 1966 à 1989. *Aquatic living Resources*, **5**, 115–26.

Laé, R. (1994) Modifications des apports en eau et impact sur les captures de poisson. In: *La Pêche dans le Delta Central du Niger* (ed. J. Quensiere), pp. 255–65. ORSTOM, Paris.

Laé, R. (1997) Estimation des rendements de pêche des lacs africains au moyen des modéles empiriques. *Aquatic Living Resources*, **10**, 83–92.

Lagardère, J.P., Bégout-Anras, M.L. & Claireaux, G. (eds) (1998) Advances in invertebrate and fish telemetry. *Hydrobiologia* 371/372 (special issue).

Lagler, K.F., Bardach, J.E., Miller, R.R. & Passino, D.R. (1977) *Ichthyology*, 2nd edn. John Wiley & Sons, New York.

Lambou, V.W. (1984) The importance of bottomland hardwood forest zones to fish and fisheries: the Atchafalaya basin, a case history. Paper presented at US Environmental Protection Agency's 'Workshop on Bottomland and Hardwood Ecosystem Characterization', 3–7 December 1984, St. Francisville, Louisiana.

Leberg, P.L. (1990) Influence of genetic variability on population growth: implications for conservation. *Journal of Fish Biology*, **37** (Suppl. A), 193–5.

Le Cren, E.D. (1958) Observations on the growth of perch (*Perca fluviatilis* L.) over twenty years with special reference to the effects of temperature and changes in population biomass. *Journal of Animal Ecology*, **27**, 287–334.

Le Cren, E.D. (1965) Some factors regulating the size of populations of freshwater fish. *Mittelung International Verein Theorical Angew Limnologie*, **13**, 88–105.

Le Cren, E.D. (1969) Estimates of fish populations and production in small stremas in England. In: *Symposium on Salmon and Trout in Streams* (ed. T.G. Northcote), pp. 269–80. Institute of Fisheries, Vancouver.

Le Cren, E.D. (1973) The population dynamics of young trout (*Salmo trutta*) in relation to density and territorial behaviour. *Rapport. R-V. Reunion du Conseil. International pour l'Exploration de la Mer*, **164**, 241–6.

Le Cren, E.D. (1987) Perch (*Perca fluviatilis*) and pike (*Esox lucius*) in Windermere from 1940 to 1985; studies in population dynamics. *Canadian Journal of Fisheries and Aquatic Sciences*, **44** (Suppl. 2), 216–28.

Lévêque, C. (1995) Role and consequence of fish diversity in functioning of African freshwater ecosystems: a review. *Aquatic Living Resources*, **8**, 59–78.

Lévêque, C., Paugy, D. & Teugels, G. (eds) (1992) *Faune des poissons d'eaux douces et saumâtres d'afrique de l'ouest*. Office pour la Recherche Scientifique et Tecnologique Outre Mer, 2 Vols.

Li, S. (1988) The principles and strategies of fish culture in Chinese reservoirs. In: *Reservoir Fishery Management and Development in Asia* (ed. S.S. De Silva), pp. 214–23. Proceedings of a Workshop held in Kathmandu, Nepal, 23–28 November 1987. IDRC, Ottawa, Canada.

Li, S. & Xu, S. (1995) *Capture and Culture of Fish in Chinese Reservoirs*. IDRC, Ottawa, Canada.

Linfield, R.S.J. (1985) An alternative concept to home range theory with respect to populations of cyprinids in major river systems. *Journal of Fish Biology*, **27** (Suppl. A), 187–96.

Liu, J.K. & Yu, Z. (1992) Water quality changes and effects on fish populations in the Hanjiang River, China following hydroelectric dam construction. *Regulated Rivers: Research and Management*, **7**, 359–68.

Lobon-Cervia, J. & Penczak, T. (1984) Fish production in the Jaruma River, Central Spain. *Holarctic Ecology*, **7**, 128–37.

Loomis, J.B. & White, D.S. (1996) Economic benefits of rare and endangered species: summary and meta-analysis. *Ecological Economics*, **18**, 197–206.

Lorda, E. & Creeco, V.A. (1987) Stock-recruitment and compensatory mortality of American shad in the Connecticut River. In: *Common Strategies of Anadromous and Catadromous Fishes* (eds M.J. Dadswell, R.J. Klauda, R.L. Saunders, R.A. Rulifson, J.E. Cooper & C.M. Moffitt), pp. 469–82. Proceedings of an International Symposium held in Boston, Massachusetts, USA, 9–13 March 1986, Vol. 1. American Fisheries Society, Bethesda, MD.

Lorenzen, K. (1995) Population dynamics and management of culture-based fisheries. *Fisheries Management and Ecology*, **2**, 61–73.

Lorenzen, K. (1996a) A simple von Bertalanffy model for density-dependent growth in extensive aquaculture, with an application to common carp (*Cyprinus carpio*). *Aquaculture*, **142**, 191–205.

Lorenzen, K. (1996b) The relationship between body weight and natural mortality in fish: a comparison of natural ecosystems and aquaculture. *Journal of Fish Biology*, **49**, 627–47.

Lorenzen, K. & Garaway, C.J. (1998) How predictable is the outcome of stocking? In: *Inland Fisheries Enhancements*, pp. 133–52. FAO Fisheries Technical Paper No. 374, FAO, Rome.

Lorenzen, K., Xu, C., Cao, F., Ye, J. & Hu, T. (1997) Analysing extensive fish culture systems by transparent population modelling: bighead carp, *Aristichthys nobilis* (Richardson 1845), culture in a Chinese reservoir. *Aquaculture Research*, **28**, 867–80.

Lorenzen, K., Juntana, J., Bundit, J. & Tourongruang, D. (1998) Assessing culture fisheries practices in small water bodies: a study of village fisheries in Northeast Thailand. *Aquaculture Research*, **29**, 211–24.

Lotrich, V.A. (1973) Growth, production and community composition of fishes inhabiting a first-, second-, third order stream of Eastern Kentucky. *Ecological Monographs*, **43**(3), 377–97.

Loubens, G. (1969) Etude de certains peuplements ichtyologiques par des pêche au poison (1ere note). *Cahiers ORSTOM (Hydrobiologie)*, **3**, 43–73.

Lowe-McConnell, R.H. (1964) The fishes of the Rupununi savanna district of British Guiana, South America. Part 1. Ecological groupings of fish species and effects of seasonal cycle on the fish. *Journal of the Linnean Society (Zoology)*, **45**, 103–44.

Lowe-McConnell, R.H. (1967) Some factors affecting fish populations in Amazonion waters. In: *Atas do simposio sobre a biota Amazonia, Conselho Nacional Pesquisas*, pp. 117–86. Consevacao de Natureza e Recursos Naturais, Rio de Janiero.

Lowe-McConnell, R.H. (1975) *Fish Communities in Tropical Freshwaters*. Longman, London.

Lowe-McConnell, R.H. (1987) *Ecological Studies in Tropical Fish Communities*, 1st edn. Cambridge University Press, Cambridge.

Lu, X. (1994) A review of river fisheries in China. *FAO Fisheries Circular No. 862*, FAO, Rome.

Lyons, J., Gonzáles-Hernández, G., Soto-Galera, E. & Guzmán-Arroyo, M. (1998) Decline of freshwater fishes and fisheries in selected drainages of west-central Mexico. *Fisheries*, **23**, 10–18.

M

MacArthur, R. (1955) Fluctuations of animal populations and a measure of community stability. *Ecology*, **36**, 533–6.

McDowall, R.M. (1994) On size and growth in freshwater fish. *Ecology of Freshwater Fish*, **3**, 67–97.

McGurk, M.D. (1986) Natural mortality of marine pelagic fish eggs and larvae: role of spatial patchiness. *Marine Ecology Progress Series*, **34**, 227–42.

Madenjian, C.P. & Ryan, P.A. (1995) Effect of gear selectivity on recommended allowable harvest with application to Lake Erie yellow perch fishery. *North American Journal of Fisheries Management*, **15**, 79–83.

Mago-Leccia, F. (1970) Estudios preliminares sobre la ecologia de los peces de los llanos de Venezuela. *Acta biogica Venezolano*, **7**, 71–102.

Mahon, R. (1981) *Patterns of fish taxocenes of small streams in Poland and Ontario: a tale of two river basins with a test for accuracy of quantitative sampling*. PhD thesis, University of Guelph, Guelph, Ontario.

Mahon, R., Balon, E.K. & Noakes, D.L.G. (1979) Distribution, community structure and production of the upper Speed River, Ontario, a preimpoundment study. *Environmental Biology of Fishes*, **4**, 219–44.

Mann, K.H. (1965) Energy transformations by fish in the River Thames. *Journal of Animal Ecology*, **34**, 253–75.

Mann, K.H. (1972) Case History of the River Thames. In: *River Ecology and Man* (eds R.T. Oglesby, C.A. Carlson & J.A. McCann), pp. 213–32. New York, Academic Press.

Mann, R.H.K. (1971) The population, growth and production of fish in four small streams in southern England. *Journal of Animal Ecology*, **40**, 155–90.

Mann, R.H.K. (1991) Growth and production. In: *Cyprinid Fishes: Systematics, Biology and Exploitation* (ed. I.J. Winfield & J.S. Nelson), pp. 456–82. Chapman & Hall, London.

Mann, R.H.K. & Mills, C.A. (1985) Variations in the sizes of gonads, eggs and larvae of the dace, *Leuciscus leuciscus*. *Environmental Biology of Fishes*, **13**, 277–87.

Mann, R.H.K. & Penczak, T. (1986) Fish production in rivers: a review. *Polish Archives of Hydrobiology*, **33**, 233–47.

Marmulla, G. & Ingendahl, D. (1996) Preliminary results of a radio telemetry study of returning Atlantic salmon (*Salmo salar* L.) and sea trout (*Salmo trutta trutta* L.) in River Sieg, tributary of River Rhine in Germany. In: *Underwater Biotelemetry*, (eds E. Baras & J.C. Philippart.), pp. 109–18. University of Liège, Liège.

Marten, G.G. & Polovina, J.J. (1982) A comparative study of fish yields from various tropcal ecosystems. In: *Theory and Management of Tropical Fisheries,* Vol. 9 (eds D. Pauly & G. Murphy), pp. 255–89. International Centre for Living Aquatic Resources Management Conference Proceedings.

May, R.M. (1971) Stability in multi-species community models. *Mathematical Bioscience*, **12**, 59–79.

McCann, K., Hastings, A. and Huxel, G.R. (1998) Weak trophic interactions and the balance of nature. *Nature* **395**, 794–98.

Meaden, G. (2001) GIS in fisheries science: foundations for a new millenium. In: *Proceedings of First International Symposium on GIS in Fishery Science* (eds T. Nishida, C.E. Hollingworth and P.J. Kailola), Seattle, Washington, USA; 2–4 March 1999. Fishery GIS Research Group. Kurofune Printing Inc., Shizuoka, Japan.

Meaden, G. & Do Chi, T. (1996) Geographical information systems: applications to marine fisheries. *FAO Fisheries Technical Paper No. 356*, FAO, Rome.

Meaden, G. & Kapetsky, J.M. (1991) Geographical information systems in inland fisheries and aquaculture. *FAO Fisheries Technical Paper No. 318*, FAO, Rome.

Mefit-Babtie (1983) Development studies. I. The Jonglei canal area: Final Report to the Government of the Democratic Republic of the Sudan, 4 Vols. Mefit-Babtie, Glasgow.

Merona, B. de & Gascuel, D. (1993) The effects of flood regime and fishing effort on the overall abundance of an exploited fish community in the Amazon floodplain. *Aquatic Living Resources*, **6**, 97–108.

Meronek, T.G., Bouchard, P.M., Buckner, E.R., et al. (1996) A review of fish control projects. *North American Journal of Fisheries Management*, **16**, 63–74.

Merron, G., Bruton, M. & La-Hausse-de-Lalouviere, P. (1993) Changes in fish communities of the Pongolo floodplain, Zululand (South Africa) before, during and after a severe drought. *Regulated Rivers: Research and Management*, **8**, 335–44.

Metcalfe, J.D. & Arnold, G.P. (1997) Tracking fish with electronic tags. *Nature (London)*, **387**, 665–6.

Minns, C.K. & Moore, J.E. (1992) Predicting the impact of climate change on the spatial pattern of freshwater fish yield capability in Eastern Canadian Lakes. *Climatic Change*, **22**, 327–46.

Minns, C.K., Meisner, J.D., Moore, J.E., Greig, L.A. & Randall, R.G. (1995) Defensible methods for pre- and post-development assessment of fish habitat in the Great Lakes. A prototype methodology for headlands and offshore structures. *Canadian Manuscript Report of Fisheries and Aquatic Sciences/Rapport Manuscrit Canadien des Sciences Halieutiques et Aquatiques*, No. 2328.

Minns, C.K., Doka, S.E., Bakelaar, C.N., Brunette, P.C. & Schertzer, W.M. (1999) Identifying habitats essential for pike, *Esox lucius* L. in the Long Point Region of Lake Erie: a suitable supply approach. *American Fisheries Society Symposium*, **22**, 363–82.

Moehl, J.F., Jr. & Davies, W.D. (1993) Fishery intensification in small water bodies: a review for North America. *FAO Fisheries Technical Paper No. 333*, FAO, Rome.

Montreuil, V., Tello, S., Maco, J. & Ismino, R. (1990) Maximum sustainable yield of commercial fisheries in the Department of Loreto, Peru. *FISHBYTE*, **8**, 13–14.

Moyle, P.B. (1993) *Fish – An Enthusiast's Guide*. University of California Press, Berkeley, CA.

Moyle, P.B. & Leidy, R.A. (1992) Loss of biodiversity in aquatic ecosystems: evidence from fish faunas. In: *Conservation Biology – The Theory and Practice of Nature Conservation, Preservation and Management* (eds P.L. Fiedler & S.K. Jain), pp. 127–69. Chapman & Hall, New York.

Moyle, P.B. & Light, T.L. (1996) Biological invasions of freshwater: empirical rules and assembly theory. *Biological Conservation*, **78**, 149–61.

Moyle, P.B. & Moyle, M.A. (1991) Introduction to fish imagery in art. *Environmental Biology of Fishes*, **31**, 5–23.

Moyle, P.B. & Moyle, P.R. (1995) Endangered fishes and economics: intergenerational obligations. *Environmental Biology of Fishes*, **43**, 29–37.

Moyle, P.B. & Randall, P.J. (1998) Evaluating the biotic integrity of watersheds in the Sierra Nevada, California. *Conservation Biology*, **12**, 1318–26.

MRAG (1992) *The CEDA Package*. Version 1.0. *User Manual*. ODA, London.

MRAG (1994a) Floodplain fisheries project; biological assessment of the fisheries. Internal Report.

MRAG (1994b) Potential yield of South Asian small reservoir fisheries. Final Report. Fisheries Management Science Programme. Overseas Development Administration.

MRAG (1995) A synthesis of simple empirical models to predict fish yields in tropical lakes and reservoirs. Marine Resources Assessment Group, London.

N

Narayan, D. (1993) Participatory evaluation: tools for managing change in water and sanitation. *World Bank Technical Paper No. 207*, Washington, DC.

Noiset, J.-L. (1994) Dynamique de populations et exploitation de trois Cichlidae (Teleostei) de la zone inondable de la riviere San Pedro (Tabasco, Mexique). Dissertation, Namur.

Novoa, D.F. (1989) The multispecies fisheries of the Orinoco river: development, present status, and management strategies. In: *Proceedings of the International Large Rivers Symposium* (ed. D.P. Dodge). *Canadian Special Publication on Fish and Aquatic Science*, **106**, 422–8.

Nyerges, T., Robkin, M. & Moore, T.J. (1997) Geographic information systems for risk evaluation:

perspectives on applications to environmental health. *Cartography and Geographic Information Systems*, **24**, 123–44.

O

Oberdorff, T., Guegan, J-F. & Hugueny, B. (1995) Global patterns of fish species richness in rivers. *Ecography*, **18**, 345–52.

O'Brien-White, S. & Thomason, C.S. (1995) Evaluating fish habitat in a South Carolina watershed using GIS. *Proceedings of the Annual Conference of Southeastern Association of Fisheries and Wildlife Administrators*, Vol. 49, pp. 153–66.

Olson, C.M. & Orr, B. (1999) Combining tree growth, fish and wildlife habitat, mass wasting, sedimentation, and hydrologic models in decision analysis and long-term forest land planning. *Forest Ecology and Management*, **114**, 339–48.

O'Neill, R.V., Hunsaker, C.T., Jones, K.B., et al. (1997) Monitoring environmental quality at the landscape scale. Using landscape indicators to assess biotic diversity, watershed integrity, and landscape stability. *BioScience*, **47**, 513–19.

Ostrom, E. (1990) *Governing the Commons: The Evolution of Institutions for Collective Action*. Cambridge University Press, Cambridge.

Ostrom, E. (1992) The rudiments of a theory of the origins, survival and performance of common-property institutions. In: *Making the Commons Work: Theory, Practice and Policy* (ed. D.W. Bromley), pp. 293-318. ICS Press, San Francisco.

Overholtz, W.J. (1989) Density-dependent growth in the Northwest Atlantic stock of Atlantic mackerel (*Scomber scombrus*). *Journal of North West Atlantic Fisheries Science*, **9**, 115–21.

Overseas Development Administration (1995) Guidance note on how to do stakeholder analysis of aid projects and programmes. Social Development Department, London. Available on http://carryon.oneworld.org/euforic/gb/stake1.htm

P

Padilla, J.E. & Charles, A.T. (1994) Bioeconomic modelling and the management of capture and culture fisheries. *NAGA*, **7**, 18–20.

Pauly, D. (1980) On the interrelationships between natural mortality, growth parameters, and mean environmental temperature in 175 fish stocks. *Journal du Conseil International pour l'Exploration de la Mer*, **39**(2), 175–92.

Pauly, D. (ed.) (1994) *On the Sex of Fish and the Gender of Scientists*. Chapman & Hall, London.

Payne, A.I. (1987) *A Survey of the Rio Pilcomayo Sabalo Fishery in July 1986*. ODA, London.

Payne, A.I. & McCarton, B. (1985) Estimation of population parameters and their application to the development of fishey management models in two African rivers. *Journal of Fish Biology*, **27** (Suppl. A), 263–77.

Penczak, T. (1981) Ecological fish production in two small lowland rivers in Poland. *Oecologia*, **48**, 107–11.

Penczak, T., Suszycka, E. & Molinsky, M. (1982) Production, consumption and energy transformations by fish populations in a small lowland river. *Ekologia Polska*, **30**, 111–37.

Penn, J.W. & Caputi, N. (1986) Spawning stock recruitment relationships environmental influences on the tiger prawn (*Penaeus esculentus*) fishery in Exmouth Gulf, Western Australia. *Australian Journal of Marine and Freshwater Research*, **37**, 491–505.

Petrere, M., Jr (1983) Yield per recruit of the tambaqui, *Colossoma macropomum* Cuvier, in the Amazonas State, Brazil. *Journal of Fish Biology*, **22**, 133–44.

Petrere, M., Jr, Welcomme, R.L. & Payne, A.I. (1998) Comparing river basins world-wide and contrasting inland fisheries in Africa and Central Amazonia. *Fisheries Management and Ecology*, **5**, 97–106.

Pinkerton, E. (1989) Co-operative management of local fisheries: new directions for improved management and community. In: *Development* (ed. E. Pinkerton). University of British Columbia Press, Vancouver.

Pinney, C. (1994) Biotic analysis of Lower Granite Reservoir drawdown using GIS. 14th Annual International Symposium of the North American Lake Management Society, Orlando, FL (USA), 31 Oct–5 Nov 1994. *Lake and Reservoir Management*, **9**(2), 104.

Pitcher, T.J. & Hart, P.J.B. (1982) *Fisheries Ecology*, 1st edn. Chapman & Hall, London.

Pitcher, T.J. & MacDonald, M.P.D. (1973) Two models for seasonal growth in fishes. *Journal of Applied Ecology*, **10**, 597–606.

Poate, C.D. & Daplyn, P.F. (1993) *Data for Agrarian Development*. Cambridge University Press, Cambridge.

Poddubnyi, A.G. (1979) The ichthyofauna of the Volga. In: *The River Volga and its Life* (ed. P.D. Mordukhai-Boltovskoi), pp. 304–39. Dr W. Junk, The Hague.

Pomeroy, R.S. & Williams, M.J. (1994) *Fisheries Co-Management: A Policy Brief*, ICLARM.

Prentice, E.F., Flagg, T.A., McCutcheon, C.S. & Brastow, D.F. (1990) PIT-tag monitoring systems for hydroelectric dams and fisheries. *American Fisheries Society Symposium*, **7**, 323–34.

Preston, F.W. (1962a) The canonical distribution of commonness and rarity. Part 1. *Ecology*, **43**, 185–215.

Preston, F.W. (1962b) The canonical distribution of commonness and rarity. Part 2. *Ecology*, **43**, 410–32.

Pretty, J.N., Guijt, I., Thompson, J. & Scoones, I. (1995) *Trainer's Guide for Participatory Learning and Action*, IIED.

Q

Quattro, J.M. & Vrijenhoek, R.C. (1989) Fitness differences among remnant populations of the endangered Sonoran topminnow. *Science*, **245**, 976–8.

Quensiere, J. (ed.) (1994) *La Peche dans le Delta Central du Niger*, pp. 255–65. ORSTOM, Paris.

Quinn, T.J. & Deriso, R.B. (1999) *Quantitative Fish Dynamics*. Oxford University Press, Oxford.

Quiros, R. (1999) The relationship between fish yield and stocking density in reservoirs from tropical and temperate regions. In: *Theroetical Reservoir Ecology and its Application* (eds J.G. Tundisi & M. Straskraba), pp. 67–83. Bakhuys.

R

Rahel, F.J., Keleher, C.J. & Anderson, J.L. (1996) Potential habitat loss and population fragmentation for cold water fish in the North Platte River drainage of the Rocky Mountains: response to climate warming. *Limnology and Oceanography*, **41**, 1116–23.

Rainboth, W.J. (1996) *FAO Species Identification Field Guide for Fishery Purposes. Fishes of the Cambodian Mekong*. FAO, Rome.

Regier, H.A. & Loftus, K.H. (1972) Effects of fisheries exploitation on salmonid communities in oligotrophic lakes. *Journal of the Fisheries Research Board of Canada*, **29**, 959–68.

Reizer, C. (1974) *Definition d'une politique d'amenagement des resources halieutiques d'un ecosys-*

teme aquatique complex per l'étude de son environnement abiotique, biotique et anthropique. Le fleuve Senegal moyen et inférieur. Doctorat en Sciences de l'environnement, Dissertation, Arlon, Fondation Universitaire.

Reynolds, J.E. & Gréboval, D.F. (1988) Socio-economic effects of the evolution of Nile perch fisheries in Lake Victoria: a review. *CIFA Technical Paper No. 17.*

Ribeiro, M.C.L.B. & Petrere, M. (1990) Fisheries ecology and management of the jaraqui (*S. insignis*) in central Amazonia. *Regulated Rivers*, **5**, 195–216.

Richardson, J.R. & Hamouda, E. (1995) GIS modelling of hydroperiod, vegetation, and soil nutrient relationships in the Lake Okeechobee marsh ecosystem. Ecological studies on the littoral and pelagic systems of Lake Okeechobee, Florida (USA). *Ergebnisse der Limnologie (Advances in Limnology)*, **45**, 95–115.

Ricker, W.E. (1954) Stock and recruitment. *Journal of the Fisheries Research Board of Canada*, **11**, 559–623.

Ricker, W.E. (1975) Computation and interpretation of biological statistics of fish populations. Department of the Environment, Fisheries and Marine Service, Canada, Bulletin 191.

Rietbergen-McCracken, J. & Narayan, D. (1998) *Participation and Social Assessment: Tools and Techniques.* World Bank, Washington, DC.

Rodriguez, M.A. & Lewis, W.M. Jr (1994) Regulation and stability in fish assemblages on neotropical lakes. *Oecologia*, **99**, 166–80.

Rogers, K.B. & Bergersen, E.P. (1996) Application of geographic information systems in fisheries: habitat use by northern pike and largemouth bass. In: *Multidimensional Approaches to Reservoir Fisheries Management* (eds L.E. Miranda & D.R. DeVries). *American Fisheries Society Symposium*, **16**, 315–23.

Ross, M.R. & Almeida, F.P. (1986) Density-dependent growth of silver hakes. *Transactions of the American Fisheries Society*, **115**, 548–54.

Rossi, L.M. (1989) Alimentacion de larvas de *Salminus maxilosus* Val., 1840 (Pisces: Characidae) *Iheringia, Seccion Zoologia*, Porto Alegre, **69**, 49–59.

Rossi, L.M. (1992) Evaluacion morfologica del aparato digestivo de postlarvas y prejuveniles de *Prochilodus lineatus* (Val., 1847) (Pisces, Curimatidae) y su relacion con dieta. *Revue d'hydrobiologie tropicale*, **28**, 159–67.

Rubec, P.J., Coyne, M.S., McMichael, R.H. Jr & Monaco, M.E. (1998) Spatial methods being developed in Florida to determine essential fish habitat. *Fisheries*, **23**(7), 21–5.

Russell, G.D., Hawkins, C.P. & O'Neill, M.P. (1997) The role of GIS in selecting sites for riparian restoration based on hydrology and land use. *Restoration Ecology* (Suppl.), **5**(4), 56–68.

Rusydi & Lampe, H.C. (1990) Economics of floating net cage common carp culture in the Saguling Reservoir, West Java, Indonesia. In: *Reservoir Fisheries and Aquaculture Development for Resettlement* (eds B. Costa-Pierce & O. Soemarwoto), pp. 218–39. ICLARM, Manila.

Rutledge, W.P., Rimmer, M.A., Barlow, C.G., Russel, D.J. & Garret, R.N. (1991) Cost benefits of stocking barramundi. *Austrasia*, **5** (April), 45–7.

Ryan, P.A. (1991) Application of surplus production models to the commercial trawl fishery for rainbow smelt (*Osmerus mordax*) in eastern Lake Erie. In: 34th Conference of the International Association for Great Lakes Research, 2–6 June 1991, Ann Arbor, MI, p. 138. Program and abstracts.

S

Salojarvi, K. (1991) Compensation in a whitefish (*Coregonus lavaretus* L. s.l.) population maintained by stocking in Lake Kallionen, northern Finland. *Finnish Fisheries Research*, **12**, 65–76.

Salojarvi, K. & Ekholm, P. (1990) Predicting the efficiency of whitefish (*Coregonus lavaretus* L. s.l.) stocking from pre-stocking catch statistics. In: *Management of Freshwater Fisheries* (eds W.L.T. van Densen, B. Steinmetz & R.H. Hughes), pp. 112–26. PUDOC, Wageningen.

Salojarvi, K. & Mutenia, A. (1994) Effects of stocking fingerlings on recruitment in the Lake Inari whitefish (*Coregonus lavaretus* L. s.l.) fishery. In: *Rehabilitation of Inland Fisheries* (ed. I. Cowx), pp. 302–13. Fishing News Books, Oxford.

Satia, B.P. (1990) *Rethinking Fisheries Management Strategies in CEDAF/IDAF Region*. FAO, Rome.

Scott, W.B. & Crossman, E.J. (1973) Freshwater fishes of Canada. *Journal of the Fisheries Research Board of Canada*, Bulletin 1849.

Selye, H. (1973) The evolution of the stress concept. *American Scientist*, **61**, 692–9.

Shapiro, J. (1980) The importance of trophic level interactions to the abundance and species coposition of algae in lakes. In: *Hypertrophic Ecosystems* (eds J. Barica & L.R. Mur), pp. 105–16. Dr W. Junk, The Hague.

Sidthimunka, A. (1970) A technical report on the fisheries survey of the Mekong River in the vicinity of the Pa Mong dam site. *Technical Paper of the Department of Fisheries of Thailand No. 8*.

Silva, S.S. de, Moreau, J., Amarasinghe, U.S., Chookajorn, T. & Guerrero, R.D. (1991) A comparative assessment of the fisheries in lacustrine inland waters in three Asian countries based on catch and effort data. *Fisheries Research*, **11**, 177–89.

Simon, T.P. & Emery, E.B. (1995) Modification and assessment of an index of biotic integrity to quantify water resource quality in great rivers. *Regulated Rivers: Research and Management*, **11**, 283–98.

Simpson, J.J. (1994) Remote sensing in fisheries: a tool for better management in the utilization of a renewable resource. *Canadian Journal of Fisheries and Aquatic Sciences*, **51**, 743–71.

Sisk, T.D., Launer, A.E., Switky, K.R. & Ehrlich, P.R. (1994) Identifying extinction threats. *BioScience*, **44**, 592–604.

Sparre, P. & Venema, S.C. (1992) Introduction to tropical fish stock assessment. *FAO Fisheries Technical Paper No. 306*, FAO, Rome.

Sparre, P. & Willman, R. (1991) Software for bio-economic analysis of fisheries: BEAM4: Analytical bio-economic simulation of space-structured multispecies and multifleet fisheries. *FAO Fisheries Technical Paper No. 310*, FAO, Rome.

St. Onge, P.D. (1995) A volume-based habitat model using GIS, bathymetric maps and field measurements: Huntsville lakes case study. In: *15th Annual International Symposium of the North American Lake Management Society*, Toronto, ON (Canada), 6–11 November 1995 (eds P.D. St. Onge, E. en Bentz, R.W. Bachmann, J.R. Jones, R.H. Peters & D.M. Soballe). *Lake and Reservoir Management*, **11**(2), 195.

Stanford, R.M., Jordan, S.W., Talhelm, D.R., Liston, C.R., Korson, C. & Steinmueller, M.H. (1982) The bioeconomic impact of impingement and entrainment on yellow perch in Lake Erie. *North American Journal of Fisheries Management*, **2**, 285–93.

Starostka, V.J. (1994) Use of the geographic information system in aquatic habitat management. *16th Northeast Pacific Pink and Chum Salmon Workshop, Juneau (USA)*, 24–26 February 1993, pp. 171–2. Alaska Sea Grant Program, Alaska University, Fairbanks.

Stocker, M., Haist, V. & Fournier, D. (1985) Environmental variation and recruitment of Pacific herring (*Clupea haraengus pallasi*) in the Strait of Georgia. *Canadian Journal of Fisheries and Aquatic Sciences*, **42**, 174–80.

Strahler, A.N. (1957) Quantitative analysis of watershed geomorphology. *Transactions of the American Geophysical Union*, **38**, 913–20.

Sugunan V.V. (1995) Reservoir fisheries of India. *FAO Fisheries Technical Paper No. 345*, FAO, Rome.

Swar, D.B. & Pradhan, B.R. (1992) Cage fish culture in the lakes of Pokhara valley, Nepal, and its impact on local fishers. *Asian Fisheries Science*, **5**, 1–13.

T

Talwaer, P.K. & Jhingran, A.G. (1991) *Inland Fishes of India and Adjacent Countries*. IBH Publishing Co., Oxford.
Tesch, F.W. (1971) Age and growth. In: *Fish Production Freshwaters* (ed. W.E. Ricker), pp. 98–130. Blackwell, Oxford.
Toepfer, C.S. (1998) Population and conservation biology of the threatened leopard darter. Dissertation, *Abstracts International Part B, Science and Engineering*, **58**, 5240.
Toews, D.R. (1979) Empirical estimates of potential fish yield for the Lake Bangweulu System, Zambia, Central Africa. *Transactions of the American Fisheries Society*, **108**, 241–52.
Townsley, P. (1997) Social issues in fisheries management. *FAO Fisheries Technical Paper No. 375*, FAO, Rome.
Travaglia, C., Kapetsky, J.M. & Righini, G. (1995) Monitoring wetlands for fisheries by NOAA AVHRR LAC thermal data. Environmental Information Management Unit, Sustainable Development Department. *Remote Sensing Centre Series No. 68*, FAO, Rome.
Tsao, E.H., Lin, Y.S., Bergersen, E.P., Behnke, R. & Chiou, C.R. (1996) A stream classification system for identifying reintroduction sites of Formosan landlocked salmon (*Oncorhynchus masou formosanus*, Jordan and Oshima. *Acta Zoologica Taiwanica, Taipei*, **7**(1), 39–59.
Turner, J.L. (1981) Changes in multi-species fisheries when many species are caught at the same time. *Committee for the Inland Fisheries of Africa (CIFA) Technical Paper*, **8**, 201–11.

U

UN (1993) *World Population Prospects. The 1992 Revision*. UN, New York, 135.
Underwood, T.J. & Bennett, D.H. (1992) Effects of fluctuating flows on the population dynamics of rainbow trout in the Spokane River of Idaho. *Northern Science*, **66**, 261–8.
United States Environmental Protection Agency (1997) *The Index of Watershed Indicators*. Office of Water, Washington, DC. EPA-841-R-97-010.
University of Michigan (1971) The fisheries of the Kafue River Flats, Zambia, in relation to the Kafue Gorge Dam. Report prepared for the FAO/UN acting as executing agency for the UNDP. *FI:SF/ZAM 11 Technical Report No. 1*. University of Michigan Press, Ann Arbor, MI.

V

Vanden Bosch, J.P. & Bernacsek, G.M. (1990) Source book for the inland fishery resources of Africa. *Committee for the Inland Fisheries of Africa (CIFA) Technical Paper*, **18**, Vols 1–3.
Vannote, R.L., Minshall, G.W., Cummins, K.W., Sedell, J.R. & Cushing, C.E. (1980) The river continuum concept. *Canadian Journal of Fisheries and Aquatic Sciences*, **37**, 130–7.
Vitousek, P.M., Mooney, H.A., Lubchenco, J. & Melillo, J.M. (1997) Human domination of Earth's ecosystems. *Science (Washington)*, **277**, 494–9.
Von Bertalanffy, L. (1934) Untersuchungen uber die gestzlichkeiten des wachstums. 1. Allgemeine Grundlagen der Theorie. *Roux'Archiv Enwichlungsmechnic Org*, **131**, 613–53.

Von Brandt, A. (1984) *Fish Catching Methods of the World*, 3rd edn. Fishing News Books, Oxford.

W

Walters, C.J. & Post, J.R. (1993) Density-dependent growth and competitive asymmetries in size-structured fish populations: a theoretical model and recommendations for field experiments. *Transactions of the American Fisheries Society*, **122**, 34–45.

Waples, R.S. (1991) Pacific salmon, *Oncorhynchus* spp., and the definition of 'species' under the Endangered Species Act. *Marine Fisheries Review*, **53**, 11–22.

Watson, D.J. & Balon, E.K. (1984) Structure and production of fish communities in tropical rainforest streams of northern Borneo. *Canadian Journal of Zoology*, **62**, 627–40.

Weatherley, A.H. & Gill, H.S. (1987) *The Biology of Fish Growth*. Academic Press, London.

Welcomme, R.L. (1967) The relationship between fecundity and fertility in the mouthbrooding cichlid fish *Tilapia leucosticta*. *Journal of the Zoological Society of London*, **151**, 453–68.

Welcomme, R.L. (1971) Evolution de la pêche intérieure, son état actuel et ses possibilités. FAO/UNDP TA, 2938.

Welcomme, R.L. (1972) An evaluation of the acadja method of fishing as practised in the coastal lagoons of Dahomey (West Africa). *Journal of Fisheries Biology*, **4**, 39–55.

Welcomme, R.L. (1976a) Some general and theoretical considerations on the fish yield of African rivers. *Journal of Fisheries Biology*, **8**, 351–64.

Welcomme, R.L. (1976b) Approaches to resource evaluation and management in tropical inland waters. IPFC/76/SYM/48.

Welcomme, R.L. (1979) *Fisheries Ecology of Floodplain Rivers*, 1st edn. Longman, London.

Welcomme, R.L. (1985) River fisheries. *FAO Fisheries Technical Paper No. 262*, FAO, Rome.

Welcomme, R.L. (1988) International introductions of inland aquatic species. *FAO Fisheries Technical Paper No. 294*, FAO, Rome.

Welcomme, R.L. (1989) Review of the present state of knowledge of fish stocks and fisheries of African rivers. In: *Proceedings of the Large River Symposium* (ed. D.P. Dodge), Vol. 106, pp. 515–32. Canadian Special Publication of Fisheries and Aquatic Sciences.

Welcomme, R.L. (1995) Relationships between fisheries and the integrity of river systems. *Regulated Rivers: Research and Management*, **11**, 121–36.

Welcomme, R.L. (1999a) Review of a model for qualitative evaluation of exploitation levels in multi-species fisheries. *Fisheries Management and Ecology*, **6**, 1–20.

Welcomme, R.L. (1999b) Institutional factors relating to aquatic genetic resources. In: *Towards Policies for Conservation and Sustainable Use of Aquatic Genetic Resources* (eds R.S.V. Pullin, D.M. Bartley & J. Kooiman). ICLARM, Penang.

Welcomme, R.L. & Bartley, D.M. (1998) An evaluation of present techniques for the enhancement of fisheries. *Journal of Fisheries Ecology and Management*, **5**, 351–82.

Welcomme, R.L. & Hagborg, D. (1977) Towards a model of a floodplain fish population and its fishery. *Environmental Biology of Fishes*, **2**, 7–24.

Welcomme, R.L. & Kapetsky, J.K. (1981) Acadjas: the brush park fisheries of Benin, West Africa. *ICLARM Newsletter*, **4**, 3–4.

Welcomme R.L. and Naeve, H. (eds) (In press) Proceedings of the EIFAC Symposium on Fisheries and Society. *Fisheries Management and Ecology*, **8**.

White, M.M. & Schell, S. (1995) An evaluation of genetic integrity of Ohio River walleye and sauger stocks. *American Fisheries Symposium*, **15**, 52–60.

Whitehead, P.J.P. (1960) The river fisheries of Kenya, Part 2. The lower Athi (Sabaki) River. *East African Agricultural and Forestry Journal*, **25**, 259–65.

Williams, R. (1971) Fish ecology of the Kafue River and floodplain environment. *Fisheries Research Bulletin, Zambia*, **5**, 305–30.

Wilson, J.A. *et al.* (1994) Chaos complexity and community management of fisheries. *Marine Policy*, **18**, 291–305.

Winemiller, K.O. (1994) Variables influencing productivity and diversity of reservoir tailwater fisheries in the upper Ohio River drainage basin. *Lake Reservoir Management*, **9**, 89–90.

Winter, J.D. (1996) Advances in underwater biotelemetry. In: *Fisheries Techniques* (eds B.R. Murphy and D.W. Willis), 2nd edn, pp. 371–95. American Fisheries Society, Bethesda, MA.

Witte, F. & Densen, W.L.T. van (1995) *Fish Stocks and Fisheries of Lake Victoria*. Samara Publishing, Cardigan.

Wootton, R.J. (1990) *Ecology of Teleost Fishes*, 1st edn. Chapman & Hall, London.

World Conservation Monitoring Centre (1998) Freshwater biodiversity: a preliminary global assessment (B. Groombridge & M. Jenkins). WCMC Biodiversity Series No. 8. WCMC–World Conservation Press, Cambridge.

Wynn, F. (1992) Controlling aquatic vegetation with triploid grass carp. *World Aquaculture*, **23**, 36–7.

Index

Access limitations to fisheries to the fishery 213–214
Acidification 224–225, 229, 273
Active methods of fishing 106
Agenda 21, 313–314
Age-structured models 140–141
Agricultural byproducts 2–3
Agriculture 220
Amazon 77–78, 79
Animal
  feeds, fish in 131
  husbandry 221
Aquaculture 14–15, 187, 188, 237–238
  balanced community of species 254–255
  distinction from inland capture fisheries 8
Aquatic vegetation 294
Area-catch studies 147
Asiatic 'lot' fisheries 214

Bag nets 103, 106
Bait fisheries 11
Baitfish, disposal of 11
Barriers 102, 103, 104
Basin planning 233–235
Berms 278, 282
Biodiversity 183, 297
  existence value of 300–301
  management for 304–307
  measures of 301–302
  productivity 297–299
  role of 297
  stability 299–300
Biodiversity Convention 327
Biomass 69–72
Blackfish 37, 38
Bottom feeders 56
Bows and arrows 116
Bypass canals 288–290

Cage culture 256
Cannibalism 52
Canning 130

Cast nets 114
Catch assessment 143–144
Catch per unit effort 144–145, 207
Catch-and-return policies 93
Catch-depletion method 148
Catches, composition of 9–10
Catchment management area (CMA) 204
Centropomidae 10
Channelisation 226
Characidae 10
Cichlidae 10
Clap nets 114–115
Clariidae 10
Classification 181
Climate change 191
Closed
  areas 213
  seasons 213
Clupeidae 10
Coastal lagoons 29, 30, 31, 40
Codes of practice 307
Co-introduction of nuisance species 239
Colastine River, Argentina 27
Collection of fish 131
Co-management in fisheries 199
  structure 200–201
Commensals 57
Competition 238
Conflict management 165–166
Conflicts of interest 13
Connectivity 229
Conservation
  objectives 12
  reserves 308
Consumption analysis 146, 168–169
Control mechanism, fish as a 11
Convention on Biodiversity 297
Convention on Biological Diversity 8, 12, 314–316
Cost effectiveness of fisheries 258–261
Cover 276, 277

Crustacea 10
Cyprinidae 10

Dam building 225–226
Dams
　removal of 290–291
　submersible 285
Demersal fish 32
Demography 327
Discharge, changes in 226–227, 230
Disease introduction 239, 252–253
Disequilibrium 252
Domestic use 220
Drawdown 227
Drying fish 130
Dynamic pool models 140–141

Echo sounding 121
Economic significance of fisheries 195–196
Ecopath models 143
Effective population size 302
Egg
　density 53
　mortality 51
Electric fishing 117
Elimination of unwanted species 253–254
Empirical regression models 138
Employment created by fisheries 13
Environment 189
　engineering 255
　data 152–153
　disturbance 238
　health 190
Environment Agency (EA) 319–320
Eutrophication 223, 228
Eutrophy 270–272
Evolutionarily significant unit (ESU) 303
Exclusive Economic Zones (EEZs) 4
Exploitation of multi-species fisheries, stages of 209
Explosives 117
Export income 14

Factors influencing fish resources 197
FAO Code of Conduct for Responsible Fisheries 8, 316–317
Fecundity 47–49
Feeding 54–55
　behaviour 59–60
　main guilds 56

Fences, fish trapping 103, 105
Fermented fish 131
Fertilisation 253
Finfish category 8–9
Fish
　abundance 146, 182
　assemblages 37
　　changes in basic characteristics of 81
　behavioural guilds 37–39
　condition 59
　counters 149
　detection, acoustic methods 149
　feed, sources of 2–3
　holes 117–119
　marketing chain 131, 132
　markets 132
　　analysis of 146
　meal 131
　oil 131
　parks 117, 118
　passes 286–287
　poison 116–117
　populations
　　size structure 40–42
　　stress, effect on 79–82, 83
　preservation 125–131
　production 167
　ramps 287–288
　role in diet 125
　size 79–80, 207
　traders 133
Fisheries
　change introduction 197
　economic policies 258–259
　effects from other users of waters 218
　intensities of 258
　in Sudan 320–321
　in United Kingdom 319–320
　in Zambia 322–323
　management
　　developing local skills 203–204
　　overarching management structure 204
　　planning and executing 251
　　roles in 201, 206
　　strategies 209–211
　needs of 196
　population dynamics 245–249
　risks 252
　types of 250

Fishermen 84–85
  food 86–88
  match 88
  recreational 88, 89
  relationships influencing 85
  specimen 88
Fishes from inland waters 33, 34, 35
Fishing
  craft 121
  gear 94
    active 106
    cost of 96
    factors influencing the choice of 94–95
    mesh selective 96, 97
    regulations and 96
    selectivity of 96, 97
    static gear 98
    types of 98–119
  regimes 92–93
  rights 214
  technology 119–121
    implications of 122
Fishing-down process 208
Flood
  control 219
  curve 291, 292
  intensity 82
  regime 292, 293, 294
  strength 82
Flood Pulse Concept 25
Flooding 27, 28
Floodplain 37–39
  fisheries simulation model (FPFMODEL) 142–143
  models 141–143
Food and Agriculture Organisation (FAO) 4
Food web 58
Forestry 220
Freezing fish 128
Fry fisheries 11
Fyke nets 103, 106

Gear
  limitations 212
  manufacturers 89
Gender analysis 162–163
Gene pool, management of 305
Genetic
  contamination 239, 252
  modification 256–258

Geographical information systems 171–181
  future use 193
  use in planning and management 191
Gill nets 98–99
Gonadosomatic index (GSI) 47
Government
  involvement in fisheries 198–199
  revenue from fisheries 13–14
Greyfish 37, 38
Gross domestic product (GDP) 14
Growth 60–61
  density-dependent 63–66
  modelling 62
Guidelines 307

Habitat
  quality 182
  rehabilitation 330
  requirements of fish 266
Habitat Quality Indices 153–154
Habitats 181, 183–184
  artificial 186
  enhancements 185
  rehabilitation and restoration 186
Harpoons 116
Harvest reserves 308–312
Herbivores 56
Heterozygosity 302
Household surveys 166–167
Human
  interventions, impact of 228, 229, 230, 231, 232
  population growth, effect on resources 1–2
Hybridisation 239
Hydrological indices (HIs) 292

Iced fish 126
Ichthyomass 77
Index of Biotic Integrity 155–156, 157, 158
Index of endemism 303
Indicators of change 305
Indices of biotic integrity (IBI) 304
Industry 221
Inland
  fisheries, regulation, historical 4–5
  waters, management strategies 15–16
Insects for the biological control 270
Instream flow methodology 154
Intermediate management areas (IMAs) 204

International initiatives 3–4
Introductions 237–238
  risk of 238–239
Inventory 181
Islands 278

Lake Chad 18, 21
Lake Chilwa, Malawi 21
Lake Nokoue, Republic of Benin 31
Lake Okeechobee 190–191
Lake Victoria 18, 55, 58, 96
Lakes 17, 22
  depression 18, 21
  dystrophic 19
  eutrophic 19
  glacial 17, 18
  high mountain 17, 19
  mesotrophic 19
  oligotrophic 19
  rift valley 17, 20, 21
  river 19
  volcanic 18
Land availability 3
Land-use patterns 190
Levees 23, 27, 226
Licences 214
Lift nets 111, 112, 113
Long lines 99

Macro-predators 57
Management of fisheries 188
Market surveys 167–168
Mark–recapture method 147
Marshes 28, 39
Maturation 43–45, 46
Maximum sustainable yield 207
Medicinal uses for fish 11
Mercenaria mercenaria 188–189
Mesh limitations 212
Meso-predators 56
Micro-predators 56
Migrations 73, 95, 181
  lateral 78–79
  longitudinal 73, 77–78
  types of 76
  vertical 79
Mining 221–222
Mitigation 262
Mobile telephones 121
Mollusca 10

Morpho-Edaphic Index (MEI) 138–139, 248
Mortality 66–69, 70, 71
Movement 72–73, 181
Multi-species stocks 141
Multi-stage channels 277

National Rivers Authority (NRA) 319–320
National trends 3
Navigation 219–220
Nile perch 58, 298–299
Non-governmental organisations (NGOs)
     200–201, 204, 206

Oligotrophy 272–273
Omnivores 56
Ophicephalidae 10
Oreochromis esculentus, disappearance of 55
Oreochromis leucostictus 96, 97
Ornamental species, fisheries for 11
Oueme river fishery, West Africa 208
Overexploitation 4
Overfishing 207, 209

Pabna Irrigation and Rural Development
     Project (PIRDP), Bangladesh 46
Parana River 78
Parana River, Argentina 26
Parasites 57
Parks 307
Participatory Monitoring and Evaluation
     164–165
Participatory Rural Appraisal 163–164
Pelagic communities 32
Percidae 10
Performance of fisheries 187
Pest control 237
Physical Habitat Simulation System 154
Planning
  inland fisheries 217
  programmes 265
Plunge basket 113, 114
Pollution 223–224, 228
Pool–riffle structures 276
Population dynamics of stocking
     245–249
Potential yield model 137
Power generation 219
Predation 238
Predator density 59–60
Predator/prey ratio 80

Processors 90
Production 72
Productivity biomass ratio 72
Protection 263
　of fish movement 286
Put, grow and take fisheries 93
Put-and-take fisheries 93

Quantitative socio-economic data 166
Quotas 215

Ramsar Convention 317–318
Recreational fisheries 10
Red River, Hanoi, Vietnam 28
Regional legislation 324
Regulation of fisheries 207
Rehabilitation 262–263
　environmental 306
　of channels 275–276, 280, 281
　of floodplains 279–280, 283
　of lakes and reservoirs 269–270
　of rivers 274–275
Re-meandering 278
Reproduction 43
　seasonality of 45–47
Reproductive behaviour 44
Reserves 307
Reservoirs 28–29, 39–40
Resource availability 1–2
Resource mapping 171
Retailers 91
Rights to fish 89–90
Risk evaluation 190
Ritual uses of fish 11
River Continuum Concept 25
River systems, profile of fish in 41
Rivers 22
　braided reach 25, 26
　floodplain 24, 26, 27
　four-phase cycle 25–27
　hierarchy of orders 22–24
　main channel 23–24, 26
　nutrient flows 24–25
　rhithronic reach 24
Rivers gradient 36–37
Rod and line 115–116

Salmon and Freshwater Fisheries Act
　319–320
Salmonidae 10

Salmonids 73–77
Salting fish 130
Sampling 169
Scoop nets 110, 112
Seasonality of fishing 123–124
Sediment 224, 228
Seine nets 108–110
Shallow bays 276
Shallows 276
Shoreline development 273
Siltation 273–274
Siluroidea 10
Size limits on fish landed 215–216
Sluice gates 290
Smoking fish 128–129
Social objectives 12–13
Spatial management units 204, 205
Spawning time 47
Species diversity 303–304
Stakeholders 160–161
State-regulated access 214
Stock assessment 135
Stocking 93, 241–245
Stock–recruitment process 50–54
　models 51
Stomach Fullness Index (SFI) 5
　7–59
Sturgeon migration 73–77
Surplus production models 140–141
Sustainability 12
Swamps 28, 39

Telemetry 149–150
Temperature changes 231
Tennessee darter 301
Terrain resource information management
　(TRIM) maps 190
Terrestrial environment 189
Total removal methods 147
Tourist industry 90
Transmitters 150–151
Transporters 91
Traps 99–102
Trawls 106–108
Trolling 115

Ubolratana reservoir 30
UNCED process 8
UNESCO Convention 317
Uptake of water 232

Vegetation within the drainage basin 267
  aquatic 268
  control of 269
  riparian 267–268
Village management areas 204
Volga River 73–77
von Bertalanffy growth function (VBGF) 62–63, 65–66

Water
  levels 80
  quality 222–223, 228, 237, 266–267
  regime management 291–292
  supply 2

Water Resources Act 1991: 319–320
Water-level control 274
Weirs 276
Whitefish 37, 38
Wildlife conservation 222
World catches from inland waters 9
World Conservation Union (IUCN) 4

Year class strength (YCS) 52
Yield prediction 135

Zonation of fish 36